£7.50.

Mr. M.C. McKENZIE.

c/o AQUILA.

Electronics IV

Electronics IV

A textbook covering the analogue (linear) content of the Level IV syllabus of the Technician Education Council

D C Green
M Tech, CEng, MIERE
Senior Lecturer in Telecommunication Engineering
Willesden College of Technology

Pitman

PITMAN BOOKS LIMITED
128 Long Acre, London WC2E 9AN

Associated Companies
Pitman Publishing New Zealand Ltd, Wellington
Pitman Publishing Pty Ltd. Melbourne

First published in Great Britain 1981
Reprinted 1983

Text set in 10/12 Linotron Times,
printed and bound in Great Britain
at The Pitman Press, Bath

0 273 01504 4

Preface

The rapid developments in the use of electronics in all kinds of fields mean that there is a continual demand for people with some knowledge of electronic circuits, devices and techniques. This book provides a comprehensive coverage of the techniques used in modern analogue equipment.

The Technician Education Council has produced a series of units at the second and third levels to introduce the student to the concepts of both analogue (or linear) and digital electronics. At these levels the approach to understanding is mainly descriptive and any analysis is fairly simple and straightforward.

Those students who attain the standard of the Technician Certificate or Diploma and who want to further their studies, then require a more analytical account of circuits, etc. The amount of desirable knowledge is very large and more than can be covered in a standard 60 hour unit, so some selection in the choice of material for a particular course is needed. With this problem in mind the TEC has produced a large "unit" covering most aspects of both digital and analogue electronics from which Colleges are invited to select material to make up a 60 hour unit.

This book has been written to cover *all* of the material specified by the *analogue* sections of this large unit. The digital sections will be the subject of the companion volume entitled *Digital Electronic Technology*.

Chapters 1 and 2 introduce the basic principles of transistors and integrated circuits. In the following chapters the applications of these devices in a wide variety of circuits are discussed. Chapters 3 to 6 inclusive discuss the ways in which transistors and integrated circuits can be used for the amplification of signals. In Chapters 4 and 5 the gain of an amplifier has sometimes been expressed as $20 \log_{10}$ (the voltage gain) dB even when the input and output impedances are (probably) not equal. This is the usual practice when considering Bode diagrams and the stability of feedback amplifiers and does lead to some simplification of the problem.

Chapters 7 and 8 are concerned with the methods employed to generate repetitive waveforms. Sinusoidal waveform generators are the subject of Chapter 7, while Chapter 8 deals with the generation of non-sinusoidal waveforms. The important topic of electrical noise is then considered in Chapter 9. The final chapter of the book is concerned with the thyristor and the triac and the ways in which these devices can be used for controlled rectification, motor speed control, and other high-power electronic applications.

The book has been written on the assumption that the reader already possesses a knowledge of Electronics, Electrical Principles and Mathematics of the standard reached by the level III units of TEC. The book should be useful to all students of analogue electronics at a level approximately that of the TEC level IV unit printed at the end of the book.

Many worked examples are provided in the text to illustrate the principles that have been discussed and each chapter concludes with a number of both short and long exercises. Answers to the numerical exercises will be found at the rear of the book.

Acknowledgement is due to the Technician Education Council for its permission to use the content of the TEC unit in an appendix to this book. The Council reserve the right to amend the content of its unit at any time.

DCG

The following abbreviations for other titles in this series are used in this text:
[EII] Electronics II
[EIII] Electronics III
[DT & S] Digital Techniques and Systems

Contents

1 Transistors

The transistor is a semiconductor device that performs many varied tasks in electronic circuitry but mainly amplification and switching. There are two kinds of transistor: the bipolar transistor and the field effect transistor. Both kinds are available both in discrete form and as a component within a monolithic integrated circuit.

The basic principles of operation of both kinds of transistor have been discussed elsewhere [EII and EIII] and will not be repeated here.

Bipolar Transistors

The **bipolar transistor** consists of a silicon crystal, or less often a germanium crystal, that has been doped to form three distinct regions within the crystal. For an n-p-n transistor, the three regions consist of two n-type regions separated by a narrow p-type region, while the p-n-p transistor has one n-type region sandwiched in between two p-type regions. The three regions are known as the emitter, base and collector, with the base region being of the opposite type to the other two. Most modern bipolar transistors are of the silicon planar type, the construction of which is shown in Fig. 1.1. Other

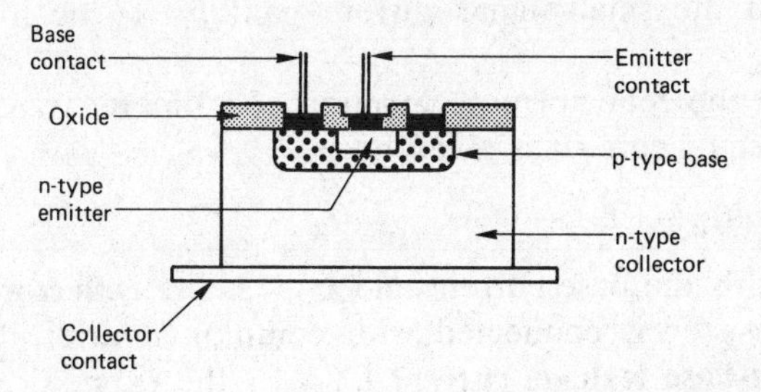

Fig. 1.1 Construction of a silicon planar transistor.

manufacturing techniques are employed for some power transistors in an attempt to obtain an improved performance in respect to the more important parameters such as voltage breakdown, and power dissipation.

D.C. Current Gain

The **d.c. current gain** h_{FE} of a bipolar transistor is the ratio of its d.c. collector current to its d.c. base current, assuming that the collector leakage current is negligibly small. The d.c. current gain is not a constant quantity but varies with both junction temperature and collector current.

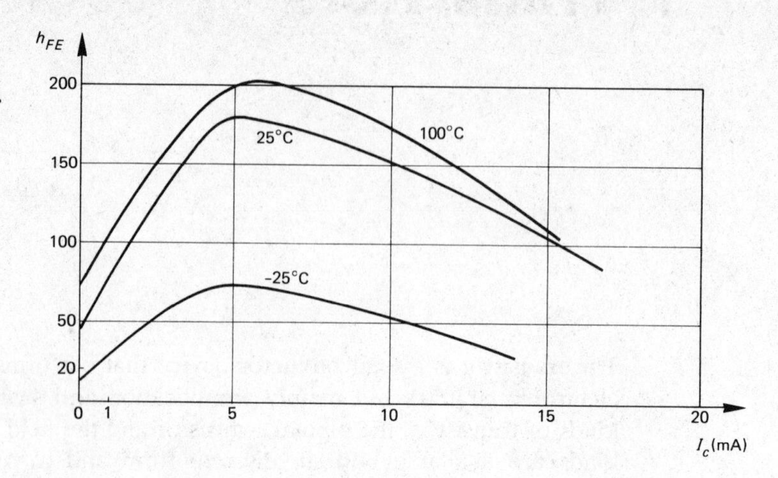

Fig. 1.2 Variation of h_{FE} with collector current and temperature.

Fig. 1.2 shows typical variations of d.c. current gain with temperature and with collector current. It is fairly obvious that if the maximum current gain is to be provided by a bipolar transistor it is necessary to bias the device to pass the appropriate value of collector current. Manufacturer's data sheets quote the maximum current gain of a transistor and the value of collector current at which it is obtained. For example, the n-p-n transistor BC 109 has a maximum current gain of 110–450 at a d.c. collector current of 2 mA. The maximum current gain quoted varies widely over the range 110–450. Such a wide variation is characteristic of many types of transistor and arises because of the difficulties associated with any attempts to closely control the manufacturing process. Usually the d.c. current gain h_{FE} is approximately equal to the small-signal current gain h_{fe} to be introduced later in this chapter.

When the collector/base junction of a bipolar transistor is reverse-biased the collector current is given by

$$I_c = h_{FE} I_b + I_{CEO} \tag{1.1}$$

where I_b is the base current and I_{CEO} is the **collector leakage current** when the transistor is connected with common emitter. I_{CEO} is related to the **common-base leakage current** I_{CBO} by the expression

$$I_{CEO} = I_{CBO}(1 + h_{FE}) \tag{1.2}$$

For most silicon transistors I_{CBO} is negligibly small—a few nanoamperes—at junction temperatures below about 120°C. This means that any rise in the collector current due to an increase in I_{CBO} because of a temperature increase is very small, and so any temperature dependence of the collector current is the result of variations in h_{FE}. The effect of an increased value of h_{FE} on the output characteristics of a transistor is to increase the spacing between the curves for each base current. With a germanium transistor I_{CBO} is *not* negligibly small—a few microamperes—and the effect of an increase

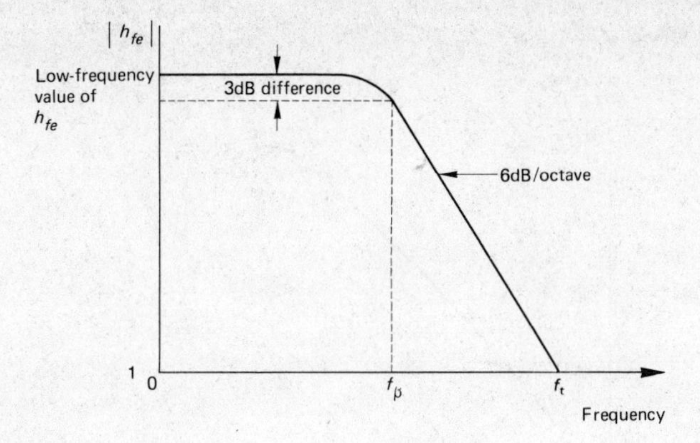

Fig. 1.3 Variation of h_{fe} with frequency.

in the junction temperature is to move the entire family of output characteristic curves upwards.

Transistor data sheets often specify I_{CBO} at room temperatures and sometimes indicate how I_{CBO} varies with change in temperature. When such information is not available it is reasonably accurate to assume that I_{CBO} doubles for every 12°C rise in temperature for a germanium transistor and for every 8°C increase for a silicon transistor.

A.C. Current Gain

The **a.c. current gain** of a bipolar transistor is the ratio $\delta I_c/\delta I_b$ for a constant value of collector/emitter voltage V_{CE}. In this book it will be denoted by the symbol h_{fe}. For collector currents much smaller than the value for maximum h_{FE}, then $h_{fe} > h_{FE}$ but the reverse is true for larger collector currents. When the collector current is somewhere in the region of the value for maximum h_{FE}, then $h_{fe} \simeq h_{FE}$.

High-frequency Performance

The current gain h_{fe} of a bipolar transistor is not the same at all frequencies but falls at the rate of 6 dB per octave at frequencies above the **cut-off frequency** f_β (see Fig. 1.3). The cut-off frequency is the frequency at which $|h_{fe}|$ has fallen by 3 dB from its low-frequency value. The frequency at which the magnitude of the current gain has fallen to unity is known as the **transition frequency** f_t. At any frequency f less than f_t the magnitude of the current gain can be determined using

$$f_t = |h_{fe}| f \tag{1.3}$$

Transistor data sheets usually specify a minimum value for f_t and very often quote a typical value as well, perhaps with a curve showing how f_t varies with collector current. A typical f_t/collector-current curve is given in Fig. 1.4. Expressions for the cut-off and transition frequencies will be derived later in this chapter (p. 27) after equivalent circuits for transistors have been discussed.

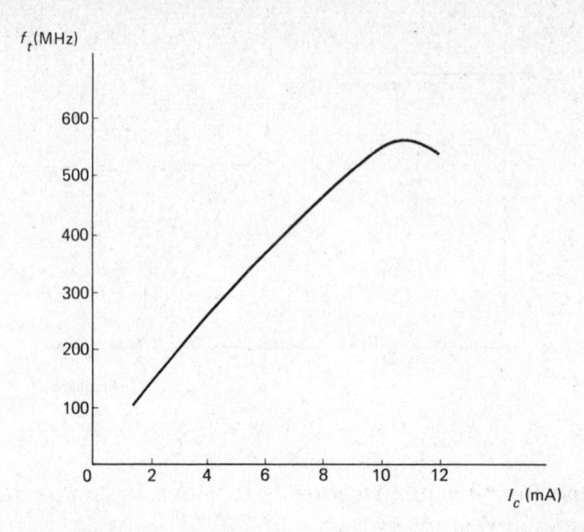

Fig. 1.4 Variation of f_t with current

Power Dissipation

An increase in the temperature of the collector/base junction will cause the collector leakage current I_{CBO} to increase. The increased collector current produces an increase in the power dissipated at the junction and this, in turn, further increases the junction temperature and so gives a further increase in I_{CBO}. The process is cumulative and, particularly in the common-emitter connection—since $I_{CEO} \gg I_{CBO}$—it may damage the device. Excessive heat within a transistor may melt soldered connections, cause the deterioration of insulation materials, and produce chemical and metallurgical changes in the semiconductor material.

The manufacturer of a transistor quotes the **maximum permissible power** that can be dissipated within the transistor without causing damage. The power dissipated in a transistor is predominantly the power that is dissipated at its collector/base junction and this, in turn, is equal to the d.c. power taken from the collector supply voltage minus the total output power (d.c. power plus a.c. power). For transistors handling small signals, the power dissipated at the collector is small and there is generally little problem. When the power dissipated by the transistor is large enough to cause the junction temperature to rise to a dangerous level, it is necessary to increase the rate at which heat is removed from the device. Power transistors are constructed with their collector terminal connected to their metallic case. To increase the area from which the heat is removed, the case of the transistor can be bolted onto a sheet of metal known as a **heat sink**. Heat will move from the transistor to the heat sink by conduction and then be removed from the heat sink by convection and radiation.

The heat dissipated at the collector/base junction flows through the **thermal resistance** θ between the junction and the external environment. Thermal resistance is analogous to electrical resistance and represents the opposition of a material to the flow of heat energy. Under steady state conditions the thermal resistance of a thermal conductor is the ratio of the temperature drop across the conductor to the heat transfer rate (in joules/sec or watts) through the conductor. The unit of thermal resistance is "degrees centigrade per watt". When a continuous power is dissipated at the collector/base junction, the heat generated will cause the junction temperature to rise until the difference between the junction and ambient temperatures is equal to the product of the thermal resistance and the power dissipated at the junction, i.e.

$$T_J - T_A = \theta_{JA} P_d \tag{1.4}$$

T_J is the **safe junction temperature**. This figure is given by the manufacturer and is usually about 85°C for a germanium transistor and about 150°C for a silicon transistor.

T_A is the ambient temperature and may be higher than the room temperature since the transistor may be installed within an enclosed space that contains one or more other heat-producing devices.

The thermal resistance consists of three parts:

1 The thermal resistance θ_{JC} between the collector/base junction and the case of the transistor.

2 The thermal resistance θ_{CS} between the transistor case and the heat sink.

3 The thermal resistance θ_{SA} between the heat sink and the environment at ambient temperature.

The conduction of heat energy is analogous to the conduction of electricity and, for calculations on heat performance, electrical analogues are used. Thus,

Heat flow ≡ current flow

Temperature difference ≡ potential difference

Thermal resistance ≡ electrical resistance

Using these analogues, the equivalent circuit for the thermal performance of a transistor can be drawn (see Fig. 1.5).

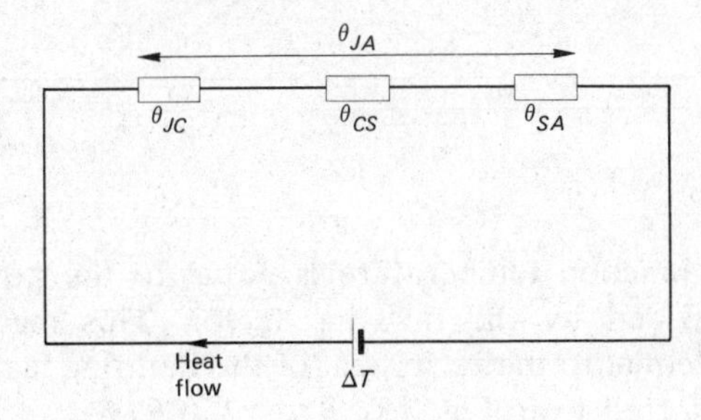

Fig. 1.5 Thermal equivalent circuit of a transistor and its heat sink.

The thermal resistance θ_{JC} between the collector/base junction and the transistor case is normally quoted by the manufacturer. The required total thermal resistance for a given power dissipation can be calculated and then a suitable heat sink can be selected with the aid of heat sink data available from makers.

Example 1.1

A transistor has a maximum collector dissipation of 6 W. The maximum junction temperature is 80°C, $\theta_{JC} = 0.9$°C/W and $\theta_{CS} = 0.25$°C/W. Calculate the necessary thermal resistance of the heat sink if the ambient temperature is 50°C.

Solution

$$\theta_{JA} = \frac{T_J - T_A}{P_d} = \frac{80 - 50}{6} = \frac{30}{6} = 5\text{°C/W}$$

Therefore,

$$\theta_{SA} = 5 - 0.9 - 0.25 = 3.85\text{°C/W} \quad (Ans)$$

The maximum power dissipation figure quoted by a manufacturer usually assumes a maximum ambient temperature of 25°C. If the ambient temperature is likely to be higher than 25°C, the maximum collector dissipation must be reduced. Sometimes the necessary derating is given as a figure, e.g. "derate 4 mW/°C for ambient temperatures above 25°C" but often a **derating curve** is given. A typical derating curve is shown in Fig. 1.6; the slope of the curve is equal to the derating factor, i.e. $-240/(85-25)$ or -4 mW/°C.

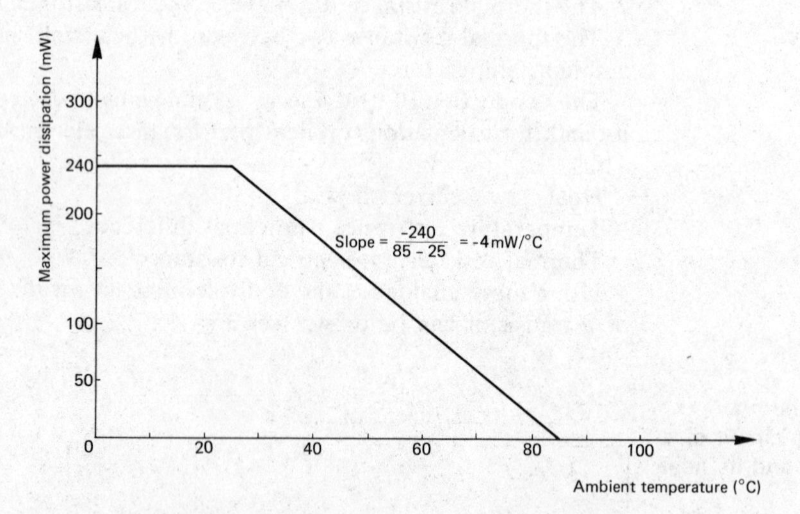

Fig. 1.6 Transistor dissipation derating curve.

The rise in junction temperature is equal to the power dissipated in the transistor divided by the derating factor. This means that the thermal resistance is equal to the reciprocal of the derating factor, and hence for the transistor referred to in Fig. 1.6, $\theta_{JA} = 250$°C/W.

Example 1.2

The following data is provided for a bipolar transistor:

Maximum power dissipation 12 W at 25°C
Maximum junction temperature 120°C
Thermal resistance 8°C/W

Fig. 1.7

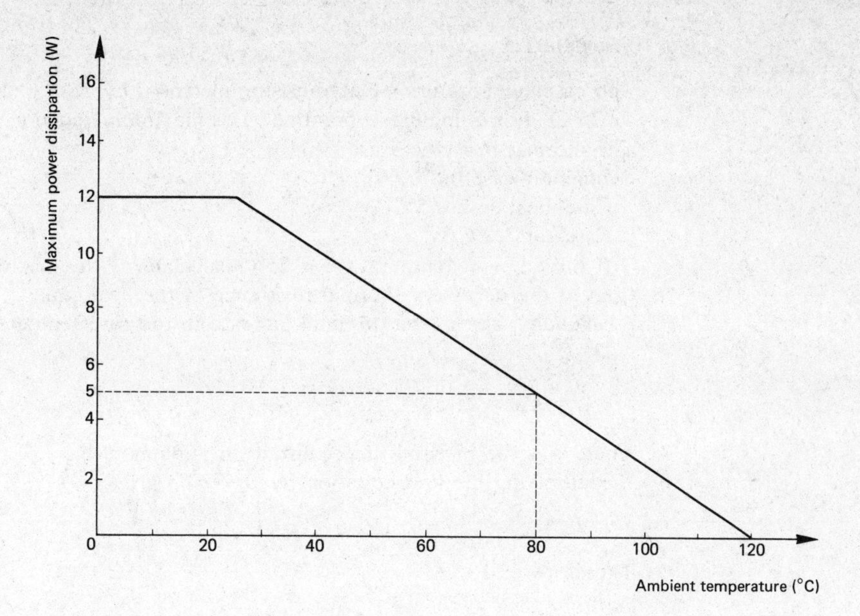

Calculate the maximum power dissipation for an ambient temperature of 80°C.
Solution

Method 1 $P_d = \dfrac{T_J - T_A}{\theta_{JA}} = \dfrac{120-80}{8} = 5\ \text{W}$ (*Ans*)

Method 2 The derating curve for the transistor falls linearly from 12 W at $T_A = 25°\text{C}$ to 0 W at $T_A = 120°\text{C}$ (see Fig. 1.7). From this graph, when $T_A = 80°\text{C}$, $P_d = 5\ \text{W}$ (*Ans*).

The maximum collector dissipation of a transistor can be marked on the output characteristics by drawing the locus of the power dissipation as the collector current varies.

Thermal Time Constant

The transistor, its case, and its heat sink must all store, or release, heat energy in order to be able to change in temperature. The effect is analogous to electrical capacitance which must store, or release, electric charge in order to change in voltage. **Thermal time constant** is the product of thermal resistance and thermal "capacitance". The electrical equivalent circuit is shown in Fig. 1.8 where C_J, C_C and C_S are respectively the thermal capacitances of the junction, case and heat sink.

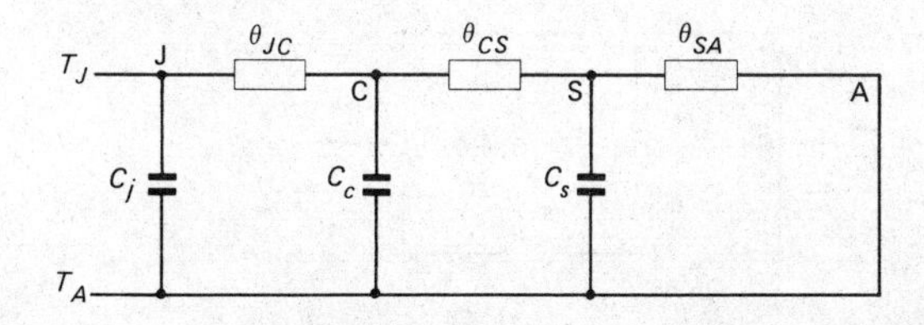

Fig. 1.8 Thermal equivalent circuit with thermal "capacitance".

Example 1.3

The case temperature of a transistor increases by 28.5°C after 1 minute, and by 42.75°C after 2 minutes operation. The maximum junction temperature is 110°C. The thermal resistances are

Junction–case 0.8°C/W
Case–heat sink 0.5°C/W
Case–air 3.0°C/W

If the ambient temperature is 25°C calculate *a*) the maximum transistor dissipation, *b*) the necessary thermal resistance of the heat sink.

Solution Let T_m be the final increase in the case temperature. Then

$$28.5 = T_m(1-e^{-1/\tau})$$
$$42.75 = T_m(1-e^{-2/\tau})$$

where τ is the thermal time constant in minutes.

Multiplying the first equation by $(1+e^{-1/\tau})$ gives

$$28.5(1+e^{-1/\tau}) = T_m(1-e^{-2/\tau}) = 42.75$$

Therefore

$$e^{-1/\tau} = \frac{42.75}{28.5} - 1 = \tfrac{1}{2}$$

and hence

$$28.5 = T_m(1-\tfrac{1}{2}) \quad \text{or} \quad T_m = 57°\text{C}$$

Therefore

$$\text{Final case temperature} = 57°\text{C} + 25°\text{C} = 82°\text{C}$$

Maximum collector dissipation

$$P_c = \frac{T_J - T_C}{\theta_{JC}} = \frac{110-82}{0.8} = 35\ \text{W}$$

Also $P_c = \dfrac{T_J - T_A}{\theta_T}$ so

$$\theta_T = \frac{T_J - T_A}{P_c} = \frac{110-25}{35} = 2.43°\text{C/W}$$

θ_T is the *total* thermal resistance between the transistor junction and the surrounding air. The thermal resistance "circuit" is shown in Fig. 1.9, from which

$$2.43 = 0.8 + \frac{3(0.5+\theta_{SA})}{3.5+\theta_{SA}}$$

$$8.4 + 2.43\theta_{SA} = 2.8 + 0.8\theta_{SA} + 1.5 + 3\theta_{SA} \qquad \theta_{SA} = 3°\text{C/W} \quad (Ans)$$

Fig. 1.9

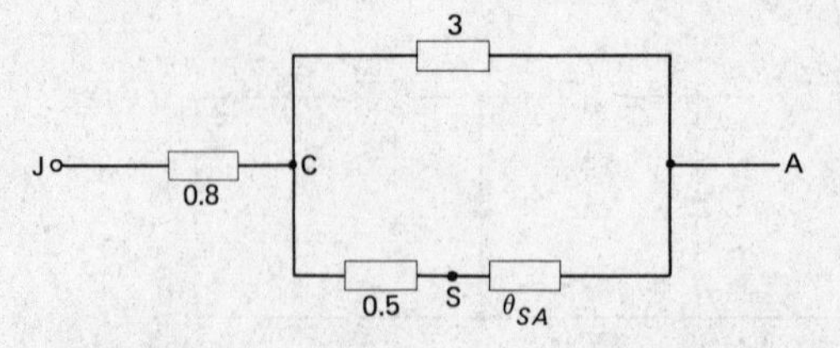

Maximum Collector Current

The maximum collector current that should be allowed to flow in a bipolar transistor is limited by the power handling capability of the internal wires and metallization of the device and the dimensions of the various regions. If the maximum safe current is exceeded, the connections may be damaged and the current gain dramatically reduced.

Maximum Collector Voltage

If the voltage applied across a p-n junction exceeds the *breakdown* value, a large current will flow which may cause damage to the transistor. The maximum safe voltages are generally quoted as between either i) the collector and base terminals with the emitter open-circuit or ii) the collector and emitter terminals with the base open-circuit. These two voltages are indicated by the labels V_{CBO} and V_{CEO} respectively. Typically, $V_{CBO} = 40$ V, $V_{CEO} = 60$ V.

There are two possible causes of the voltage breakdown of a bipolar transistor.

1 Punch-through Increasing the collector/emitter voltage V_{CE} of a transistor will *increase* the reverse-bias of the collector/base junction and the forward-bias of the emitter/base junction and in so doing will *reduce* the effective width of the base region. Eventually, the base width will approach zero and the current flowing will increase to a large value since the transistor has effectively become a low resistance connected across the power supply. The heat developed by this large current will damage the transistor.
2 Avalanche multiplication When the reverse-bias voltage across a p-n junction is sufficiently large, a charge carrier crossing the junction may collide with an atom and in so doing remove an electron from a covalent bond. The hole-electron pair thus formed exists in a region of high electric field strength, and so the two charge carriers are rapidly swept away. In their passage through the material these secondary charge carriers may, in turn, also collide with atoms to produce further hole-electron pairs. This process is cumulative and will lead to a rapid increase in the collector current if the voltage is high enough.

For a given transistor, voltage breakdown might be due to either punch-through or avalanche multiplication, depending on the structure of the device.

The Safe Operation of Power Transistors

The main ratings of a power transistor are based upon the maximum permissible values of collector current, collector voltage, power dissipation and junction temperature. The **safe operating area** (SOAR) of a bipolar transistor is a plot of the log of collector current to a base of the log of the collector/emitter voltage on which four limits are marked. Fig. 1.10 shows a typical SOAR diagram. Three of the boundaries to the diagram have

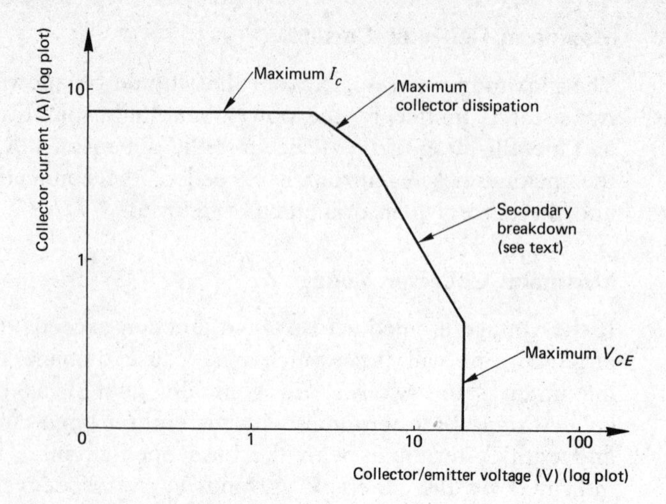

Fig. 1.10 SOAR diagram for a transistor.

already been discussed; the fourth, namely **secondary breakdown**, occurs if the collector current is allowed to still further increase after avalanche breakdown has taken place. Secondary breakdown results in an excessive temperature rise somewhere within the transistor and will usually destroy the device. When designing a circuit, each of the four SOAR boundaries should be regarded as a limit which must not be exceeded.

Example 1.4

A transistor has a maximum collector current of 10 A and maximum collector/emitter voltage V_{CEO} of 80 V. The maximum power dissipated at an ambient temperature of 25°C is 40 W. Draw the SOAR diagram. Assume that the maximum collector current at $V_{CE} = 80$ V before the onset of secondary breakdown is 0.3 A.

Solution Since the maximum collector dissipation is 40 W the relationship between I and V is

I_C	(A)	1	2	4	8	10
V_{CE}	(V)	40	20	10	5	4

Plotting these figures on log/log axes gives a straight line with a slope of −1 (Fig. 1.11).

The maximum collector current at $V_{CE} = 80$ V is 0.3 A; this point is marked as X in Fig. 1.11. A straight line must now be drawn, with a slope greater than unity, to connect point X to the maximum dissipation line. The slope of this line must be such that the reduction in collector/emitter voltage is large enough to prevent secondary breakdown. A suitable line is shown in the figure.

Fig. 1.11

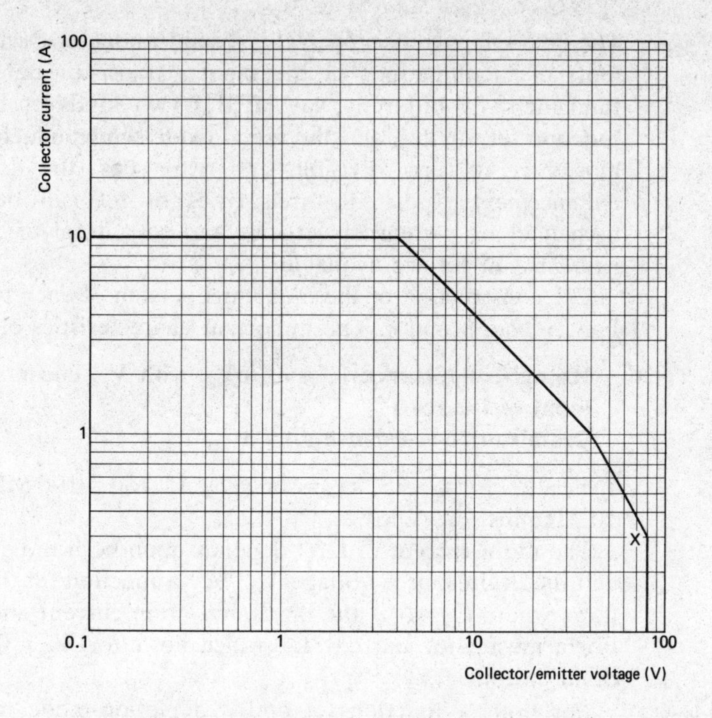

Transistor Capacitances

Each of the p-n junctions in a bipolar transistor has a self-capacitance, the value of which is dependent upon both the magnitude and polarity of the junction voltage. This capacitance represents the charge stored when a voltage is applied.

When a transistor is operated in its active region, i.e. neither saturated nor cut-off, there is an emitter diffusion capacitance which is proportional to the d.c. collector current flowing. The capacitance $C_{b'e}$ of the base/emitter junction is the sum of this diffusion capacitance and the space charge capacitance. Typically $C_{b'e}$ is about 20 pF†.

The collector/base junction also possesses some capacitance and, since the junction is reverse-biased, $C_{b'c}$ is smaller than $C_{b'e}$. Typically $C_{b'c}$ is about 5 pF.

The output capacitance C_{ob} of the transistor is sometimes quoted in manufacturers' literature and this capacitance depends upon both $C_{b'e}$ and $C_{b'c}$ (see p. 24) but is predominantly due to $C_{b'c}$. Typically C_{ob} is about 6 pF.

† The reason for the use of the suffix b' to represent the base will become apparent on page 24.

Field Effect Transistors

The field effect transistor (fet) is a semiconductor device which can perform most of the functions of the bipolar transistor but which operates in a fundamentally different way [EIII]. Two kinds of fet are available: the **junction fet**, or jfet, and the **metal oxide semiconductor fet**, or mosfet. The mosfet is, in turn, divisible into two types: the *depletion* type and the *enhancement* type. All three types of fet can be obtained in either n-channel or p-channel versions and so a total of *six* different types of *small-signal* fet are available.

The construction of the n-channel version of each type of fet is shown by Figs. 1.12*a*, *b* and *c*. The important characteristics of a fet are its

Mutual conductance $g_m = \delta I_d/\delta V_{gs}$ with V_{ds} constant
Input resistance R_{IN}
Drain/source resistance $r_{ds} = \delta V_{ds}/\delta I_d$

Typically, $g_m = 4$ mS, $r_{ds} = 100$ kΩ and $R_{IN} = 100$ MΩ plus for a jfet and 10^{10} Ω plus for a mosfet.

The **drain current** of a fet depends upon both the gate/source voltage V_{gs} and the drain/source voltage V_{ds}. For a junction fet the drain current which flows when $V_{gs} = 0$ is the maximum drain current and it is labelled as I_{dss}. For a mosfet the current I_{dss} which flows for $V_{gs} = 0$ is not the maximum drain current.

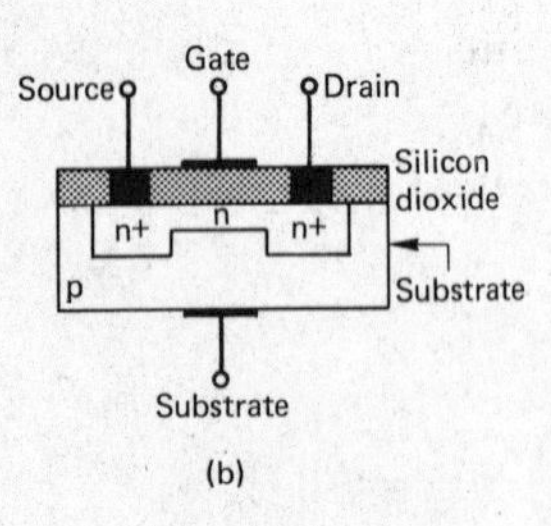

For both a junction fet and a depletion-mode mosfet, two important parameters are

a) the **gate pinch-off voltage** V_p
b) the drain current I_{dss} which flows when $V_{gs} = 0$.

When a jfet is operated in its pinch-off region, i.e. as an amplifying device,

$$I_d = I_{dss}(1 - |V_{gs}/V_p|)^2 \tag{1.5}$$

The mutual conductance g_m is given by $\delta I_d/\delta V_{gs}$, hence differentiating equation (1.5)

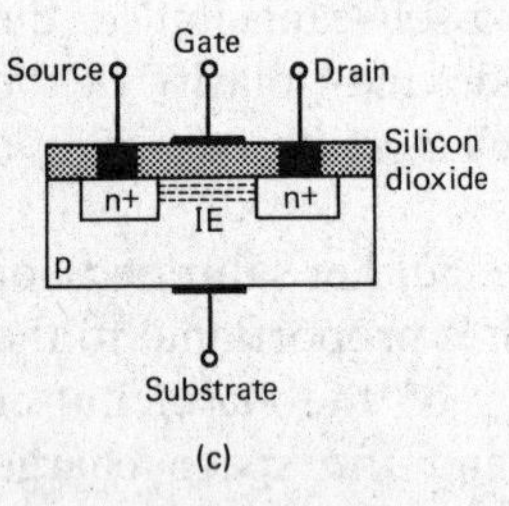

$$g_m = dI_d/dV_{gs} = \frac{-2I_{dss}}{V_p}(1 - |V_{gs}/V_p|)^2 \tag{1.6a}$$

When V_{gs} is zero

$$g_m = g_{mo} = -2I_{dss}/V_p \tag{1.6b}$$

Typically, $I_{dss} = 1$ mA and $V_p = -2$ V whence $g_{mo} = 1$ mS.

The mutual conductance varies linearly with the gate/source voltage but it is always very much smaller than the mutual conductance of a bipolar transistor, e.g. 1 mS as opposed to 38 mS for 1 mA drain or collector current. This means that the fet will generally give a much lower voltage gain than is available from a bipolar transistor.

Fig. 1.12 The construction of (*a*) an n-channel jfet, (*b*) an n-channel depletion mode mosfet, (*c*) an n-channel enhancement mode mosfet. IE = induced electrons forming a virtual channel when the gate voltage is positive.

For an enhancement-type mosfet the drain current I_d is related to the gate/source voltage V_{gs} by

$$g_m = 2I_d/(V_{gs} - V_{th}) \tag{1.7}$$

where V_{th} is the gate/source voltage at which the induced channel just becomes evident. V_{th} is often known as the *threshold voltage.* Typical values for V_{th} are 1 V to 5 V; when the gate voltage is less than V_{th} there is no channel and $I_d = 0$.

Effects of Temperature

The velocity with which majority charge carriers travel through the channel is dependent upon both the drain/source voltage and the temperature of the fet. An increase in the temperature reduces the velocity of the channel charge carriers and this appears in the form of a reduction in the drain current which flows for given gate/source and drain/source voltages.

A further factor that may also affect the variation of drain current with change in temperature is the barrier potential across the gate-channel p-n junction. An increase in temperature will cause the barrier potential to fall and this, in turn, will reduce the width of the depletion layer for a given gate/source voltage. The channel resistance will fall and so the drain current will increase.

The two effects tend to vary the drain current in opposite directions and as a result the overall variation can be quite small. Indeed, if the gate/source voltage is made equal to the pinch-off voltage *plus* 0.63 V then zero temperature coefficient is obtained. In general, the overall result is that the drain current decreases with increased temperature. This is the opposite of the variation in collector current with temperature experienced by a bipolar.

Power Dissipation

The **total power** dissipated within a field effect transistor is the product of the drain current I_d and the drain/source voltage V_{ds}. The manufacturer of a device specifies a maximum safe value to the power dissipation which should not be exceeded.

For a small-signal fet the allowable power dissipation is in the region of 200–300 mW.

Maximum Drain/Source Voltage

If the drain/source voltage of a field effect transistor is increased to too large a value, a large increase in the drain current will suddenly take place. There are two contributory sources for this: *a*) avalanche breakdown of the reverse-biased drain-substrate p-n junction, and *b*) punch-through across the channel. The avalanche breakdown effect is similar to that already described for the bipolar transistor.

Punch-through occurs when the reverse bias voltage at the drain-substrate p-n junction is large enough for the depletion layer thus formed to widen to such an extent that it reaches the source. The source-substrate p-n junction will then break down and the drain and source terminals of the fet are effectively short-circuited together. The manufacturer of a device quotes the maximum drain/source voltage that can be applied without danger of breakdown taking place.

For most small-signal fets, the maximum drain/source voltage is 25–30 V.

Reverse Gate/Source Voltage

There is always a maximum value to the reverse gate/source voltage which can be applied without damaging the device. For a junction field effect transistor, the gate-channel junction must be held in the reverse-biased condition or else a gate current will flow and the input resistance will fall to a low value, probably damaging the device. On the other hand the reverse bias voltage must not be too large or the junction will break down. The gate-channel junction of either type of mosfet can be either forward-biased or reverse-biased but again damage can be caused if too large a gate/source voltage is applied. Typically, $V_{gs(max)}$ is 30 V.

FET Capacitances

A field effect transistor has capacitances between its gate and source terminals, and between its gate and drain terminals, labelled respectively as C_{gs} and C_{gd}. Data sheets usually quote values for capacitances C_{iss} and C_{rss}, where C_{iss} is the common-source short-circuit input capacitance and is numerically equal to the sum of C_{gs} and C_{gd}, and C_{rss} is the common-source reverse transfer capacitance which is equal to C_{gd}. Because of the difference in construction these capacitances are smaller in the depletion type mosfet than in the enhancement type. Typically, a field effect transistor has an input capacitance of 5–7 pF.

The VMOS Power Field Effect Transistor

All three types of field effect transistor so far described are constructed in such a way that the current flows *horizontally* through the channel from source to drain. As a result the maximum possible current density is limited and it is difficult to remove heat so that the devices are unable to dissipate very much power. An alternative structure which is more suitable for large current and power ratings is known as a *vertical structure metal oxide semiconductor power field effect transistor* or **vmosfet**. The construction of a vmosfet is shown by Fig. 1.13. The n^+ substrate has a n^- epitaxial layer diffused onto it and then a lightly-doped p-type region is made. Next, another n^+ region is diffused before the V-shaped groove is etched into the n^+ and p regions and partly into the n^- epitaxial layer. A layer of silicon oxide is grown over the surface of the semiconductor before parts of the

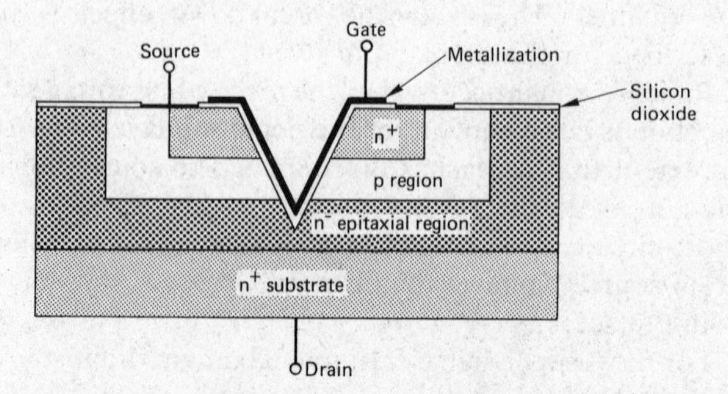

Fig. 1.13 Construction of a vmosfet.

oxide are etched away and the resulting gaps are metallized to provide the gate and source connections.

Essentially the device is of the enhancement type; with the drain held positive with respect to the source, a positive voltage applied to the gate will induce n-type channels onto both surfaces of the body adjacent to the gate. Electrons are then able to flow via the two channels from source to n^- epitaxial layer and then via the n^+ substrate to the drain. The junction between the p-type and n-type regions is reverse-biased and produces a depletion layer which mainly lies within the n^- epitaxial layer and this results in an increased drain/source breakdown voltage. As with the more conventional type of mosfet the magnitude of the drain current is determined by the gate/source voltage and Fig. 1.14 shows a typical set of drain characteristics.

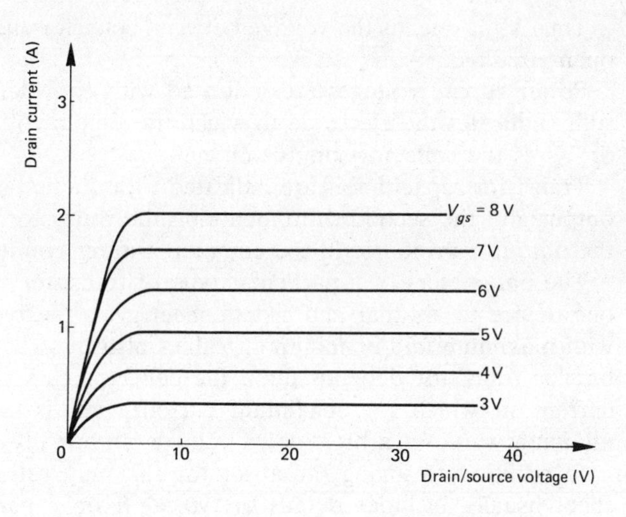

Fig. 1.14 Vmosfet drain characteristics.

The vmosfet possesses a number of advantages over the more usual horizontal channel fets. These are briefly as follows:

1 The current density in the device can be much greater. There are two contributory factors to this: firstly the channel length, which is determined by the diffusion depth, can be precisely controlled and this permits the use of channels with a large width/length ratio, and, secondly, two parallel channels are formed.
2 The self-capacitance of the device is less.
3 The saturation resistance is smaller.
4 Partly because of **1** and partly because the drain can be connected to a heat sink, the maximum permissible power dissipation within the device is much larger. In addition the vmosfet shares the advantages of the conventional mosfet over the bipolar transistor [EIII].

Data Sheets

The selection of a transistor for a particular application can best be made by consulting the data sheets provided by the manufacturer. A great deal of information about a device is provided by a data sheet and to make full use of it is essential to understand the meanings of the various symbols which are employed.

The system of symbols employed uses the usual symbols *I*, *V*, *P*, etc, to represent current, voltage, power, and so on. Each symbol has a number of suffixes which are written in a definite order; thus

First subscript — denotes the electrode at which the current or voltage is measured,
Second subscript — denotes the reference electrode,
Third subscript — indicates whether the third electrode is open, or short-circuited, or connected via a specified resistance *R* to the reference electrode.

Thus V_{CBO} means the voltage between collector and base with the emitter open-circuited.

Power supply voltages are indicated with two identical suffixes. the first suffix indicates the electrode to which the supply voltage is applied, e.g. E_{cc} or V_{cc} is the collector supply voltage.

Transistor capacitances are indicated with the first suffix denoting input or output and the second suffix denoting the transistor connection, i.e. C_{oe} is the output capacitance in the common-emitter connection.

The parameters of a particular type of transistor vary considerably from one device to another and a data sheet gives the typical value, very often with maximum and/or minimum values also quoted. The current gain of a bipolar transistor depends upon the collector current and so the collector current at which the maximum current gain is obtained is given. The minimum value of a breakdown voltage is normally given.

In addition to giving the absolute maximum ratings of a device a data sheet usually includes details of typical *h* or *y* parameters and provides graphs of such things as output or drain characteristics, input and transfer characteristics, the variation of *h* or *y* parameters with collector or drain current or voltage, or with frequency.

Information may also be provided about the noise performance, the power dissipation derating and the high-frequency characteristics.

Shortened versions of transistor data are given in catalogues supplied by distributors and other retailers of semiconductors and Table 1.1 shows some examples of such bipolar transistor data.

Table 1.1 Bipolar transistor data

Type	*Material*	P_{max} (mW)	$I_{c(max)}$ (mA)	$V_{CEO(max)}$ (V)	$V_{CBO(max)}$ (V)	$V_{EBO(max)}$ (V)	*Typical* h_{FE} *at* I_c(mA)	*Typical* f_t(MHz)
AD 161	Ge npn	4000	3000	20	32	10	150 at 500	3.0
AD 162	Ge pnp	6000	3000	−20	−32	−10	150 at 500	1.5
BC 107	Si npn	300	100	45	50	6	240 at 2	300
BC 108	Si npn	300	100	20	30	5	240 at 2	300

Equivalent Circuits

An equivalent circuit is one that behaves electrically in exactly the same way as the device whose a.c. performance it represents. The equivalent circuit is only valid over a limited frequency range and provided the device is, or can reasonably be supposed to be, linear in its operation. The values of the components of the equivalent circuit are only valid at the specified collector or drain current and frequency, and provided the signal amplitude is small.

Bipolar Transistor Equivalent Circuits

Several possible equivalent circuits for the bipolar transistor can be drawn but only a few are still used. For audio-frequency work, the *h-parameter* equivalent circuit is generally employed although the *hybrid-π* circuit is also sometimes used. At higher frequencies the hybrid-π and the y-parameter equivalent circuits are available, while for very high frequency work s-parameters provide the greatest accuracy. In this chapter, the h-parameter, the hybrid-π and the y-parameter equivalent circuits will be discussed, and a discussion of the s-parameter approach to circuit calculation will be left for a subsequent volume.

h-parameters

The performance of a common-emitter transistor may be specified in terms of a general four-terminal network that is itself defined in terms of the

ig. 1.15 Representa-ion of a transistor as "black box"

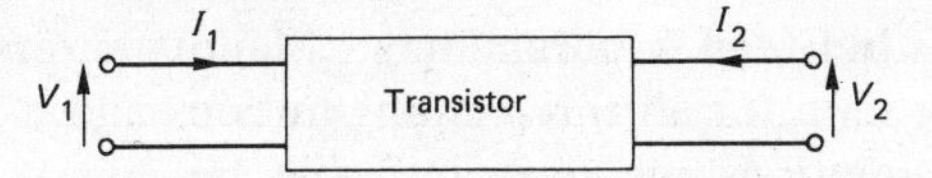

currents and voltages existing at its input and output terminals (Fig. 1.15). If the input current I_b and the output voltage V_{ce} are taken as the independent variables, then the output current I_c and the input voltage V_{be} are given by these equations:

$$V_{be} = h_{ie}I_b + h_{re}V_{ce} \tag{1.8}$$

$$I_c = h_{fe}I_b + h_{oe}V_{ce} \tag{1.9}$$

Equation (1.8) states that the input voltage V_{be} is equal to a parameter h_{ie} times the input current I_b plus another parameter h_{re} times the output voltage V_{ce}. The right-hand side of the equation must have the dimensions of voltage, and hence h_{ie} must be an impedance and h_{re} must be dimensionless. Similarly, the right-hand side of equation (1.9) must have the dimensions of a current; therefore h_{fe} is dimensionless and h_{oe} is an admittance.

The a.c. equivalent circuit that is suggested by equations (1.8) and (1.9) is shown in Fig. 1.16; the circuit is equally valid for the common-base and common-collector configurations, the configuration concerned being indicated by a different second suffix, b or c. The h-parameters of a transistor are defined in Table 1.2.

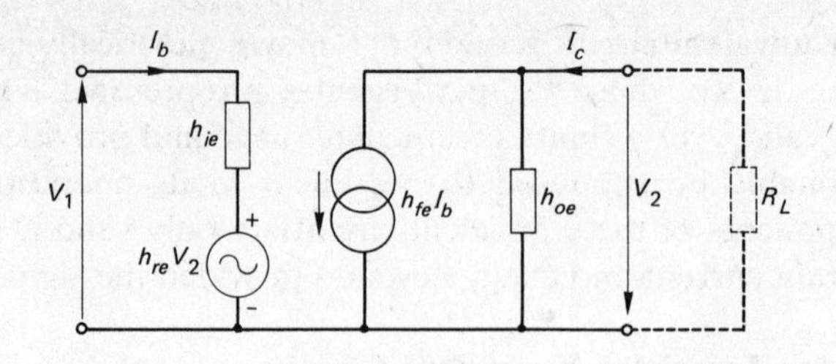

Fig. 1.16 *h*-parameter equivalent circuit of a transistor.

Table 1.2 *h*-parameters of a transistor

h-parameter Common base	Common emitter	Common collector	*Definition*
$h_{ib}=\dfrac{V_{eb}}{I_e}$	$h_{ie}=\dfrac{V_{be}}{I_b}$	$h_{ic}=\dfrac{V_{bc}}{I_b}$	Input impedance with output terminals short-circuited (ohms)
$h_{rb}=\dfrac{V_{eb}}{V_{cb}}$	$h_{re}=\dfrac{V_{be}}{V_{ce}}$	$h_{rc}=\dfrac{V_{bc}}{V_{ec}}$	Reverse voltage ratio with input terminals open-circuited (dimensionless)
$h_{fb}=\dfrac{I_c}{I_e}$	$h_{fe}=\dfrac{I_c}{I_b}$	$h_{fc}=\dfrac{I_e}{I_b}$	Forward current gain with output terminals short-circuited (dimensionless)
$h_{ob}=\dfrac{I_c}{V_{cb}}$	$h_{oe}=\dfrac{I_c}{V_{ce}}$	$h_{oc}=\dfrac{I_e}{V_{ec}}$	Output admittance with input terminals open-circuited (siemens)

Relationships between *h*-parameters Manufacturers do not normally publish transistor data for all three configurations and it is sometimes necessary to know the relationships given in Table 1.3.

Table 1.3 *h*-parameter relationships

Common base	*Common emitter*	*Common collector*
h_{ib}	$h_{ie}=\dfrac{h_{ib}}{1+h_{fb}}$	$h_{ic}=\dfrac{h_{ib}}{1+h_{fb}}$
h_{rb}	$h_{re}=\dfrac{h_{ib}h_{ob}}{1+h_{fb}}-h_{rb}$	$h_{rc}\approx 1$
h_{fb}	$h_{fe}=\dfrac{-h_{fb}}{1+h_{fb}}$	$h_{fc}=\dfrac{-1}{1+h_{fb}}$
h_{ob}	$h_{oe}=\dfrac{h_{ob}}{1+h_{fb}}$	$h_{oc}=\dfrac{h_{ob}}{1+h_{fb}}$

The *h*-parameters of a transistor are not constant but vary with change both in temperature and in collector current. Fig. 1.17 shows how the four *h*-parameters may vary for a typical transistor. Typical values for the four *h*-parameters are

$$h_{ie}=1000\ \Omega \qquad h_{re}=3\times10^{-4} \qquad h_{oe}=300\times10^{-6}\ \text{S} \qquad h_{fe}=250$$

Fig. 1.17 Variation of transistor *h*-parameters with (*a*) collector current, (*b*) temperature

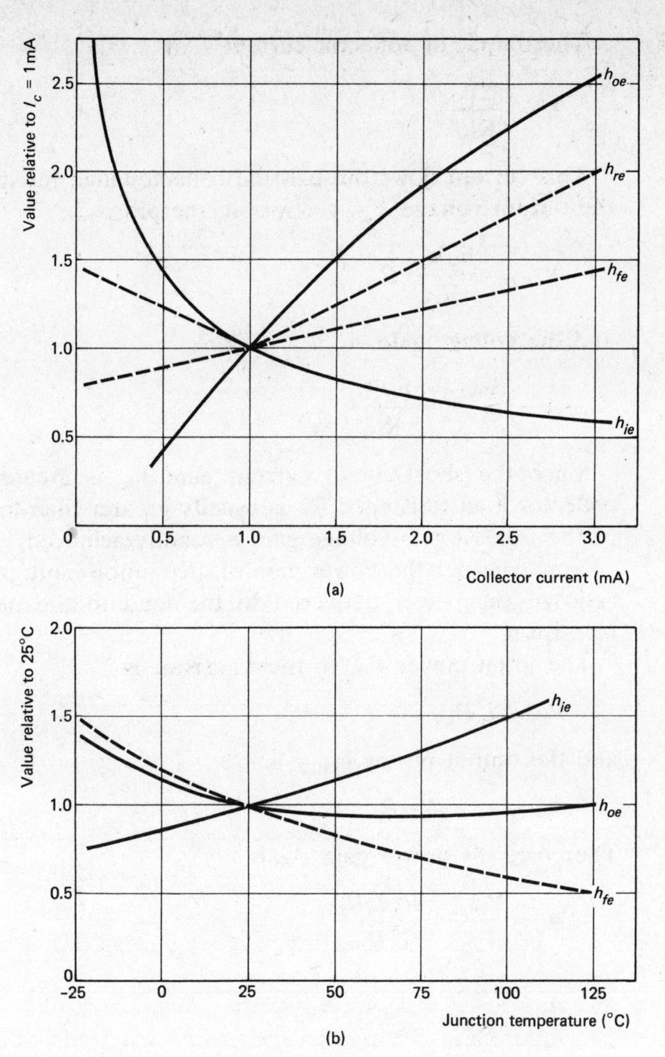

Voltage Gain and Power Gain of a Transistor

To obtain expressions for the voltage gain and the power gain of a transistor connected with common emitter: let the r.m.s. voltage and internal resistance of the voltage source applied to the input terminals of the transistor be E_S volts and R_S ohms respectively, and let the input resistance of the transistor be R_{IN} ohms.

Then, the input current to the transistor, I_b, is

$$I_b = \frac{E_S}{R_s + R_{IN}}$$

and the voltage V_{IN} appearing across the transistor input terminals is

$$V_{IN} = I_b R_{IN} \quad \text{or} \quad I_b = \frac{V_{IN}}{R_{IN}}$$

The output or collector current I_c is

$$I_c = \frac{h_{fe} V_{IN}}{R_{IN}}$$

This current flows through the collector load resistance R_L and develops the output voltage V_{OUT} across it, therefore

$$V_{OUT} = \frac{h_{fe} V_{IN}}{R_{IN}} R_L$$

and the **voltage gain** A_v is

$$A_v = \frac{V_{OUT}}{V_{IN}} = \frac{h_{fe} R_L}{R_{IN}}$$

Since the short-circuit current gain h_{fe} is greater than unity and the collector load resistance R_L is usually greater than the input resistance R_{IN} of the transistor, a voltage gain is readily achieved.

Now consider the power gain of a common-emitter transistor. This is the ratio of the power delivered to the load to the power delivered to the transistor.

The input power P_{IN} to the transistor is

$$P_{IN} = I_b^2 R_{IN}$$

and the output power P_{OUT} is

$$P_{OUT} = (h_{fe} I_b)^2 R_L$$

Therefore the **power gain** A_p is

$$A_p = \frac{P_{OUT}}{P_{IN}} = \frac{h_{fe}^2 I_b^2 R_L}{I_b^2 R_{IN}}$$

$$A_p = \frac{h_{fe}^2 R_L}{R_{IN}} = A_v A_i$$

Again, since R_L is greater than R_{IN} a power gain is possible.

Example A transistor is connected with common-emitter in a circuit and has a collector load resistance of 2000 Ω. The short-circuit current gain of the transistor is 100 and its input resistance is 1000 Ω. Calculate the voltage and power gains of the transistor.

Solution

From the above equation

$$\text{Voltage gain} = \frac{h_{fe} R_L}{R_{IN}} = \frac{100 \times 2000}{1000} = 200 \quad (Ans)$$

Also

$$\text{Power gain} = \frac{h_{fe}^2 R_L}{R_{IN}} = 200 \times 100 = 20\,000 \quad (Ans)$$

Determination of Gain and Input/Output Resistance using *h*-parameters

The voltage gain A_v and power gain A_p of a single-stage transistor amplifier are easily determined if the current gain A_i and input resistance R_{IN} are known since, as in the previous section,

$$A_v = A_i R_L / R_{IN} \quad \text{and} \quad A_p = A_v A_i$$

The input resistance of a bipolar transistor is the ratio of the input voltage to the input current, i.e.

$$R_{IN} = V_{be}/I_b \ \Omega$$

Now an increase in the collector current will produce a larger voltage drop across the load resistance R_L and so the collector/emitter voltage will fall. Thus $V_{ce} = -I_c R_L$.

Substituting into equation (1.9),

$$I_c = h_{fe} I_b - h_{oe} I_c R_L$$

$$I_c(1 + h_{oe} R_L) = h_{fe} I_b$$

$$I_c = \frac{h_{fe} I_b}{1 + h_{oe} R_L}$$

so that

$$A_i = I_c / I_b = \frac{h_{fe}}{1 + h_{oe} R_L} \tag{1.10}$$

Equation (1.8) can be written as

$$V_{be} = h_{ie} I_b - h_{re} I_c R_L$$

$$= h_{ie} I_b - \frac{h_{re} h_{fe} R_L I_b}{1 + h_{oe} R_L}$$

So that the input resistance is

$$R_{IN} = V_{be} / I_b = h_{ie} - \frac{h_{re} h_{fe} R_L}{1 + h_{oe} R_L} \tag{1.11}$$

An expression for the **voltage gain** can be obtained by substituting both (1.10) and (1.11) into the expression

$$A_v = A_i R_L / R_{IN}$$

This gives

$$A_v = \frac{h_{fe} R_L}{1 + h_{oe} R_L} \times \frac{1 + h_{oe} R_L}{h_{ie} + h_{ie} h_{oe} R_L - h_{re} h_{fe} R_L}$$

$$= \frac{h_{fe} R_L}{h_{ie} + h_{ie} h_{oe} R_L - h_{re} h_{fe} R_L} \tag{1.12}$$

Very often, h_{re} is negligibly small, and then equations (1.11) and (1.12) reduce to

$$R_{IN} = h_{ie} \qquad (1.11A)$$

$$A_v = \frac{h_{fe}R_L}{h_{ie}(1 + h_{oe}R_L)} \qquad (1.12A)$$

Example 1.5

A transistor connected with common emitter has $h_{ie} = 800\ \Omega$, $h_{oe} = 50 \times 10^{-6}$ S and $h_{fe} = 55$. Calculate the voltage and power gains obtained with a collector load resistor of 200 Ω. Find also the percentage error if h_{oe} is neglected.

Solution

$$A_i = 55/(1 + 50 \times 10^{-6} \times 2 \times 10^3) = 50$$

$$R_{IN} = 800\ \Omega$$

Therefore,

$$A_v = 50 \times 2 \times 10^3/800 = 125 \quad (Ans)$$

$$A_p = 125 \times 50 = 6250 \quad (Ans)$$

If h_{oe} is neglected, $A_i = h_{fe} = 55$ and

$$A_v = 55 \times 2 \times 10^3/800 = 137.5$$

$$A_p = 137.5 \times 55 = 7562.5$$

The percentage error in the calculated voltage gain is

$$\frac{137.5 - 125}{125} \times 100 = +10\% \quad (Ans)$$

and the percentage error in the power gain is

$$\frac{7562.5 - 6250}{6250} \times 100 = +21\% \quad (Ans)$$

These errors may seem rather large but they are of the same order as the probable errors caused by resistor tolerances (±5% or ±10%) and also the transistor parameters have wide tolerances.

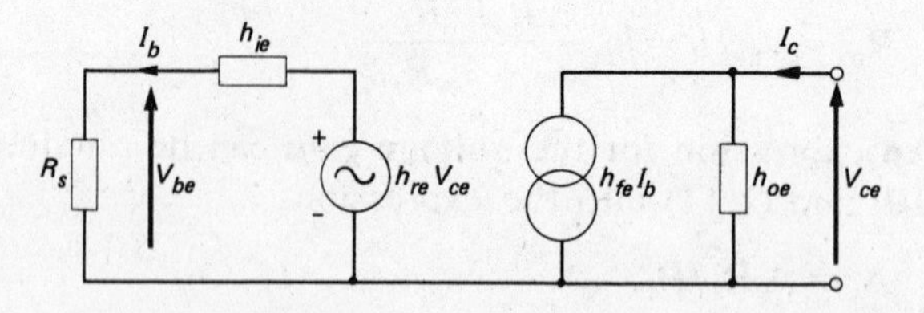

Fig. 1.18 Calculation of R_{OUT}.

The **output resistance** R_{OUT} is the ratio V_{ce}/I_c with zero input voltage. From Fig. 1.18

$$V_{be} = -I_bR_S$$

Therefore, equation (1.8) becomes

$$-I_bR_S = h_{ie}I_b + h_{re}V_{ce} \quad \text{and} \quad I_b = \frac{-h_{re}V_{ce}}{h_{ie} + R_S}$$

Then, equation (1.9) becomes

$$I_c = \frac{-h_{fe}h_{re}V_{ce}}{h_{ie}+R_S} + h_{oe}V_{ce}$$

$$R_{OUT} = V_{ce}/I_c = \frac{h_{ie}+R_S}{h_{oe}h_{ie}+h_{oe}R_S - h_{fe}h_{re}} \qquad (1.13)$$

Note that the output resistance of a bipolar transistor depends upon the value of the source resistance. However, if h_{re} is negligibly small,

$$I_c = h_{oe}V_{ce} \quad \text{and} \quad R_{OUT} = \frac{1}{h_{oe}}$$

The h-parameters are really the slopes of the various transistor characteristic curves which can be drawn. Thus, h_{oe} is the slope $\delta I_c/\delta V_{ce}$ of an output characteristic curve, h_{ie} is the slope $\delta V_{be}/\delta I_b$ of an input characteristic curve. Similarly h_{fe} and h_{re} are the slopes of the forward current transfer I_c/I_b and the reverse voltage transfer V_{be}/V_{ce} curves of the transistor. Most often, though, h_{fe} would be determined from the output characteristics [EII].

Although the h-parameters can be determined from the static characteristics of the transistor it is usually more convenient to measure them directly. Simple circuits that can be used for this purpose are given in Fig. 1.19*a* and *b*.

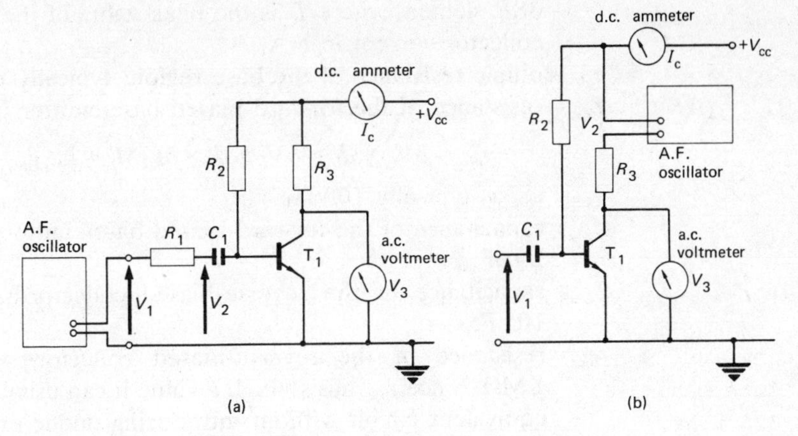

Fig. 1.19 Measurement of *h*-parameters (*a*) h_{ie} and h_{fe} (*b*) h_{re} and h_{oe}.

To measure h_{fe} and h_{ie}, use Fig. 1.19*a*, set I_c to the required value and then measure V_1, V_2 and V_3. Then

$$I_b = (V_1 - V_2)/R_1 \quad \text{and} \quad I_c = V_3/R_3$$

$$h_{fe} = I_c/I_b \qquad h_{ie} = V_2/I_b$$

h_{oe} and h_{re} are measured using Fig. 1.19*b*. Measure V_1, V_2 and V_3. Then

$$I_c = V_3/R_3 \qquad h_{oe} = I_c/V_2 \qquad h_{re} = V_1/V_2$$

In both circuits R_3 must be only a 100 Ω or so, so that the output circuit can be considered to be approximately a short circuit to a.c.

Hybrid-π Equivalent Circuit

The h-parameter equivalent circuit is normally only used at audio frequencies where the self-capacitances of the transistor have negligible effect on the performance of the device. At higher frequencies the effects of the transistor capacitances cannot be ignored and so these capacitances must appear in any equivalent circuit of the transistor. One circuit, which includes the transistor capacitances in a way which takes account of the manner in which the current gain of the transistor varies with frequency, is known as the **hybrid-π circuit** and is shown in Fig. 1.20. The point labelled b' is supposed to be in the middle of the base region. The meanings of each of the components shown is as follows:

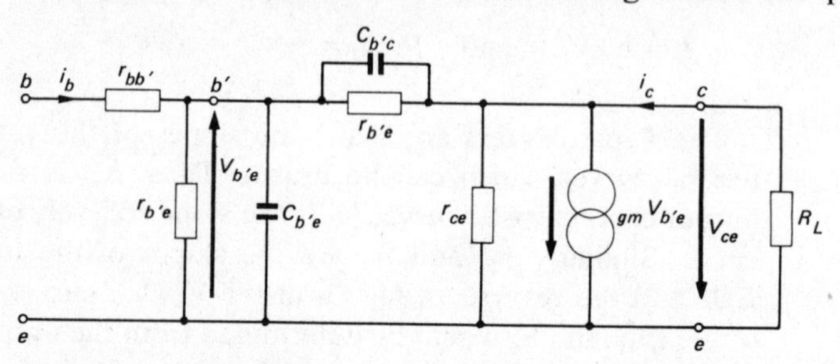

Fig. 1.20 Hybrid-π equivalent circuit of a transistor.

g_m = mutual conductance $\delta I_c/\delta V_{be}$ of the transistor. At 25°C, g_m is equal to $38I_c$ siemen, where I_c is the peak value of the a.c. component of the collector current in mA.

$r_{bb'}$ = ohmic resistance of the base region: typically $r_{bb'} = 200\ \Omega$.

$r_{b'e}$ = resistance of the forward-biased base/emitter junction

$$r_{b'e} = \delta V_{b'e}/\delta I_b = \delta V_{b'e}/\delta I_c \times \delta I_c/\delta I_b = h_{fe}/g_m$$

$r_{b'e}$ is typically 1000 Ω.

$C_{b'e}$ = capacitance of the forward-biased base/emitter junction, and is about 20 pF.

$C_{b'c}$ = capacitance of the reverse-biased collector/base junction, typically 10 pF.

$r_{b'c}$ = resistance of the reverse-biased collector/base junction, typically 1 MΩ. Since $r_{b'c}$ has so high a value it can usually be omitted from the equivalent circuit without introducing undue error.

r_{ce} = output resistance of the transistor when the base/emitter terminals are short-circuited and is typically about 40 kΩ.

It is always possible to convert from one equivalent circuit to another and Table 1.4 gives the relationships in a h-parameter to hybrid-π conversion.

Table 1.4

Equivalent circuit	*Parameter*			
h	h_{ie}	h_{re}	h_{fe}	h_{oe}
hybrid-π	$r_{bb'} + r_{b'e}$	$r_{b'e}/r_{b'c}$	$g_m r_{b'e}$	$1/r_{ce}$

Calculation of Gain Using the Hybrid-π Equivalent Circuit

At low frequencies where the capacitances $C_{b'e}$ and $C_{b'c}$ can be neglected

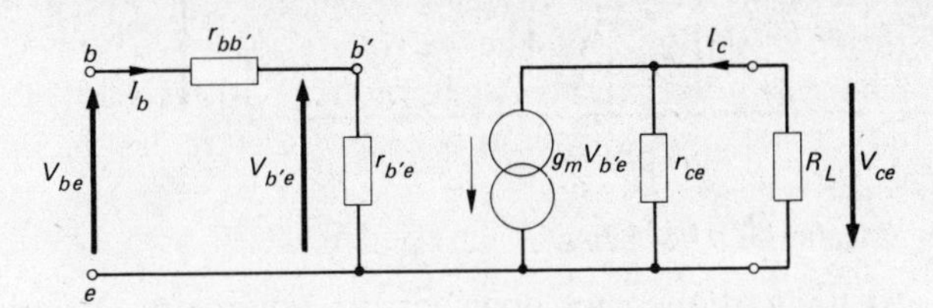

Fig. 1.21 Simplified hybrid-π equivalent circuit of a transistor.

and removed from the equivalent circuit (Fig. 1.21), a voltage V_{be} applied between the base and emitter produces

$$V_{b'e} = V_{be} r_{b'e}/(r_{bb'} + r_{b'e})$$

If the source resistance R_S is taken into account (i.e. add to $r_{bb'}$)

$$V_{b'e} = \frac{E_S r_{b'e}}{R_S + r_{bb'} + r_{b'e}}$$

The output voltage V_{ce} is

$$V_{ce} = \frac{g_m V_{be} r_{b'e}}{r_{bb'} + r_{b'e}} \times \frac{R_L r_{ce}}{R_L + r_{ce}}$$

Usually, $R_L \ll r_{ce}$ and then the voltage gain A_v is

$$A_v = V_{ce}/V_{be} = \frac{g_m r_{b'e} R_L}{r_{bb'} + r_{b'e}} \simeq g_m R_L \tag{1.14}$$

At high frequencies the reactances of $C_{b'e}$ and $C_{b'c}$ are small enough to affect the current and voltage gains of the transistor. The effect of the capacitance $C_{b'c}$ linking the input and output circuits of the transistor is to considerably increase the input capacitance of the device.

The current i flowing through $C_{b'c}$ is

$$\begin{aligned} i &= j\omega C_{b'c}(V_{b'e} - V_{ce}) \\ &= j\omega C_{b'c}(V_{b'e} - [-g_m R_L V_{b'e}]) \\ &= j\omega C_{b'c} V_{b'e}(1 + g_m R_L) \end{aligned}$$

Thus, the admittance seen looking into the left-hand side of $C_{b'c}$ is

$$Y_{IN} = i/V_{b'e} = j\omega C_{b'c}(1 + g_m R_L)$$

The effective input capacitance C_{IN} of the transistor is the sum of $C_{b'e}$ and the capacitance represented by Y_{IN}, i.e.

$$C_{IN} = C_{b'e} + C_{b'c}(1 + g_m R_L) \tag{1.15}$$

The increase in the input capacitance brought about in this way is known as the **Miller Effect**. The simplified high-frequency equivalent circuit of the transistor is shown by Fig. 1.22. From this circuit the low-frequency **voltage gain** A_v is unchanged at

Fig. 1.22 High-frequency hybrid-π equivalent circuit.

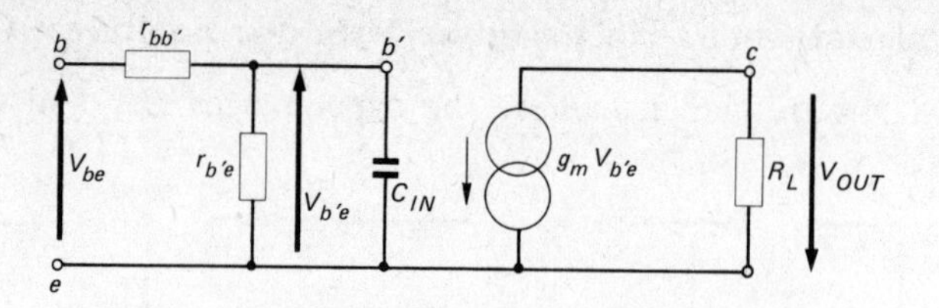

$g_m r_{b'e} R_L/(r_{bb'} + r_{b'e})$

At high frequencies however the voltage $V_{b'e}$ is given by

$$V_{b'e} = \frac{V_{be} r_{b'e}/(1 + j\omega C_{IN} r_{b'e})}{r_{bb'} + r_{b'e}/(1 + j\omega C_{IN} r_{b'e})}$$

$$= \frac{V_{be} r_{b'e}}{r_{bb'} + r_{b'e} + j\omega C_{IN} r_{b'e} r_{bb'}} \quad (1.16)$$

$$V_{ce} = g_m V_{b'e} R_L$$

Hence

$$A_{v(HF)} = V_{ce}/V_{be} = \frac{g_m r_{b'e} R_L}{r_{bb'} + r_{b'e} + j\omega C_{IN} r_{b'e} r_{bb'}} \quad (1.17)$$

Equation (1.17) can be rewritten as

$$A_{v(HF)} = \frac{g_m r_{b'e} R_L}{r_{bb'} + r_{b'e}} \times \frac{1}{1 + \dfrac{j\omega C_{IN} r_{b'e} r_{bb'}}{r_{bb'} + r_{b'e}}}$$

or

$$A_{v(HF)} = \frac{A_{v(MF)}}{1 + j\omega\tau} \quad (1.18)$$

where $\tau = C_{IN} r_{b'e} r_{bb'}/(r_{bb'} + r_{b'e})$.

It is clear from equation (1.18) that the high-frequency voltage gain of a transistor falls with increase in frequency and has fallen by 3 dB from its low-frequency value at the frequency at which $\omega\tau$ is unity.

Example 1.6

A common-emitter amplifier uses a bipolar transistor having the following data:

$r_{bb'} = 80\ \Omega$ $\quad C_{b'e} = 100\ \text{pF}$ $\quad r_{b'e} = 1200\ \Omega$,

$r_{b'c} = 2.5\ \text{M}\Omega$ $\quad C_{b'c} = 2\ \text{pF}$ $\quad r_{ce} = 60\ \text{k}\Omega$

If $g_m = 38$ mS determine the value of purely resistive load for which the circuit has a 3 dB bandwidth of 3 MHz.

Solution From equation (1.18)

$$\left|\frac{A_{v(HF)}}{A_{v(MF)}}\right| = \frac{1}{\sqrt{2}} = \frac{1}{\sqrt{[1 + \omega_{3\text{dB}}^2 \tau^2]}}$$

or $\tau = 1/\omega_{3\text{dB}}$

Hence,

$$\frac{C_{IN}r_{b'e}r_{bb'}}{r_{bb'}+r_{b'e}}=\frac{1}{2\pi\times3\times10^{6}}=53.052\times10^{-9}$$

$$C_{IN}=\frac{53.052\times10^{-9}\times(80+1200)}{80\times1200}=707.36\text{ pF}$$

Therefore

$$707.36\times10^{-12}=100\times10^{-12}+2\times10^{-12}(1+38\times10^{-3}R_L)$$

$$\frac{607.36\times10^{-12}}{2\times10^{-12}}=1+38\times10^{-3}R_L$$

$$R_L=\frac{302.68}{38\times10^{-3}}=7965\ \Omega\quad(Ans)$$

High-frequency Current Gain of a Bipolar Transistor

The short-circuit current gain of a bipolar transistor falls at the higher frequencies because of its inherent self-capacitances. The meanings of the terms *cut-off* and *transition* frequencies have already been explained (page 3) and the hybrid-π equivalent circuit will now be used to derive expressions for these frequencies.

Consider the simplified equivalent circuit given in Fig. 1.22 again. To obtain an expression for the **short-circuit current gain** h_{fe} of the transistor, set R_L to zero. Then, from equation (1.15)

$$C_{IN}=C_{b'e}+C_{b'c} \tag{1.19}$$

Now

$$I_b=V_{b'e}\left[\frac{1}{r_{b'e}}+j\omega(C_{b'e}+C_{b'c})\right]\quad\text{and}\quad I_c=g_mV_{b'e}$$

Therefore,

$$h_{fe}=I_c/I_b=\frac{g_m}{\frac{1}{r_{b'e}}+j\omega(C_{b'e}+C_{b'c})}=\frac{g_mr_{b'e}}{1+j\omega r_{b'e}(C_{b'e}+C_{b'c})} \tag{1.20}$$

The transition frequency f_t is the frequency at which the magnitude of h_{fe} has fallen to unity. Hence,

$$1\simeq g_m/\omega_t(C_{b'e}+C_{b'c})$$

or $$f_t=\frac{g_m}{2\pi(C_{b'e}+C_{b'c})} \tag{1.21}$$

h_{fe} has fallen by 3 dB from its low-frequency value $h_{fe(LF)}$ at the *cut-off frequency* f_β.

$$\left|\frac{h_{fe}}{h_{fe(LF)}}\right|=\frac{1}{\sqrt{2}}=\frac{1}{\sqrt{[1+\omega_\beta^2(C_{b'e}+C_{b'c})^2r_{b'e}^2]}}$$

Thus, $1 = \omega_\beta(C_{b'e} + C_{b'c})r_{b'e}$ and

$$f_\beta = \frac{1}{2\pi(C_{b'e} + C_{b'c})r_{b'e}} \tag{1.22}$$

Multiplying equation (1.22) by the low-frequency value of h_{fe}, i.e. $g_m r_{b'e}$ gives

$$h_{fe(LF)} \times f_\beta = f_t$$

and this confirms that f_t is the short-circuit gain bandwidth product of the transistor and thus equation (1.3) can be used to find h_{fe} at any frequency.

y-parameter Equivalent Circuit

The use of the h-parameter equivalent circuit is restricted to the lower frequencies at which the effects of transistor capacitances can be neglected. At higher frequencies the hybrid-π circuit can be used but also available is the **y-parameter equivalent circuit**. There are two reasons for the use of y-parameters at higher frequencies:

1 The y-parameters are easily and directly measured.
2 They are well suited to nodal analysis.

Referring to Fig. 1.15 the a.c. performance of a transistor can be described by the equations

$$I_b = y_{ie}V_{be} + y_{re}V_{ce} \tag{1.23}$$

$$I_c = y_{fe}V_{be} + y_{oe}V_{ce} \tag{1.24}$$

It will be noticed that the input and output currents are expressed in terms of the input and output voltages; this means that all four parameters have the dimensions of *admittance*.

$y_{fe} = I_c/V_{be}$ with $V_{ce} = 0$ and is the short-circuit forward transfer admittance.
$y_{ie} = I_b/V_{be}$ with $V_{ce} = 0$ and is the short-circuit input admittance.
$y_{re} = I_b/V_{ce}$ with $V_{be} = 0$ and is the short-circuit reverse transfer admittance.
$y_{oe} = I_c/V_{ce}$ with $V_{be} = 0$ and is the short-circuit output admittance.

The a.c. equivalent circuit described by these equations is shown by Fig. 1.23.

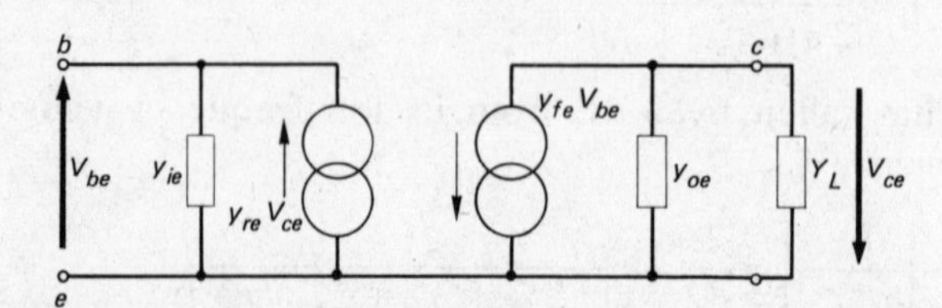

Fig. 1.23 *y*-parameter equivalent circuit of a transistor.

Calculation of Gain using y-parameters

The **voltage gain** $A_v = V_{ce}/V_{be}$ of a transistor is easily determined by writing

$$V_{ce} = -I_c R_L = -I_c/Y_L$$

and substituting into equation (1.24).

$$-V_{ce}Y_L = y_{fe}V_{be} + y_{oe}V_{ce}$$
$$-V_{ce}(Y_L + y_{oe}) = y_{fe}V_{be}$$

Therefore,

$$A_v = V_{ce}/V_{be} = -y_{fe}/(Y_L + y_{oe}) \qquad (1.25)$$

To determine an expression for the input admittance $Y_{IN} = I_b/V_{be}$, equation (1.24) is first rearranged to yield

$$-V_{ce}Y_L = y_{fe}V_{be} + y_{oe}V_{ce}$$
$$V_{ce} = -V_{be}y_{fe}/(Y_L + y_{oe})$$

Substituting into equation (1.23)

$$I_b = y_{ie}V_{be} - y_{re}y_{fe}V_{be}/(Y_L + y_{oe})$$

and so

$$Y_{IN} = I_b/V_{be} = y_{ie} - \frac{y_{re}y_{fe}}{y_L + y_{oe}} \qquad (1.26)$$

An expression for the **current gain** can be similarly determined but, alternatively, use can be made of the relationship

$$A_v = A_i R_L/R_{IN}$$

Rearranging,

$$A_i = A_v R_{IN}/R_L = A_v Y_L/Y_{IN}$$

Hence,

$$A_i = I_c/I_b = \frac{-y_{fe}Y_L}{(Y_L + y_{oe})\left(y_{ie} - \dfrac{y_{re}y_{fe}}{Y_L + y_{oe}}\right)} = \frac{-y_{fe}Y_L}{y_{ie}Y_L + y_{ie}y_{oe} - y_{re}y_{fe}} \qquad (1.27)$$

The output admittance of the transistor is $Y_{OUT} = I_C/V_{ce}$ siemen.
From Fig. 1.24 $V_{be} = -I_b R_S = -I_b/Y_S$
Substituting into equation (1.23)

$$-Y_S V_{be} = y_{ie}V_{be} + y_{re}V_{ce}$$

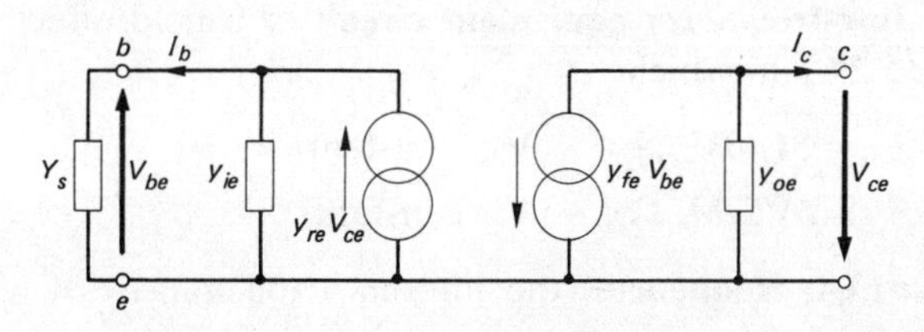

Fig. 1.24 Calculation of Y_{OUT}.

$$-V_{be}(Y_S + y_{ie}) = y_{re}V_{ce}$$

$$V_{be} = -\frac{y_{re}V_{ce}}{Y_S + y_{ie}}$$

Substituting for V_{be} into equation (1.24) gives

$$I_c = y_{oe}V_{ce} - \frac{y_{fe}y_{re}V_{ce}}{Y_S + y_{ie}}$$

and

$$Y_{OUT} = I_c/V_{ce} = y_{oe} - \frac{y_{fe}y_{re}}{Y_S + y_{ie}} \tag{1.28}$$

Example 1.7

A common-emitter transistor amplifier has a collector load resistor of 6.8 kΩ. The y-parameters of the transistor are

$$y_{ie} = 480 \times 10^{-6}\ \text{S} \qquad y_{fe} = 0.04\ \text{S}$$

$$y_{oe} = 40 \times 10^{-6}\ \text{S} \qquad y_{re} = -2 \times 10^{-6}\ \text{S}.$$

If the source resistance is 500 Ω calculate *a*) the input resistance *b*) the output resistance and *c*) the voltage gain.

Solution

a) From equation (1.26)

$$Y_{IN} = 480 \times 10^{-6} + \frac{0.04 \times 2 \times 10^{-6}}{\frac{1}{6.8 \times 10^3} + 40 \times 10^{-6}} = 9.077 \times 10^{-4}\ \text{S}$$

and $R_{IN} = 1/Y_{IN} = 1102\ \Omega$ (*Ans*)

b) From equation (1.28)

$$Y_{OUT} = 40 \times 10^{-6} + \frac{0.04 \times 2 \times 10^{-6}}{\frac{1}{500} + 480 \times 10^{-6}} = 7.226 \times 10^{-5}\ \text{S}$$

and $R_{OUT} = 1/Y_{OUT} = 13\,839\ \Omega$ (*Ans*)

c) From equation (1.25)

$$A_v = \frac{0.04}{\frac{1}{6.8 \times 10^3} + 40 \times 10^{-6}} = 213.8 \quad (Ans)$$

Field Effect Transistors

The **low-frequency equivalent circuit** of a field effect transistor is shown in Fig. 1.25*a* in which

$$g_m = \delta I_d/\delta V_{gs}\ \text{S} \qquad (V_{ds}\ \text{constant})$$

$$r_{ds} = \delta V_{ds}/\delta I_d\ \Omega \qquad (V_{gs}\ \text{constant})$$

At **high frequencies** the internal capacitances of a fet can no longer be

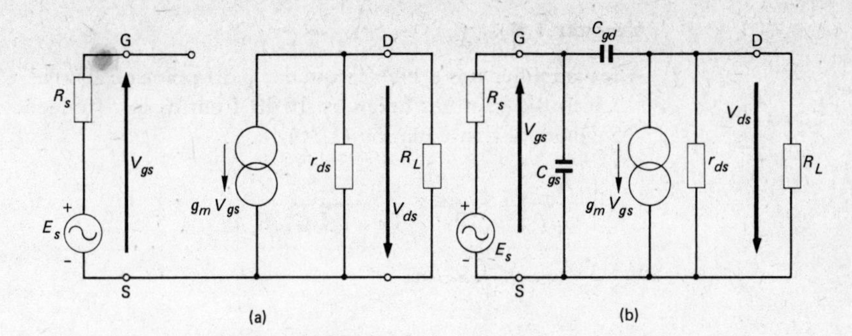

Fig. 1.25 Equivalent circuit of a fet: (*a*) low-frequency, (*b*) high-frequency.

neglected and the equivalent circuit that must be used is given by Fig. 1.25*b*. Typical values for these components are $C_{gs} = 4$ pF, $C_{gd} = 1$ pF, and $r_{ds} =$ 100 kΩ.

At low frequencies the voltage gain of a fet is simply

$$A_v = V_{ds}/V_{gs} = g_m V_{gs} R_L / V_{gs} = g_m R_L \quad (1.29)$$

(assuming that $r_{ds} \gg R_L$).

As the frequency is increased, the voltage gain V_{ds}/E_S falls because of the effect of the internal capacitances C_{gs} and C_{gd}. The input capacitance C_{IN} of the fet is

$$C_{IN} = C_{gs} + C_{gd}(1 + g_m R_L)$$

and hence

$$V_{gs} = \frac{E_S \times 1/j\omega C_{IN}}{R_S + 1/j\omega C_{IN}} = \frac{E_S}{1 + j\omega C_{IN} R_S}$$

and

$$V_{ds} = \frac{g_m R_L E_S}{1 + j\omega C_{IN} R_S}$$

so that the voltage gain is

$$A_v = V_{ds}/E_S = \frac{g_m R_L}{1 + j\omega C_{IN} R_S} \quad (1.30)$$

This analysis neglects the capacitance C_{ds} between the drain and the source but the error involved is small since the input time constant is predominant.

The voltage gain falls by 3 dB from its low-frequency value at the frequency f_{3dB} which makes $1 = \omega C_{IN} R_S$ or

$$f_{3dB} = 1/2\pi C_{IN} R_S \quad (1.31)$$

The gain-bandwidth product is

$$A_{v(LF)} \times f_{3dB} = \frac{g_m R_L}{2\pi C_{IN} R_S} \quad (1.32)$$

Example 1.8

A fet amplifier has a high-frequency 3 dB point of 30 kHz. Calculate the frequency at which the gain has fallen by 10 dB from its low-frequency value.

Solution From equation (1.30)

$$\left|\frac{A_v}{A_{v(LF)}}\right| = \frac{1}{\sqrt{10}} = \frac{1}{\sqrt{[1+\omega_{10dB}^2 C_{IN}^2 R_S^2]}}$$

and $3 = \omega_{10dB} C_{IN} R_S$ or

$$f_{10dB} = \frac{3}{2\pi C_{IN} R_S} = 3\times 30\text{ kHz} = 90\text{ kHz} \quad (Ans)$$

Bias Circuits

The **operating point** of a bipolar transistor must be chosen to satisfy one or more of the following factors

1 Transistors of the same type, even when made in the same batch, very often possess widely differing h_{FE} values, commonly over a 1–4 or even greater range.
2 The values of h_{FE}, I_{CEO} and V_{BE} are all temperature dependent.
3 The current gain depends upon the collector current and the maximum gain is obtained at a collector current specified by the manufacturer.
4 For maximum output voltage, the operating point must be at the middle of the d.c. load line so that the maximum possible swings of current and voltage can be obtained.
5 The input resistance is a function of the collector current and for a high impedance a low value of current is necessary.
6 The transition frequency f_t also depends upon the magnitude of the current collector.
7 As the collector current is reduced so the unwanted noise voltages generated within a transistor reduce in magnitude.

The selected operating point must be specified by applying a *bias* current to the transistor. An amplifier stage will be designed to have the chosen d.c. collector current assuming the *nominal* value of h_{FE} for the transistor. The **bias circuit** should operate to ensure that approximately the same collector current will flow if a transistor having either the maximum or the minimum h_{FE} should be used instead. This means that the bias circuit must also provide **d.c. stabilization** to keep the collector current at more or less the chosen value, even though h_{FE}, I_{CEO} and V_{BE} may vary considerably.

1 The most commonly employed bias circuit is the **potential divider** arrangement shown in Fig. 1.26. The circuit operates to provide d.c. stabilization in the following way. The base/emitter voltage V_{BE} of the transistor is equal to the voltage

$$V_B = V_{cc}R_2/(R_1+R_2)$$

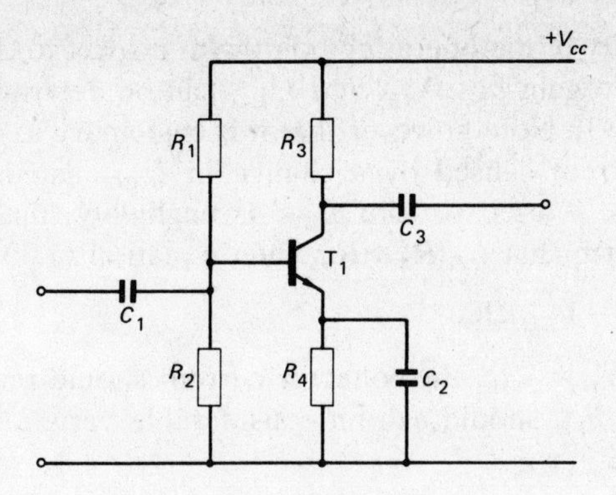

Fig. 1.26 Potential divider bias.

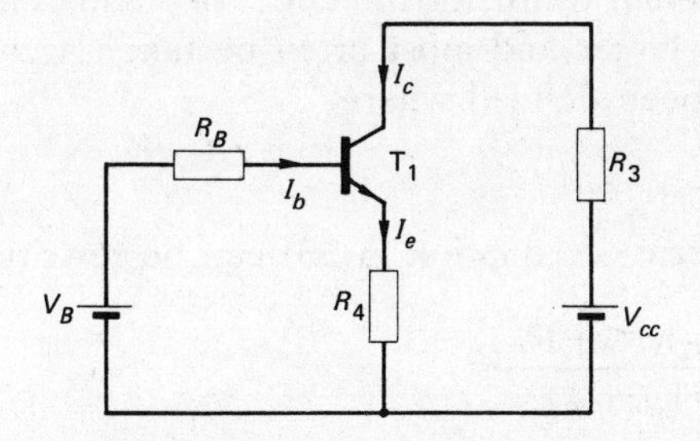

Fig. 1.27 Thevenin equivalent circuit of the bias circuit.

dropped across R_2 minus the voltage $V_E = I_E R_4$ across R_4. If the collector current should increase for *any* reason, the voltage across R_4 will rise and in so doing will reduce the forward bias voltage V_{BE} of T_1. This will reduce the collector current and so tend to oppose the initial increase in the current. Signal-frequency currents flowing in the emitter resistance will apply *negative feedback* to the circuit. If this is not required, an emitter decoupling capacitor must be provided.

For analytical purposes the circuit shown in Fig. 1.26 can be simplified by the application of Thevenin's theorem to the left of the base terminal of T_1. The Thevenin equivalent circuit thus obtained is shown in Fig. 1.27 in which

$$V_B = V_{cc} R_2/(R_1 + R_2) \quad \text{and} \quad R_B = R_1 R_2/(R_1 + R_2)$$

Hence, applying Kirchhoff's second law

$$V_B = I_b R_B + V_{BE} + I_E R_4$$

$$V_B - V_{BE} = I_b(R_B + R_4) + I_c R_4$$

But $I_c = h_{FE} I_b + I_{CEO}$, therefore

$$V_B - V_{BE} = \frac{(I_c - I_{CEO})}{h_{FE}}(R_B + R_4) + I_c R_4$$

$$= \frac{I_c}{h_{FE}}(R_B + R_4[1 + h_{FE}]) - \frac{I_{CEO}}{h_{FE}}(R_B + R_4)$$

$$I_c = \frac{(V_B - V_{BE})h_{FE}}{R_B + R_4(1 + h_{FE})} + \frac{I_{CEO}(R_B + R_4)}{R_B + R_4(1 + h_{FE})} \qquad (1.33)$$

The stability of the operating point with respect to changes in one or more of the current gain h_{FE}, V_{BE} and I_{CEO} can be determined by differentiating equation (1.33). Note however that it is customary to consider the change in collector current caused by a change in I_{CBO} and *not* I_{CEO}. Further, if a silicon device is used so that I_{CEO} is negligibly small and the component values are such that $h_{FE}R_4 \gg R_B$, then equation (1.33) reduces to

$$I_c = (V_B - V_{BE})/R_4$$

Provided $V_B \gg V_{BE}$, the collector current should remain sensibly constant even though h_{FE} should exhibit considerable variation.

2 For a germanium transistor the effect of changes in the collector leakage current is much larger and must often be taken account of, and a **stability function** S has been defined where

$$S = \delta I_c / \delta I_{CBO} \tag{1.34}$$

The second term of equation (1.33) can be rewritten as

$$\frac{I_{CBO}(1+h_{FE})(R_4+R_B)}{R_B+R_4(1+h_{FE})}$$

and differentiating this with respect to I_{CBO} gives

$$S = \frac{(1+h_{FE})(R_4+R_B)}{R_B+R_4(1+h_{FE})} \tag{1.35}$$

Ideally S should be zero but, in practice, it is made as small as possible. For this, R_4 should be as large, and R_B as small, as possible. Unfortunately increasing R_4 also increases the d.c. voltage developed across the emitter resistance and thereby limits the maximum possible output voltage. Also, reducing the value of R_B increases the shunting effect that the bias resistors have upon the signal path. As a result the design of a bias circuit must be a compromise between conflicting factors.

Example 1.9

A single-stage transistor amplifier of the type shown in Fig. 1.26 has the following component values: $R_1 = 12\ \text{k}\Omega$, $R_2 = 6.8\ \text{k}\Omega$, $h_{FE} = 140$, $R_4 = 1000\ \Omega$. Calculate its stability factor $\delta I_c/\delta I_{CBO}$ and determine the change in its collector current when I_{CBO} changes by 20 nA. Assume all other parameters are constant in value.

Solution The total base resistance is

$$R_B = 12 \times 10^3 \times 6800/(12 \times 10^3 + 6800) = 4340\ \Omega$$

From equation (1.35)

$$S = \frac{141(4340+1000)}{4340+141\times 1000} = 5.18 \quad (Ans)$$

Change in collector current $= 5.18 \times 20 = 104$ nA (*Ans*)

3 It is possible to similarly derive equations for the stability obtained when h_{FE} varies, but the main cause of h_{FE} variations is transistor substitution and then the change is usually too large for differentiation to yield accurate results. An alternative method is best used which is also valid when two or more of the three potential variables change simultaneously.

Suppose that both h_{FE} and I_{CBO} vary but V_{BE} remains constant. From Fig. 1.27,

$$V_B = I_b R_B + V_{BE} + (I_c + I_b) R_4$$

$$I_b = \frac{V_B - V_{BE} - I_c R_4}{R_B + R_4}$$

Suppose that h_{FE} varies from h_{FE1} to h_{FE2} while I_{CEO} changes from I_{CEO1} to I_{CEO2} causing I_c to alter from I_{c1} to I_{c2}. Then

$$I_{b1} = \frac{V_B - V_{BE} - I_{c1} R_4}{R_B + R_4}$$

and $$I_{b2} = \frac{V_B - V_{BE} - I_{c2} R_4}{R_B + R_4}$$

Therefore,

$$I_{b1} - I_{b2} = (I_{c2} - I_{c1}) R_4/(R_B + R_4)$$

or $$I_{c2} - I_{c1} = \frac{(I_{b1} - I_{b2})(R_B + R_4)}{R_4} \quad (1.36)$$

Clearly the same result would be obtained if the changes in collector current were due only to a change in h_{FE}, probably the result of device replacement.

Example 1.10

A transistor has a current gain of $h_{FE} = 100$ and is used in a potential divider circuit in which $R_1 = 120\ \text{k}\Omega$ and $R_2 = 18\ \text{k}\Omega$. The collector current is 1.2 mA at an ambient temperature of 20°C. The collector leakage current I_{CBO} is 10 nA at 20°C and doubles in value for every 10°C rise in temperature. The current gain h_{FE} increases by 5% for every 10°C rise in temperature. Calculate the emitter resistance which will ensure that the collector current will not exceed 1.22 mA at an ambient temperature of 50°C. Assume V_{BE} is constant at 0.7 V.

Solution

At 20°C, $h_{FE1} = 100$, $I_{CEO1} = 101 \times 10\ \text{nA} = 1.01\ \mu\text{A}$, $I_{c1} = 1.2\ \text{mA}$.

At 50°C, $h_{FE2} \simeq 116$, $I_{CBO2} = 80\ \text{nA}$ so $I_{CEO2} = 80\ \text{nA} \times 117 = 9.36\ \mu\text{A}$.

Now $I_c = h_{FE} I_b + I_{CEO}$ so that

$$1.2 \times 10^{-3} = 100 I_{b1} + 1.01 \times 10^{-6} \quad \text{or} \quad I_{b1} = 11.99\ \mu\text{A}$$

$$1.22 \times 10^{-3} = 116 I_{b2} + 9.36 \times 10^{-6} \quad \text{or} \quad I_{b2} = 10.44\ \mu\text{A}$$

Also, $R_B = 120 \times 10^3 \times 18 \times 10^3/(120 + 18) \times 10^3 = 15.65\ \text{k}\Omega$.

From equation (1.36)

$$\frac{R_4}{R_4 + 15.65 \times 10^3} = \frac{(11.99 - 10.44) \times 10^{-6}}{1.22 \times 10^{-3} - 1.2 \times 10^{-3}} = 7.75 \times 10^{-2}$$

$R_4 = 1315\ \Omega$ (*Ans*)

This is the smallest value of R_4 which should be used and probably the nearest preferred value, i.e. 1500 Ω would be used.

Fig. 1.28

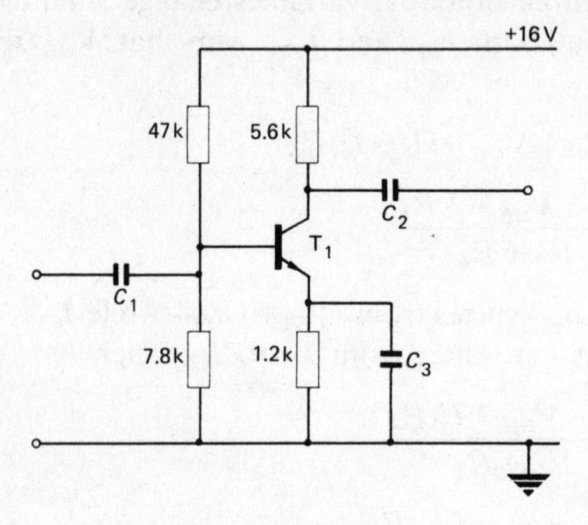

Example 1.11

For the circuit shown in Fig. 1.28 calculate the collector current and the collector/emitter voltage if $V_{BE} = 0.6$ V, $h_{FE} = 120$ and $I_{CBO} = 10$ nA.

Solution Applying Thevenin's theorem to the base circuit,

$$V_B = \frac{16 \times 7.8}{7.8 + 47} = 2.28 \text{ V}$$

and the total base resistance is

$$R_B = \frac{7.8 \times 47}{7.8 + 47} = 6.69 \text{ k}\Omega$$

Hence

$$2.28 = I_b \times 6.69 \times 10^3 + 0.6 + 1.2 \times 10^3 \times 121 I_b$$

$$I_b = \frac{1.68}{(6.69 + 1.2 \times 121) \times 10^3} = 11\ \mu\text{A}$$

Therefore,

$$I_c = 120 \times 11 \times 10^{-6} + 121 \times 10 \times 10^{-9} = 1.33 \text{ mA} \quad (Ans)$$

$$V_{CE} = 16 - 1.33 \times 10^{-3} \times 5.6 \times 10^3 - 1.33 \times 10^{-3} \times 1.2 \times 10^3 = 7.0 \text{ V} \quad (Ans)$$

Design of the Bias Circuit

In the design of a potential divider bias circuit there are a greater number of variables than there are equations and so some element of judgement is necessary. This means that there are a number of different ways in which a bias circuit can be designed. One method has been touched upon with Example 1.9 although no indication was given as to the origin of the values of the base circuit potential divider resistors.

Applying Kirchhoff's second law to the collector/emitter circuit of Fig. 1.26,

$$V_{cc} = V_{CE} + I_c R_3 + (I_c + I_b) R_4$$

$$V_{cc} - V_{CE} \simeq I_c (R_3 + R_4)$$

The operating point of the transistor is specified by the choice made for $V_{cc} - V_{CE}$. For maximum voltage output the operating point should lie at the middle of the d.c. load line and then $V_{CE} = V_{cc}/2$. Choosing this point,

$$V_{cc}/2 = I_c (R_3 + R_4)$$

The value of I_c can be determined by drawing a suitable d.c. load line on the output characteristics of the transistor. Then

$$R_3 + R_4 = V_{cc}/2I_c$$

The division of resistance between R_3 and R_4 must now be determined. The higher the value of R_4 the better will be the d.c. stabilization of the circuit but, on the other hand, the maximum possible output voltage will be reduced. A reasonable choice is to allow $V_{cc}/10$ to be developed across R_4, then $R_4 = V_{cc}/10I_c$ Ω. The base voltage of the transistor is now

$$V_B = V_{BE} + I_c R_4$$

The current flowing through R_2 must be several times larger than the base current I_b so that V_B remains sensibly constant as required for good d.c. stabilization. Choose I_{R2} as

$$10 I_b \simeq 10 I_c / h_{FE}$$

where h_{FE} is the nominal value of h_{FE}. Then

$$V_{cc} - V_B = (I_{R2} + I_b) R_1$$

$$R_1 = (V_{cc} - V_B)/(I_{R2} + I_b)$$

and $R_2 = V_B / I_{R2}$.

Example 1.12

Design a potential divider bias circuit for a transistor with $h_{FE} = 100$ and $V_{BE} = 0.7$ V if the collector current is to be 1.2 mA at an ambient temperature of 20°C and the collector supply voltage is 20 V.

Solution Choose the collector/emitter voltage to be $V_{cc}/2 = 10$ V. Then

$$R_3 + R_4 = 10/1.2 \times 10^{-3} = 8.33 \text{ k}\Omega$$

If $V_{cc}/10$ or 2 V are to be dropped across R_4 then

$$R_4 = 2/1.2 \times 10^{-3} = 1667\ \Omega = 1800\ \Omega \text{ preferred value}$$

Now $R_3 = 8333 - 1667 = 6667\ \Omega = 6800\ \Omega$ preferred value

The base voltage is equal to

$$0.7 + 1.2 \times 10^{-3} \times 1800 = 2.86 \text{ V}$$

and the base current is

$1.2 \times 10^{-3}/100 = 12\ \mu A$

The current in R_2 should be at least 10 times greater than the base current, say 120 μA. Hence

$$R_1 = (20 - 2.86)/132 \times 10^{-6}\ k\Omega = 129.8 = 120\ k\Omega \text{ preferred value}$$

Lastly,

$$R_2 = 2.86/120 \times 10^{-6} = 23.8\ k\Omega = 22\ k\Omega \text{ preferred value}$$

Notice that the values of R_1 and R_2 obtained are nearly the values used in Example 1.10. The emitter resistance of 1800 Ω is higher than minimum value necessary to keep the collector current within the limits quoted in that example; with $R_4 = 1800\ \Omega$ the collector current will only increase from 1.2 mA to 1.2154 mA when the quoted increase in h_{FE} and I_{CBO} occur.

Other methods of biasing bipolar transistors are also available; some of these have been discussed elsewhere [EIII]; others are used in conjunction with integrated circuits.

Integrated Circuit Bias Methods

The transistors within an integrated circuit must also be biased to a required operating point but the bias circuits used with discrete components are not used because

a) The emitter decoupling capacitance needed is much larger than can be provided in an i.c.
b) Transistors and diodes are cheaper to provide than resistors.
c) If all the bias elements are fabricated in the same chip they will have the same coefficient of temperature.

One method of biasing an integrated bipolar transistor is shown in Fig. 1.29. Transistor T_1 has its base and collector terminals connected directly together and it is therefore operated as a diode. The diode is connected in parallel with the base and emitter terminals of T_2 which is connected as an emitter follower. This connection ensures that the base/emitter voltages of the two *identical* transistors are equal and this means that they must conduct equal currents. Hence,

$$I_{c1} = I_{c2} = (V_{cc} - V_{BE})/(R_1 - I_{b1} - I_{b2})$$

or $\quad I_{c2} \simeq V_{cc}/R_1$

since I_{b1} and I_{b2} are equal

The magnitude of the collector current I_{c2} is controlled by the choice of V_{cc} and/or R_1. The disadvantage of this bias circuit is that the input signal is applied across a low input impedance.

An improved bias circuit is shown by Fig. 1.30. R_3 and R_4 are of equal value. T_1 and T_2 are also identical and so $I_{b1} = I_{b2}$. To determine I_{c2} apply Kirchhoff's law to the circuit:

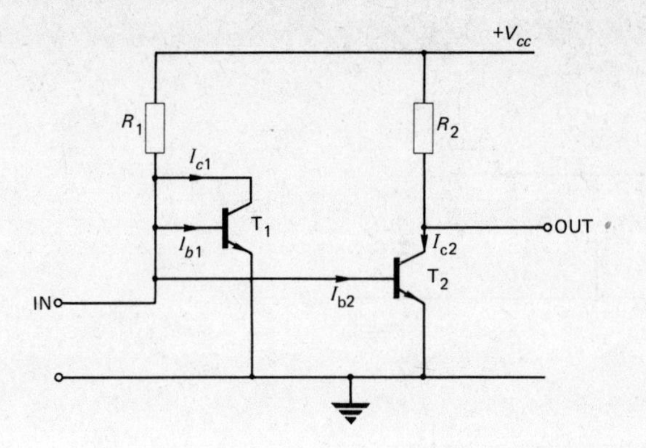

Fig. 1.29 Biasing an integrated circuit bipolar transistor.

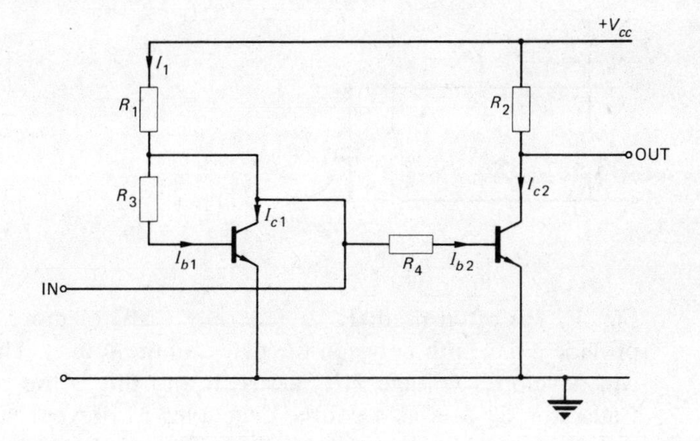

Fig. 1.30 An improved i.c. bias circuit.

$$V_{cc} = I_1 R_1 + V_{CE1}$$
$$= I_{c1} R_1 + I_b(2R_1 + R_3) + V_{BE1}$$
$$= I_{c1} R_1 + I_b(2R_1 + R_3) + V_{BE1}$$

$$I_{c1} = I_{c2} = \frac{V_{cc} - V_{BE1} - (2R_1 + R_3)I_b}{R_1} \tag{1.37a}$$

Since $V_{cc} \gg V_{BE1}$ and $V_{cc} \gg (2R_1 + R_3)I_b$

$$I_{c2} \simeq V_{cc}/R_1 \tag{1.37b}$$

which is, of course, the same expression as was obtained for the previous circuit. The chosen operating point is independent of temperature variations provided the transistors are closely matched, and this is easily arranged since they are both fabricated in the same chip.

Bias for Field Effect Transistors

The n-channel jfet and the n-channel depletion-mode mosfet are operated with the gate biased negatively with respect to the source, and the circuit of

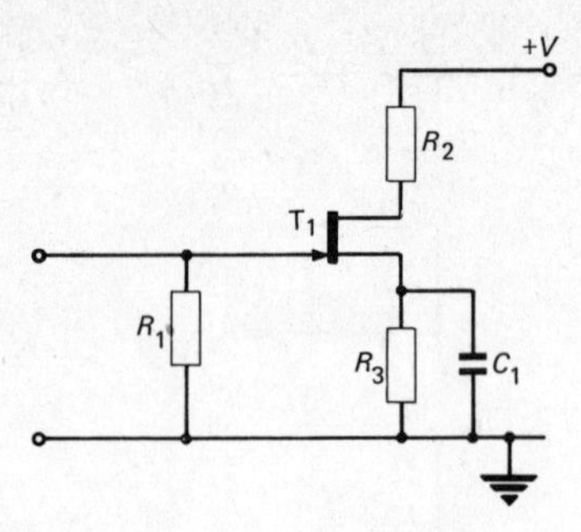

Fig. 1.31 Junction fet and depletion mode mosfet source bias.

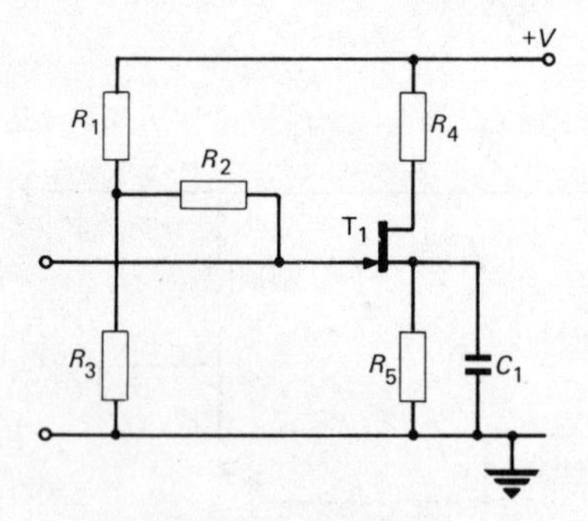

Fig. 1.32 Junction fet and depletion mode mosfet potential divider bias.

Fig. 1.31 is often used. R_1 is generally 1 MΩ or more and its function is to provide a d.c. path between the gate and the source. The drain current in R_3 will develop a voltage $I_d R_3$ across it and this is the required bias voltage. Capacitor C_1 acts as a source decoupler to prevent n.f.b. being applied to the circuit. This arrangement provides adequate d.c. stability for most small-signal stages. Devices of the same type are, just like bipolar transistors, subject to wide spreads in their parameters and it may often be necessary to use the more effective circuit of Fig. 1.32.

For a specified drain current I_d the required V_{gs} can be found from the drain characteristics or from equation (1.5) or equation (1.8).

Example 1.13

In Fig. 1.31, $R_2 = 10\ \text{k}\Omega$, $R_1 = 1\ \text{M}\Omega$, $R_3 = 560\ \Omega$, $V_{dd} = 24$ V, $I_{dss} = 2$ mA and $V_p = -2.0$ V. If I_d is to be 1 mA, calculate *a*) V_{gs} and *b*) V_{ds}.

Solution

a) From equation (1.5) $1 \times 10^{-3} = 2 \times 10^{-3}(1 + V_{gs}/2.0)^2$

$V_{gs} = -0.6\ \text{V}$ (*Ans*)

b) $V_{ds} = 24 - 1 \times 10^{-3}(10 \times 10^3 + 560) = 13.44\ \text{V}$ (*Ans*)

Fig. 1.32 is very similar to the bipolar transistor circuit of Fig. 1.26 but includes an extra resistor R_2. This resistor (not always fitted) is there to increase the input impedance of the circuit from

$$R_1R_3/(R_1 + R_3) \quad \text{to} \quad R_2 + R_1R_3/(R_1 + R_3)$$

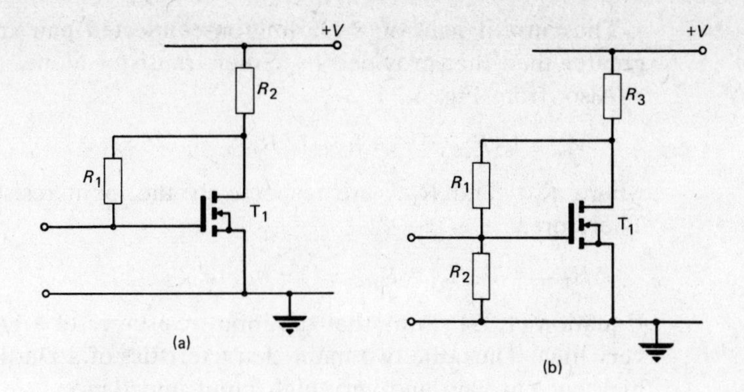

ig. 1.33 Enhancement mode mosfet bias.

or to approximately R_2 (since R_2 can be 1 MΩ or more).

An n-channel enhancement mosfet must be operated with its gate positive with respect to its source and Figs. 1.33*a* and *b* show two possible circuits. Fig. 1.33*a* is only suitable if the bias point $V_{gs} = V_{ds}$ is required.

he Darlington ircuit

The **Darlington** connection, illustrated by Fig. 1.34, has two transistors with their base and collector terminals commoned to produce an element which can often be used to replace a single transistor.

From Fig. 1.34,

$$I_{c1} = h_{fe1} I_{b1} \quad \text{and} \quad I_{e1} = I_{b2} = I_{b1}(1 + h_{fe1})$$

Also

$$I_{c2} = h_{fe2} I_{b2} = h_{fe2} I_{b1}(1 + h_{fe1})$$

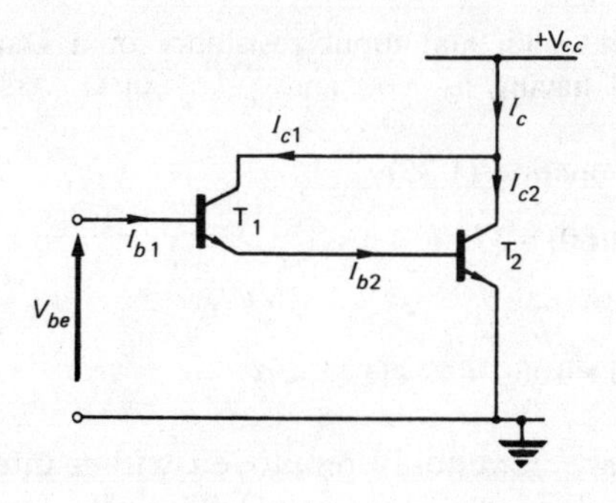

g. 1.34 Darlington onnection.

The *total* collector current I_c is the sum of the individual collector currents, i.e.

$$I_c = h_{fe1} I_{b1} + h_{fe2} I_{b1}(1 + h_{fe1})$$

and the short-circuit current gain is

$$A_i = I_c / I_{b1} = h_{fe1} + h_{fe2}(1 + h_{fe1}) \qquad (1.38)$$

The current gain of a Darlington-connected pair of transistors is much greater than that provided by either transistor alone.

Also, from Fig. 1.34,

$$V_{be} = I_{b1}R_{IN1} + (1 + h_{fe1})I_{b1}R_{IN2}$$

where R_{IN2} and R_{IN2} are respectively the input resistances of T_1 and T_2. Therefore,

$$R_{INT} = V_{be}/I_{b1} = R_{IN1} + (1 + h_{fe1})R_{IN2} \quad (1.39)$$

Equation (1.39) shows that the input resistance of a Darlington pair can be very high. Thus, the two main characteristics of a Darlington pair are a very high current gain and very high input impedance.

The emitter current of T_1 must always be larger than the collector leakage current of T_2 and, to ensure this, it is often necessary to connect an emitter resistor in the position shown in Fig. 1.35.

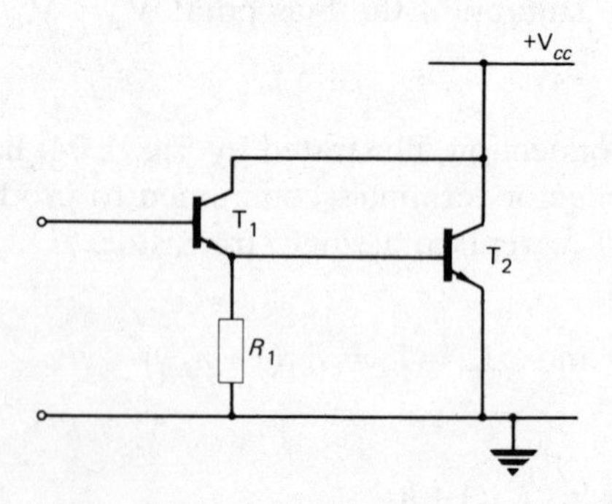

Fig. 1.35 Use of an emitter resistor in a Darlington pair.

Example 1.14

Calculate the current gain and input resistance of a Darlington pair if identical transistors are used having $h_{fe} = 60$ and $h_{ie} = 1000\ \Omega$. Assume h_{re} and h_{oe} can be neglected.

Solution From equation (1.38)

$$A_i = 60 + 60(1 + 60) = 3720 \quad (Ans)$$

From equation (1.39)

$$R_{IN} = 1000 + 61 \times 1000 = 62\ \text{k}\Omega \quad (Ans)$$

Darlington pairs are commonly employed within integrated circuits and as the output devices in audio-frequency push-pull power amplifiers. With the latter, T_1 is usually a low-power high-current-gain type and T_2 is the power transistor proper which passes a large current. Difficulties can arise with the circuit if the collector current of the second transistor is low and its gain fairly large. Suppose, for example, that T_2 has $h_{fe} = 150$ and $I_c = 2$ mA. Then

$$I_{b2} \simeq I_{e1} \simeq I_{c1} = 2 \times 10^{-3}/150 = 13.33\ \mu\text{A}$$

and for such a low value of collector current the gain of T_1 will be sma

This would mean that much of the advantage of the Darlington connection would be lost. Darlington pairs are available as discrete packages, e.g. the MJ 1001 is a n-p-n pair.

Current Sources

There are many applications in electronics where a source of **constant current** is required as opposed to the more common requirement for a constant voltage source. The term constant current means that the current provided to a load remains more or less constant as the load resistance is varied. This means that the ideal constant current source will have an infinite output impedance but practical sources are of very high impedance.

Fig. 1.36 Fet current source.

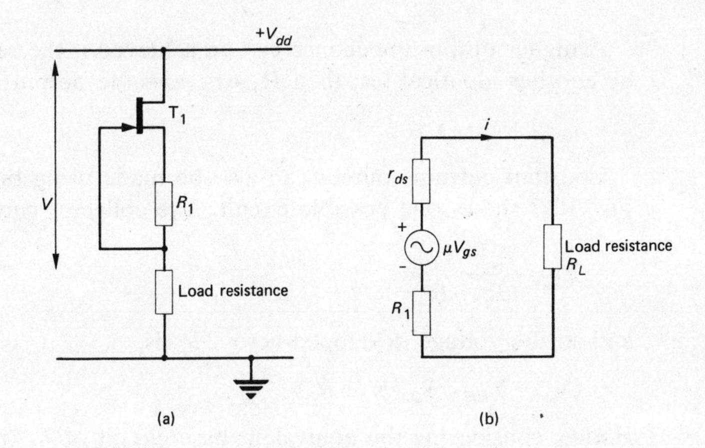

One of the simplest methods of providing a constant current source is shown by Fig. 1.36*a* in which a jfet has its gate connected to its source via a resistor R_1. With the fet biased beyond pinch-off, its drain current is almost constant, at a value set by its gate/source bias voltage V_{gs}, regardless of the load impedance. The bias voltage is developed across R_1 by the drain current and this allows the constant current to be set to be the required value by the suitable choice of R_1.

The output impedance of the circuit is easily determined with the aid of the equivalent circuit of the source (Fig. 1.36*b*) in which $\mu = g_m r_{ds}$. From the figure,

$$i = \frac{\mu V_{gs}}{r_{ds} + R_1 + R_L} = \frac{\mu(V - iR_1)}{r_{ds} + R_1 + R_L}$$

$$i(r_{ds} + R_1 + R_L) = \mu V - \mu i R_1$$

$$i = \mu V / [r_{ds} + R_1(1+\mu) + R_L]$$

Assuming that $r_{ds} + R_1(1+\mu) \gg R_L$ a constant current will be provided. The effective output impedance is

$$\frac{\mu V}{i} = r_{ds} + R_1 + g_m r_{ds} R_1 \qquad (1.40)$$

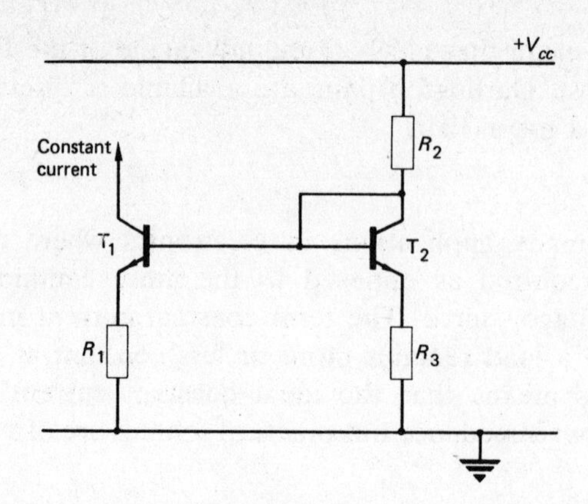

Fig. 1.37 Bipolar transistor current source.

A higher output impedance can be achieved if the resistor R_1 is replaced by another identical fet; then $R_1 = r_{ds}$ and the output impedance becomes

$$r_{ds}(2 + g_m r_{ds})$$

Constant current sources can also be made using bipolar transistors and Fig. 1.37 shows one possible circuit. The collector current passed by T_2 is

$$I_{c2} = \frac{V_{cc} - V_{BE2}}{R_2 + R_3}$$

and so the voltage developed across R_3 is

$$(V_{cc} - V_{BE2})R_3/(R_2 + R_3)$$

Hence, considering the equivalent base circuit of T_1 (refer to Fig. 1.26)

$$\frac{(V_{cc} - V_{BE2})R_3}{R_2 + R_3} + V_{BE2} = V_{BE1} + (I_{b1} + I_{c1})R_1$$

Now, $I_{c1} \gg I_{b1}$ and $V_{BE2} \simeq V_{BE1}$ and so

$$I_{c1} = \frac{(V_{cc} - V_{BE2})R_3}{R_1(R_2 + R_3)} \quad (1.41$$

Transistor T_1 has negative feedback applied to it and this increases its output impedance to a high value. Two other constant current sources are given in Fig. 1.38; their operation is left as an exercise for the reader.

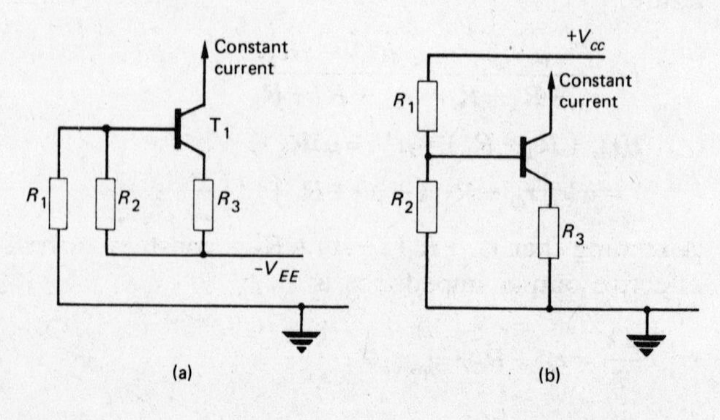

Fig. 1.38 Two other constant current sources.

Exercises 1

1.1 A power transistor has a maximum collector dissipation of 4 watts for an ambient temperature of 30°C. If the maximum junction temperature is 170°C and the thermal resistance θ_{JS} between junction and heat sink is 4°C/W, calculate the maximum possible thermal resistance for the heat sink.

1.2 A transistor is to dissipate 12 W power at an ambient temperature of 45°C with a maximum collector temperature of 90°C. The thermal resistance between junction and case is 0.9°C/W and between case and air is 6°C/W. Calculate the thermal resistance of the required heat sink. Find also the maximum power which can be dissipated if a mica washer of thermal resistance 1.8°C/W is inserted between the case of the transistor and the heat sink.

1.3 Design a potential divider bias circuit for a transistor if $h_{FE} = 100$ and $V_{BE} = 0.6$ V and the collector supply voltage is 20 V and the required collector current is 3 mA.

1.4 Derive an expression for the output resistance of the current source shown in Fig. 1.39. Calculate its value if T_1 has $g_m = 1.2$ mS and $r_{ds} = 85$ kΩ and T_2 has $g_m = 1$ mS and $r_{ds} = 90$ kΩ.

Fig. 1.39

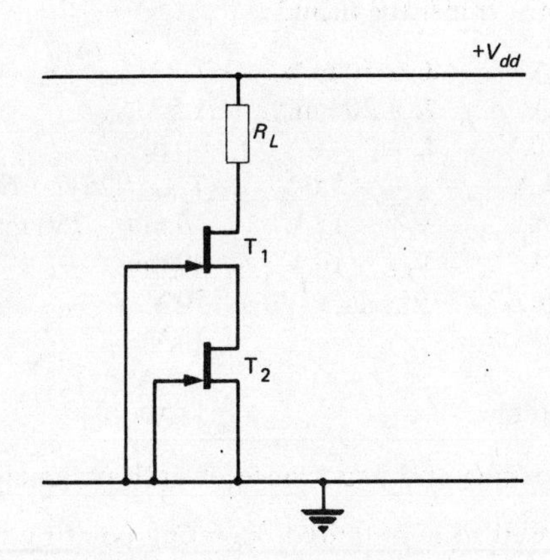

1.5 Define the four h-parameters and draw the h-parameter equivalent circuit of a bipolar transistor having a resistive load R_L. A transistor has the following parameters: $h_{ie} = 600$ Ω, $h_{fe} = 100$, $h_{oe} = 20 \times 10^{-6}$ S, and $h_{re} = 2 \times 10^{-4}$. Calculate the voltage gain of the transistor when $R_L = 3.3$ kΩ. Determine also the percentage error involved in neglecting i) h_{re} only, ii) h_{re} and h_{oe}.

1.6 Define each of the four y-parameters and draw the y-parameter equivalent circuit of a bipolar transistor having a resistive load R_L. A common-emitter transistor has a 7.8 kΩ load and the following y-parameters: $y_{ie} = 500 \times 10^{-6}$ S, $y_{oe} = 50 \times 10^{-6}$ S, $y_{fe} = 0.025$ S, $y_{re} = 0$. Calculate i) the input and output impedances, ii) the voltage gain of the circuit.

1.7 Design a potential divider bias circuit for an a.f. transistor stage if $I_c = 1$ mA, $V_{cc} = 12$ V, $V_{CE} = 5.5$ V, $h_{FE} = 120$ and $V_{BE} = 0.6$ V. Calculate the values of I_c and V_{BE} if h_{FE} increases to 150.

1.8 Explain the factors that cause the current gain of a bipolar transistor to fall at hig frequencies.

A transistor has a current gain h_{fe} of 100 at audio frequencies. The 3 dB cut-c frequency is 400 kHz. Calculate the current gain to be expected at i) 800 kHz, 2 MHz, iii) 4 MHz.

1.9 What is meant by the Miller effect? Derive an expression for the input capacitance a field effect transistor.

A fet has $C_{gs} = 5$ pF, $C_{gd} = 1.5$ pF and $g_m = 2.4$ mS and a drain/load resistor 4.7 kΩ. Calculate the high frequency at which the voltage gain of the circuit h fallen by 3 dB on its low-frequency value if the source resistance is 1000 Ω.

1.10 A bipolar transistor has $r_{bb'} = 80\ \Omega$, $r_{b'e} = 700\ \Omega$, $C_{b'e} = 50$ pF and $C_{b'c} = 5$ pF, ar $g_m = 40$ mS. Calculate the value of collector load resistor to give the voltage ga an upper 3 dB frequency of 5 MHz. Find also the low-frequency gain.

1.11 A potential divider bias circuit has been designed for use with a bipolar transist having a nominal h_{FE} of 80. The effective base resistance $R_1R_2/(R_1+R_2)$ is 6 kΩ ar the collector current is 1.5 mA. Determine the minimum value of emitter resistan needed to limit the collector current to 1.55 mA when a transistor having th maximum h_{FE} of 100 is used.

1.12 The specifications of a transistor include:

$V_{BE(SAT)}$	$I_c = 0.5$ A	$I_b = 50$ mA	1.2 V_{max}	
	$I_c = 2.0$ A	$I_b = 200$ mA	1.5 V_{max}	
C_{TC}	$V_{CB} = 5.0$ V	$I_e = 0$	60 pF	
f_t	$I_c = 0.25$ A	$V_{CE} = 5.0$ V	$T_{amb} = 25$°C	60 MHz min
h_{FE}	$I_c = 0.5$ A	$V_{CE} = 12$ V	78 min	250 max
	$I_c = 2.0$ A	$V_{CE} = 10$ V	40 min	
$V_{CBO(max)}$	$I_c \leq 1.0$ mA		70 V	
$V_{CEO(max)}$			45 V	
$I_{c(max)}$			6.0 A	
P_{total}	$T_{amb} \leq 60$°C		11 W	

Explain fully the meaning and importance of each parameter.

1.13 A junction fet is biased to its required operating point by the circuit shown in Fi 1.40. Determine the quiescent values of the gate/source voltage, the drain curre and the drain/source voltage if $I_{dss} = 4$ mA, $V_p = -3.8$ V, and the quiesce drain/earth voltage is to be 10 V.

Fig. 1.40

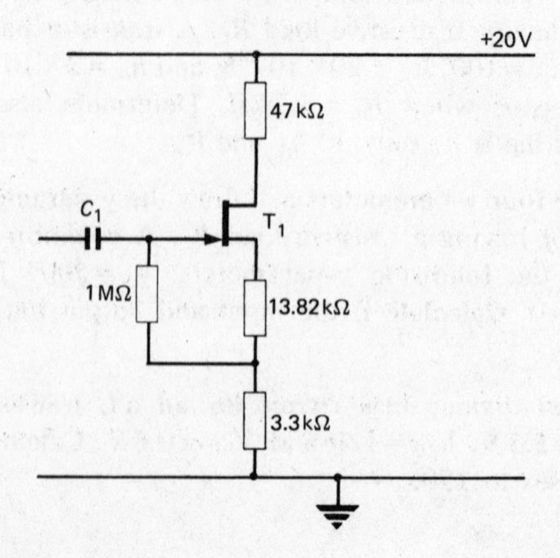

Short Exercises

1.14 Draw the h-parameter equivalent circuit of a transistor and define each parameter. State the limitations on its use.

1.15 An amplifier has input impedance of $1000\angle 45^\circ\ \Omega$, an output impedance of $50\angle 10^\circ\ \Omega$, and a current gain of $150^\circ\angle 0^\circ$ at a particular frequency. Calculate its voltage gain when the load impedance is $200\angle 0^\circ\ \Omega$.

1.16 Define the h-parameters of a bipolar transistor and relate them to the static characteristics of the device.

1.17 Draw the hybrid-π equivalent circuit of a bipolar transistor and relate each resistance to the parameters of the transistor.

1.18 A bipolar transistor has $h_{ie} = 1100\Omega$, $h_{fe} = 180$ and $h_{oe} = 80\ \mu S$. Calculate values for the hybrid-π equivalent circuit components.

1.19 A bipolar transistor has $h_{ie} = 1000\ \Omega$, $h_{fe} = 150$; calculate its peak a.c. collector current.

1.20 The f_t of a bipolar transistor is 500 MHz. Calculate its current gain at i) 100 MHz, ii) 50 MHz.

1.21 A transistor has $f_t = 600$ MHz and $g_m = 30$ mS. Determine the sum of its capacitances $C_{b'e}$ and $C_{b'c}$.

1.22 A transistor has the following data: $I_{c(max)} = 10$ A, $V_{CEO(max)} = 75$ V, maximum power dissipation = 50 W. Explain why the product $I_{c(max)} \times V_{CEO(max)}$ can be so many times greater than 50 W.

1.23 A common-emitter transistor amplifier stage has a collector load resistor of 4.7 kΩ. The transistor has $h_{oe} = h_{re} = 0$, $h_{ie} = 1$ kΩ (min) to 4 kΩ (max), $h_{fe} = 150$. Calculate the maximum and the minimum voltage gains of the circuit.

1.24 Show that $I_{CEO} = I_{CBO}(1 + h_{FE})$

1.25 For the circuit in Fig. 1.41 calculate the value of R_1 to give a drain current of 0.5 mA. $I_{dss} = 1$ mA and $V_p = -1$ V.

Fig. 1.41

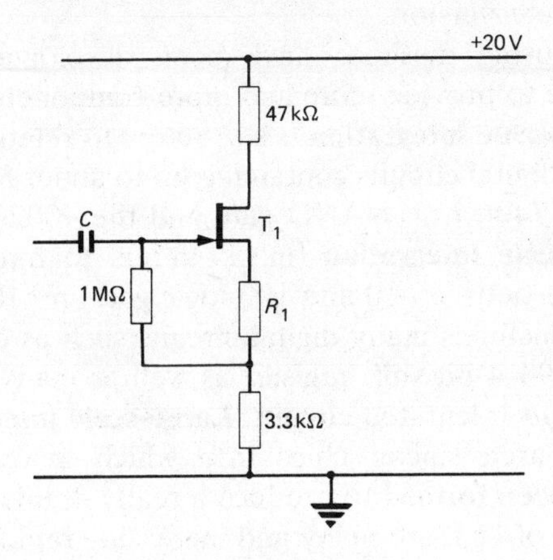

1.26 Calculate the current gain, input resistance and voltage gain of a Darlington pair if, for T_1, $h_{ie} = 1000\ \Omega$, $h_{fe} = 120$, and, for T_2, $h_{ie} = 300\ \Omega$, $h_{fe} = 80$. The load resistance is 1 kΩ.

2 Integrated Circuits

Introduction

The methods used to make a planar bipolar transistor or a field effect transistor [EIII] can be extended to allow a complete circuit to be fabricated within a single silicon chip. Most of the components required by the circuit are formed at the same time and the device is known as a **monolithic integrated circuit** since only the one chip is used. Other kinds of integrated circuit are also available and are used to a much lesser extent; these i.c.s are known as *thick-film* and *thin-film* circuits. The monolithic i.c. is very much the most common type and the general term integrated circuit is generally taken to mean this kind of device. Certainly this will be the case throughout the remainder of this book.

The use of integrated circuits offers a number of advantages over the use of discrete circuitry. These advantages are: greatly reduced size, lower costs, greater reliability, and the ability to perform complex circuit functions economically. The reduction in the physical dimensions of circuits occurs because a complex circuit can be enclosed within a volume of comparable dimensions to a single discrete transistor. The cost of an integrated circuit depends upon both its complexity and the quantity manufactured but in many instances the cost of an i.c. is no greater than that of a single transistor. Even when several integrated circuits are interconnected to produce a more complex circuit, the overall costs are still much lower than for the all-discrete version.

As manufacturing methods have been developed and refined it has become possible to provide more and more components within a single chip. The term **small-scale integration** (s.s.i.) refers to relatively simple integrated circuits such as digital circuits containing up to about 8 logic gates. Examples of s.s.i. are the 7400 t.t.l. NAND gate and the 4002 cmos NOR gate. The term **medium-scale integration** (m.s.i.) refers to integrated circuits which have somewhere between 10 and 100 logic gates (or the equivalent circuits). This definition includes many digital circuits such as counters and registers, e.g. the t.t.l. 7494 4-bit shift register as well as many of the more complex analogue or *linear* integrated circuits. **Large-scale integration** (l.s.i.) refers to relatively large-area silicon chips into which many hundreds of circuit elements have been formed to produce a really dramatic reduction in circuit costs. Examples of l.s.i. are many and increasing rapidly; e.g. semiconductor memories, microprocessors, frequency synthesizers and t.v games. A single l.s.i. chip may replace 100 or more discrete and s.s.i./m.s.i. devices to give a large reduction in costs, physical dimensions and power dissipation.

Monolithic Integrated Circuits

The fabrication of an integrated circuit component is achieved by a sequential series of oxidizing, etching and diffusing [EIII]. Bipolar and field effect transistors, diodes and small-valued resistors and capacitors can all be formed in this way but high values of resistance and capacitance cannot be achieved.

A thin wafer, about 0.02 cm thick, is sliced from a rod of p-type silicon and will have a surface area of about 26 cm^2. Since a complete integrated circuit may only occupy a surface area of about 0.06 cm^2, a large number of *identical* circuits can be simultaneously formed in the one wafer. The principle is illustrated by Fig. 2.1, although to simplify the drawing fewer chips have been shown.

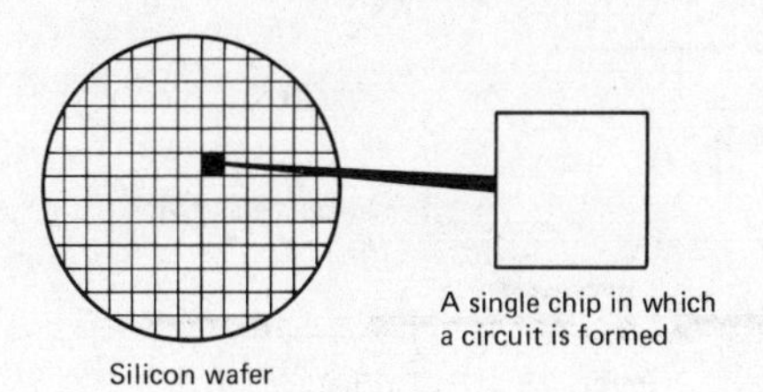

Fig. 2.1 Showing how a silicon wafer is divided into a number of chips.

Each individual p-type silicon chip acts as a substrate into which the various components making up the circuit can be formed. The components are simultaneously formed by the diffusion of impurity elements into selected parts of the chip.

When very precise depths of doping are required, a more accurate method of placing the doping atoms into the semiconductor is needed and a technique such as *ion implantation* is often used. A high-energy beam of boron ions (p-type) or phosphorous (n-type) is directed into the substrate at the point where the precise doping is required. The depth at which the ions are implanted depends upon the voltage used to accelerate the ions and the density of the doping achieved depends upon the number of ions in the beam. By controlling both of these variables any desired doping can be obtained with great accuracy. The disadvantage of ion implantation is the increased costs incurred.

Since the p-type substrate is an electrical conductor it is necessary to ensure that each of the formed components is insulated from the substrate and from each other, otherwise the various components will be coupled together by the resistance of the substrate. There are four ways in which the necessary isolation can be obtained:

a) **Reverse-biased p-n junctions** Several n-type regions, equal in number to the number of components to be formed in the chip, are diffused into the p-type substrate (Fig. 2.2). Each of the n-type regions will be isolated from the substrate if the junction is maintained in the reverse-biased condition by connecting the substrate to a potential which is more negative than any other part of the circuit.

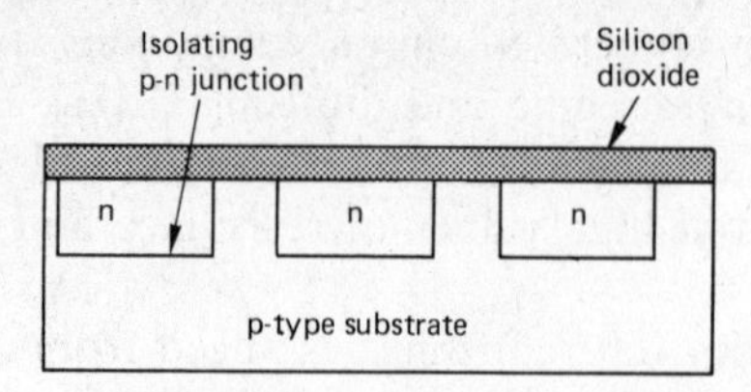

Fig. 2.2 A method of isolating the components in an integrated circuit.

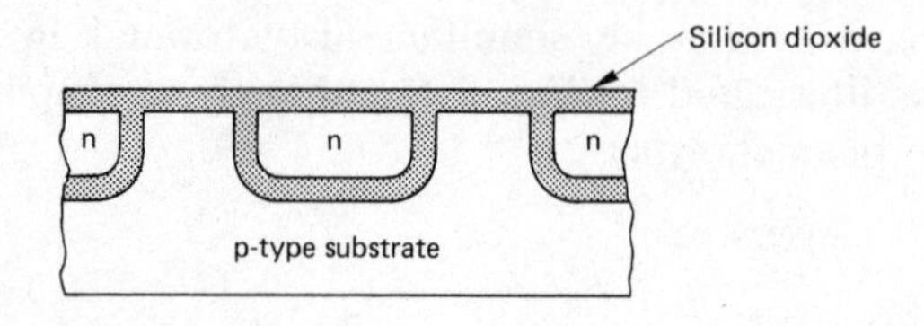

Fig. 2.3 Dielectric isolation.

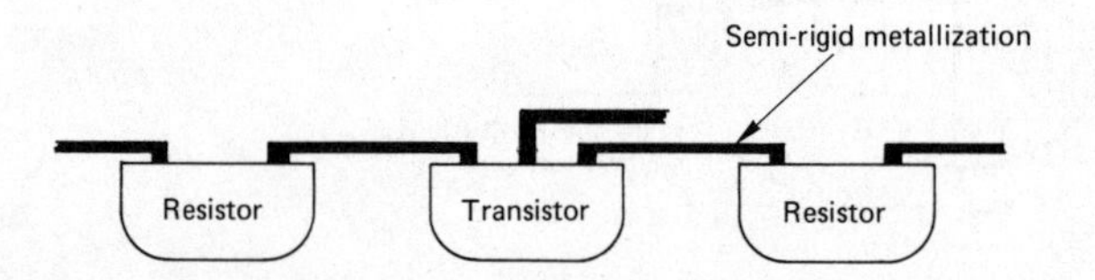

Fig. 2.4 Beam lead bonding.

b) **Dielectric isolation** The use of p-n junctions to provide component isolation leads to various unwanted capacitances and leakage paths within the integrated circuit which will adversely affect its high frequency/speed performance. These unwanted effects can be considerably reduced by the use of the dielectric isolation illustrated by Fig. 2.3.

The islands in which the components will be diffused are electrically isolated from one another by a layer of silicon dioxide. The isolating layers have a much smaller capacitance per unit area than a reverse-biased p-n junction and so the unwanted capacitances are reduced.

c) **Beam lead bonding** The components are formed in the substrate and then a heavier than normal interconnecting metallization pattern is deposited on top. All the silicon that is not used for an element is then removed by further etching so that the circuit is reduced to a number of separate elements isolated from one another by air. The elements are both connected together according to the dictates of the circuit *and* supported by the semi-rigid metallization pattern (see Fig. 2.4).

d) **Collector diffusion isolation** Since this is essentially an alternative method of forming a bipolar transistor it will be considered in the next section.

Integrated Bipolar Transistor

The most commonly used element in integrated circuits is the n-p-n bipolar transistor, the construction of which is shown in Fig. 2.5 (n^+ denotes a region of greater conductivity). The buried layer consists of an n^+ low-resistance region that is effectively in parallel with the n-type collector region and so reduces its series resistance. The series resistance of the

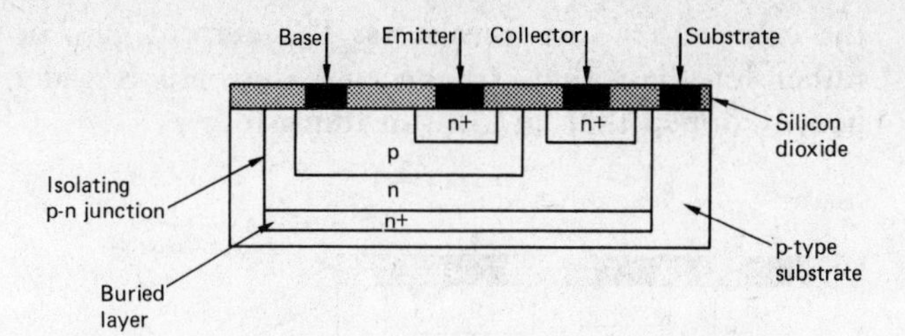

Fig. 2.5 Integrated n-p-n transistor.

collector cannot be reduced by using a lower resistivity material for the collector region since this would reduce the breakdown voltage of the collector/base junction. Typically, an h_{fe} of about 100 is obtained.

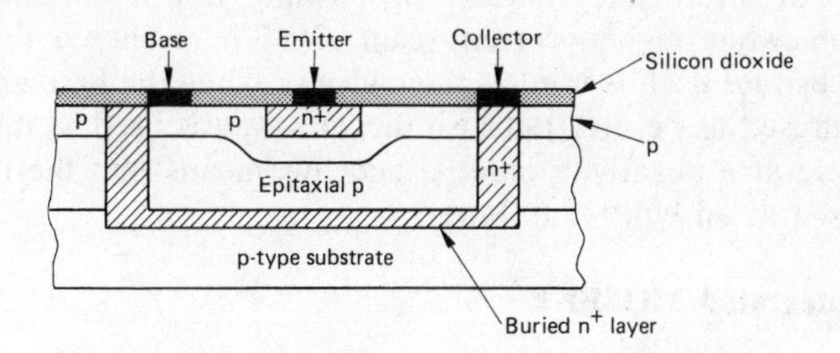

Fig. 2.6 Integrated n-p-n transistor using c.d.i.

Collector diffusion isolation (c.d.i.) is a simpler process that occupies a smaller chip area per transistor and is therefore particularly important for l.s.i. devices. The construction is shown in Fig. 2.6 from which it can be seen that the buried n^+ region is used as the collector region. Onto this is grown a narrow p-type region which forms the base region. Then a single n^+ diffusion produces both the collector and the isolation, since this diffusion penetrates through the p-type region and links up with the collector region. A thin p-type layer is deposited on top, before, lastly, the n-type emitter is formed. The c.d.i. process produces a transistor with a high current gain over a wide range of collector currents.

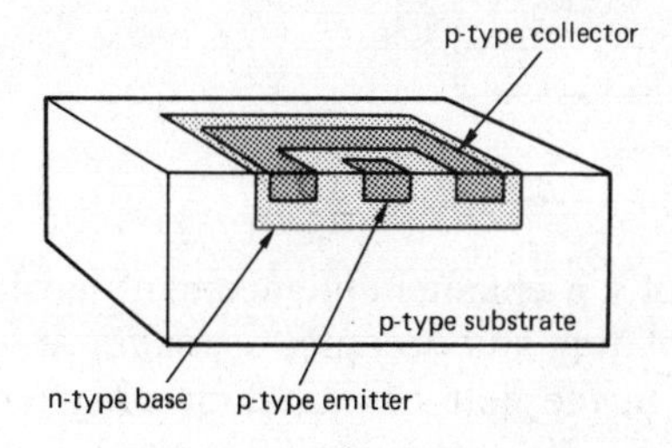

Fig. 2.7 Lateral p-n-p transistor.

Most integrated circuits use a p-type substrate and n-p-n transistors wherever possible but sometimes a p-n-p transistor is also necessary. When the collector current to be passed is fairly small a *lateral* p-n-p transistor is used (Fig. 2.7). The current in this kind of transistor flows laterally, or parallel to the surface, from emitter to collector rather than vertically as is the case with n-p-n transistors. The current gain of a lateral transistor is rather low, less than 10, because the base is wider and the emitter less heavily doped than in a n-p-n transistor.

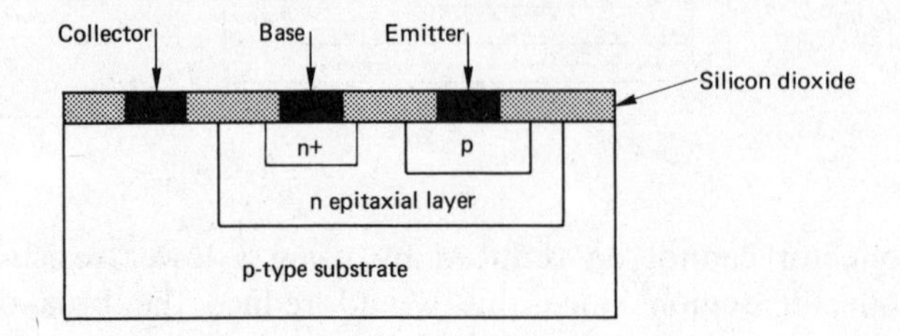

Fig. 2.8 Substrate p-n-p transistor.

An alternative method of forming a p-n-p transistor which offers a somewhat higher current gain, 20–30, is shown by Fig. 2.8. Now the substrate itself is used as the collector while the base and emitter regions are diffused as before. Because the substrate is used as the collector it must be held at a negative potential and this means that the transistor can only be used as an emitter follower.

Integrated MOSFET

Many linear integrated circuits, particularly operational amplifiers, require a very high input impedance and this can often best be provided by a fet input stage. The fet is either a junction device or it is a p-channel enhancement-mode mosfet. The fet is formed using the same diffusion as that for the n-p-n bipolar transistors in the chip, although this presents some problems because the fet requires much lower doping levels than the bipolar devices. This difficulty can be overcome if ion implantation is used. The p-channel enhancement mosfet is normally employed because it is easier and cheaper to fabricate than any of the other types.

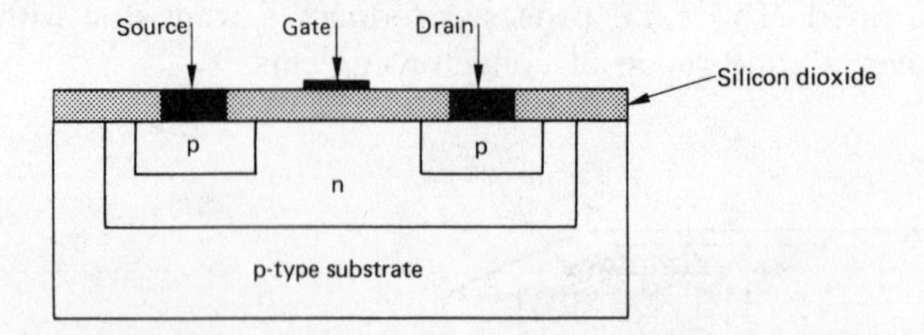

Fig. 2.9 Integrated p-channel enhancement-mode mosfet.

The construction of a p-channel enhancement mosfet is shown in Fig. 2.9. The device is self-isolating and occupies a smaller area of the chip than does a bipolar transistor. In the field of digital circuitry mosfets are employed in three different logic families. A **pmos** integrated circuit consists exclusively of p-channel enhancement mosfets and provides a high circuit density at low cost and is used for l.s.i. devices. Sometimes depletion-mode p-channel mosfets are also used. The advantage of pmos circuitry is its simplicity but this is in many cases outweighed by its slow speed of operation in comparison with nmos circuits. An **nmos** circuit contains only n-channel mosfets;

these are mainly of the enhancement type, depletion-type elements being mainly used as resistors. Unfortunately, nmos devices are more difficult to fabricate. Applications of nmos include memories and microprocessors.

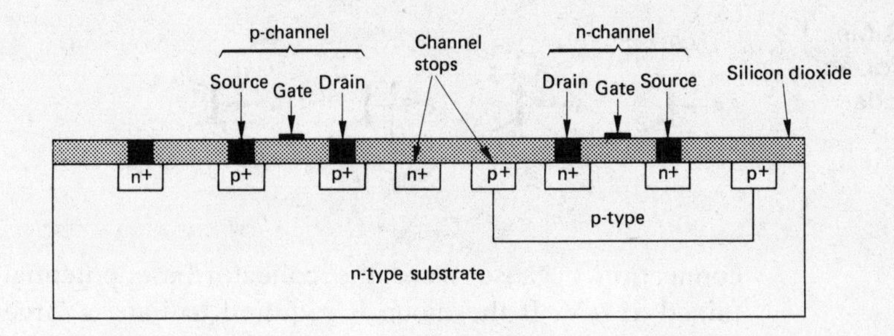

Fig. 2.10 Cmos construction.

One of the most popular logic families, that includes many of the more simple digital circuits, is the cmos family. **Complementary mos** employs pairs of p-channel and n-channel enhancement-mode fets and gives logic circuits whose main merit is a very low power dissipation. A cmos device is fabricated onto an n-type substrate and Fig. 2.10 shows how the fets are formed. It can be seen that the n-channel devices are formed within a p-type isolation region. The p^+ and n^+ regions, labelled as *stops,* are necessary to avoid an unwanted channel being induced between adjacent transistors. Because of the need to provide these stops cmos is more extravagant in its use of chip area than either nmos or pmos.

A modern development is the use of sapphire substrates into which silicon islands are formed. The islands are isolated from one another by the non-conducting substrate and within each island a mosfet is formed. *Silicon-on-sapphire* gives a very good packing density with the minimum unwanted capacitance but the technique is expensive.

Integrated Junction FET

The junction field effect transistor is used mainly in some linear integrated circuits for its high input impedance. Fig. 2.11 shows the construction of an integrated n-channel jfet.

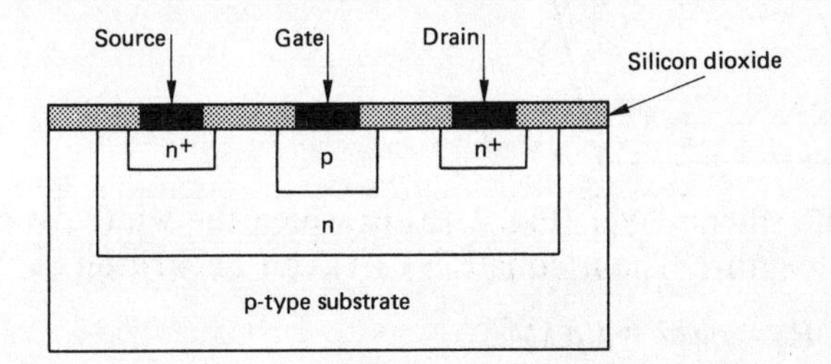

Fig. 2.11 Integrated n-channel jfet.

Integrated Diode

A diode is formed from one of the transistor connections shown in Fig. 2.12. For each of the connections a, b and c, forward-biasing the diode will result in currents flowing in the substrate. This does not occur with the fourth

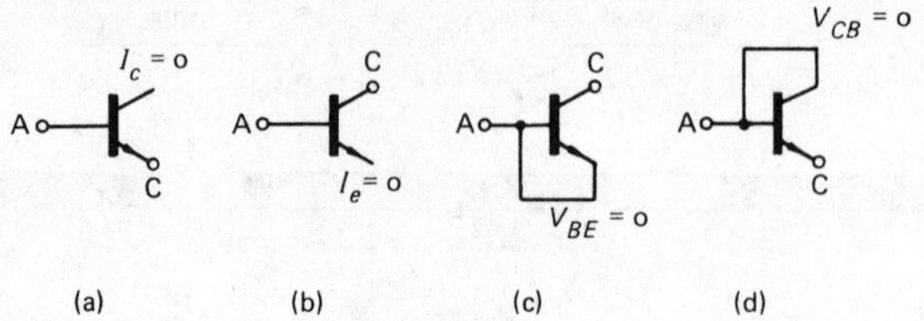

Fig. 2.12 Possible connections for an integrated diode.

connection because there the collector/base potential difference is maintained at 0 V. If the diode is required to have a large breakdown voltage, either connection b or c is best since they each involve a collector/base junction. This has a breakdown voltage at least 30 V compared with about 10 V for an emitter/base junction. Should the minimum leakage current be of prime importance, either connection a or d should be used. Connection b provides the lowest forward resistance and the lowest minority carrier storage time.

Integrated Resistors

Integrated resistors are made using a thin layer of p-type silicon that is diffused at the same time as either the base or the emitter of a bipolar transistor. The choice depends upon the required resistance value, the tolerance and the temperature coefficient.

The resistance of a silicon layer depends upon the length l, area a, and the resistivity ρ of the layer according to

$$R = \rho l / a \ \Omega \tag{2.1}$$

The area a of a layer is the product of the width W and the depth d, thus

$$R = \rho l / W d$$

It is usual to express the resistance in terms of the resistance of a square

Fig. 2.13

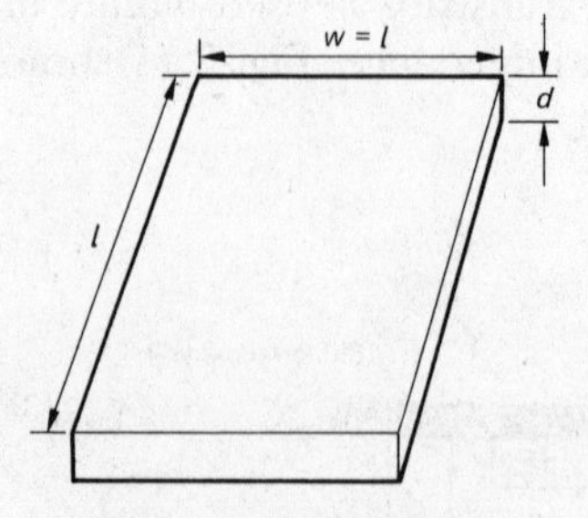

of the silicon layer (Fig. 2.13) in which the width W of the layer is equal to the length l. Then, equation (2.1) can be written as

$$R_S = \rho l / l d = \rho / d \ \Omega/\square \tag{2.2}$$

The resistance R_S is now the resistance between the opposite sides of a square and it is known as the **sheet resistance**, measured in a unit known as the *ohm per square*. The sheet resistance depends only upon the resistivity of the silicon layer and not upon the dimensions of the square. The resistivity of the layer is determined by the number of charge carriers that are diffused into the layer and the depth to which they penetrate. For a base-diffused resistor, $R_S = 100\ \Omega/\square$–$300\ \Omega/\square$, while for an emitter-diffused resistor, $R_S = 2\ \Omega/\square$–$10\ \Omega/\square$. A required resistance value is obtained by the suitable choice of both the length and the width of the resistive path. The sheet resistance can be increased by increasing the length of the path and decreased by increasing the width of the path.

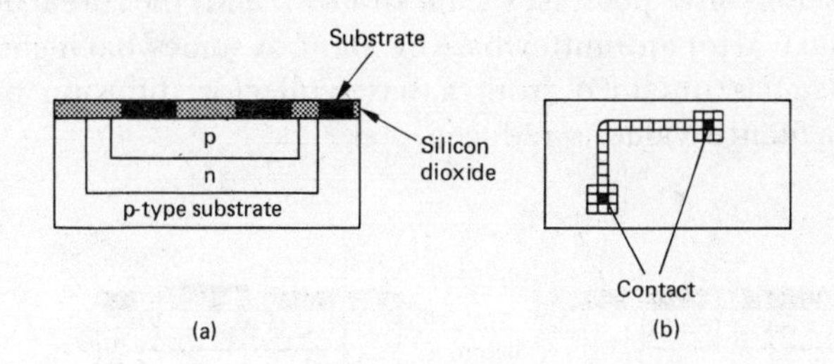

Fig. 2.14 Side and top views of an integrated resistor.

The constructional details of a **base-diffused resistor** are given in Fig. 2.14*a* and *b*. Fig. 2.14*a* shows that the resistive path is formed by a p-type region that joins together the resistor contacts. Fig. 2.14*b* shows the top view of the resistor and indicates how a required resistance value may be obtained by connecting a number of squares in series. The resistor can follow any path which will best utilize the surface area of the chip. The practical range of resistances is from about 50 Ω to 30 kΩ. When a lower resistance is needed, emitter diffusion is necessary.

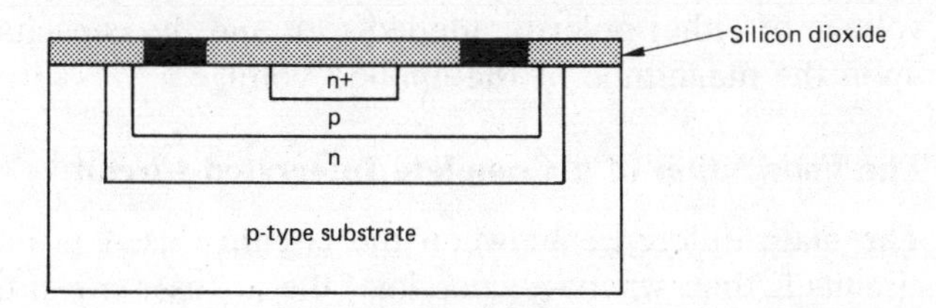

Fig. 2.15 Pinch resistor.

When a high resistance value must be provided, an alternative construction known as a **pinch resistor** must be used (see Fig. 2.15). An n^+ diffusion into the p-region reduces the area of the resistive path and so increases its resistance. It is now possible to obtain a sheet resistance of some $5\ k\Omega/\square$ to $15\ k\Omega/\square$, but large tolerances and a poor temperature coefficient cannot be avoided.

Integrated resistors cannot be made to closer tolerances than about ±10%. However, when two resistors are simultaneously diffused the *ratio* of their two values *can* be kept within about ±2% tolerance.

Integrated Capacitors

Integrated capacitors are fabricated in two different ways: either the capacitance of a reverse-biased p-n junction can be utilized, or the capacitance can be provided by a layer of silicon separating two conducting areas. The p-n junction is formed at the same time as either the emitter/base or the collector/base junction of a transistor. Provided the p-n junction is held in its reverse-biased condition, a capacitance of about 0.2 pF/mil^2 can be obtained. Since the chip area available for a capacitor is limited, the maximum capacitance which can be provided is about 100 pF. The disadvantages of this type of capacitor are: the capacitance is low and is voltage-dependent; p-n junction isolation is difficult because the isolating junction also possesses capacitance; and the breakdown voltage is only about 6 V for an emitter/base element. A somewhat higher value of breakdown voltage is obtained from a base/collector diffusion but then the possible capacitance value is reduced.

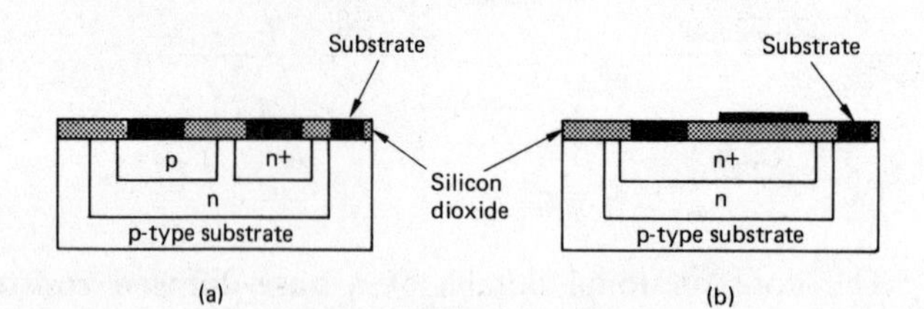

Fig. 2.16 Integrated mos capacitor.

Fig. 2.16*b* shows how a **mos capacitor** is fabricated. One electrode of the capacitor is provided by an aluminium layer that is deposited onto the top of the silicon layer and the other electrode is provided by the diffused n$^+$ region. The capacitance depends upon the thickness of the silicon dioxide layer and the area of the aluminium plate and is about 0.3 pF/mil^2. Capacitances of up to a few hundred pF are possible with a breakdown voltage of 60–100 V and a low temperature coefficient. The mos capacitor can have a voltage of either polarity applied to it and the capacitance does not depend upon the magnitude of the applied voltage.

The Fabrication of a Complete Integrated Circuit

The main difference between the circuitry used in integrated and discrete circuits is that, whenever possible, the i.c. uses n-p-n transistors rather than other components since they are the cheapest.

In the fabrication of a complete integrated circuit all the components, active and passive, required to make up the circuit are formed simultaneously. The components are then interconnected as required by means of an aluminium pattern that is deposited on top of the chip. Circuits are then separated into individual chips and sealed within a package. A number of packages are available such as dual-in-line, TO circular, and, less frequently, flat and quad-in-line. Manufacturers place a letter after the device type number to indicate the package used but unfortunately the letters used have not been standardized.

A wide variety of integrated circuits are available and new ones are continually being introduced. Most i.c.s can be placed into one of two general categories, namely linear and digital, although some devices like timers and analogue-to-digital converters straddle the boundary line.

When an integrated circuit is used in an equipment, a number of external components are also needed. Large values of resistance and capacitance must be externally provided as must variable components such as volume and tuning controls. Inductors cannot be produced within an i.c. and must always be externally fitted.

Linear Integrated Circuits

The term linear is used to refer to integrated circuits which mainly respond to analogue signals rather than digital signals. The most commonly employed linear integrated circuit (l.i.c.) is the **operational amplifier**. The operational amplifier is a high-gain high-input-impedance low-output-impedance device that is used for a wide variety of applications, including amplifiers and waveform generators. Other widely used types of l.i.c. are timers, voltage regulators, and audio power amplifiers.

Digital Integrated Circuits

With digital circuits, the signals to be processed are **digital** in nature and so the active elements are operated as *two-state* devices. The fundamental building blocks of most digital circuitry are the various kinds of logic gate, e.g. AND, OR, NAND and NOR, and the bistable multivibrator or flip-flop. The commonly employed logic families are the various forms of transistor-transistor logic (t.t.l.)—with the low-power Schottky version expected to dominate in the near future (a new development being Advanced Schottky)—and cmos. (Some manufacturers have already dropped t.t.l. versions other than the low-power Schottky.) The t.t.l. and cmos families are widely used for random logic applications, and for s.s.i.

Film Integrated Circuits

A film integrated circuit is one in which a pattern of resistors and capacitors and their interconnections is formed onto an insulating substrate, such as glass or ceramic. Any other components that are wanted are added later and soldered into position. There are two kinds of film integrated circuit known as *thick film* and *thin film*. The term *hybrid* is applied when thick and thin films are mixed and/or transistor/monolithic i.c.s are added. A film integrated circuit is generally used when it is required to minimize the use of discrete components but when for some reason or the other a monolithic i.c. cannot be used.

Thick film integrated circuits are produced by screen printing various kinds of pastes or inks onto a ceramic substrate and then firing the substrate in an oven. The printing of components and interconnections is carried out by the use of a stainless steel wire screen which has about 100 meshes per centimetre and is coated with a photo-sensitive emulsion. The screen is mounted above and parallel to the substrate. The screen and the substrate are painted with the appropriate paste; a conductive paste for connections, a resistive paste for resistors, and a non-conductive paste for cross-overs and capacitor dielectrics.

The value of resistance formed is determined by the paste which sets the Ω/□ value, and by the dimensions of the resistive path. A resistor can normally be fabricated to a tolerance of ±15% but if a closer tolerance is needed a resistance value can always be increased by trimming the width of the resistive path. Capacitors are usually of the multi-layer ceramic type for values such as 10 pF to 0.1 μF, while larger values can be obtained if tantalum is used as the dielectric material. *Small* values of inductance can be made by using a spiral path. Active devices and any other components required are then connected to the appropriate points in the conductive pattern. Discrete devices are available in miniature packages or if the space is available conventional devices can be used.

A **thin film integrated circuit** consists of a very thin film deposited by either *vacuum evaporation* or *sputtering* onto a glass or glazed aluminium oxide substrate. The deposited materials are gold, tantalum, nichrome or aluminium for conductors, and either silicon or tantalum oxide for dielectrics.

The advantages claimed for film circuits are lower costs, smaller size and greater reliability than the discrete alternatives. Thin films are more expensive than thick films but can be made with closer tolerances and smaller temperature coefficients and with an extremely close packing density.

In general, **hybrid** technology cannot compete with monolithic i.c.s and is only considered if i) a suitable monolithic i.c. is not available and the expected volume is not enough to justify the design costs for a new custom-built monolithic i.c., or ii) the required performance is in excess of that provided by a monolithic i.c.; this refers in particular to power handling and temperature stability.

Exercises 2

2.1 Explain what is meant by each of the following and give a possible application: i) thick film circuit, ii) thin film circuit.

Calculate the dimensions of a thick film resistor of 3 kΩ and 100 mW dissipation if the sheet resistance is 400 Ω/□ and the maximum power dissipation is 1 W per cm^2.

2.2 Explain the meaning of each of the following terms: i) monolithic integrated circuit, ii) diffusion, iii) beam lead bonding, and iv) metallization.

2.3 Describe how an enhancement-type mosfet and a junction fet can be fabricated in a monolithic integrated circuit. Explain why high values of resistance and capacitance are not formed within an i.c.

2.4 Fig. 2.17 shows the basic circuit of a t.t.l. NAND gate. Draw, using side and plan views, the layout of this circuit when it is formed in a monolithic integrated circuit. Discuss briefly why the reliability of an integrated circuit is greater than its discrete component version.

Fig. 2.17

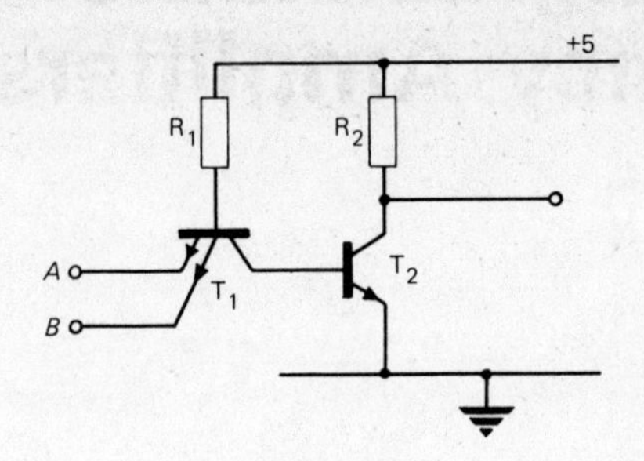

Fig. 2.18

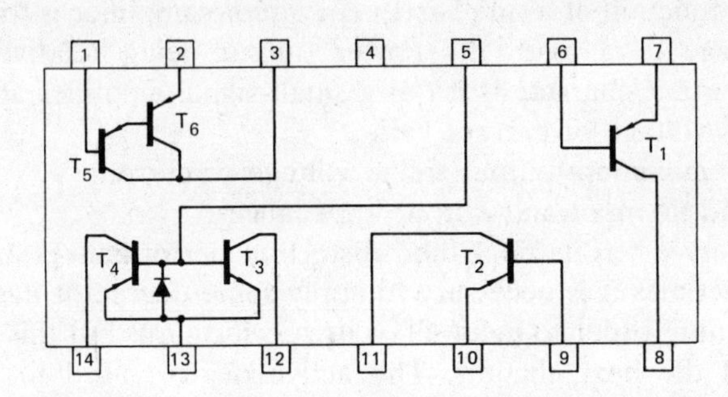

2.5 Describe, with the aid of diagrams, the four ways in which an integrated diode can be formed from a transistor p-n junction. Say which of the four methods is generally employed and draw a side view of your choice in an integrated circuit.

2.6 The pin connections of the CA 3084 transistor array are shown in Fig. 2.18. The array provides *a*) two matched transistors T_1 and T_2, *b*) a current-mirror pair T_3 and T_4, and *c*) a Darlington pair T_4 and T_5. Draw a circuit to illustrate a possible application for the first and third pairs.

Short Exercises

2.7 Sketch the four basic diode connections for monolithic integrated circuits. Which of them has the lowest forward voltage drop and which has the highest breakdown voltage?

2.8 Calculate the resistance of an i.c. resistor 5 mils long and 1 mil wide if its sheet resistance is 300 $\Omega/\square$.

2.9 List the relative merits of monolithic and film integrated circuits.

2.10 List the various ways in which an integrated bipolar transistor can be isolated from the substrate. Why is such isolation necessary and why is it not required for a fet?

2.11 Draw side-by-side the constructions of a discrete and an integrated bipolar transistor and then point out the differences between them.

2.12 What are the main differences between a junction and a mos capacitor in a monolithic i.c.? Give typical values for each.

3 Small-signal Audio-frequency Amplifiers

Introduction

The function of a small-frequency audio amplifier is to deliver a current or a voltage to a load, the power output being relatively unimportant. The important characteristics of a small-signal amplifier are

its voltage or current gain

its maximum output signal voltage or current

and its input and output impedances.

Very often its noise and distortion performances are of importance too. Sometimes it is necessary to apply some degree of *negative feedback* to the circuit in order to meet all of its specifications but this topic will be deferred until the next chapter. The active devices used to provide the required amplification may be either a bipolar transistor or one of the three kinds of field effect transistor. Alternatively an integrated circuit may be employed. Generally small-signal integrated amplifiers use one of the many kinds of operational amplifier which are available but for low-noise applications it may be necessary to use a purpose-built low-noise integrated a.f. preamplifier.

The term *small-signal* implies that the swings of collector or drain current or voltage are small enough for the parameters of the device to be considered as essentially constant. This means that the determination of the gain, etc. can be carried out using one of the a.c. equivalent circuits discussed in Chapter 1. Alternatively, the gain of a circuit can be determined by drawing a load line on the output or the drain characteristics of the device [EIII].

Single-stage Amplifiers

1 The circuit of a **single-stage bipolar transistor amplifier** is shown by Fig. 3.1. Potential-divider bias is employed with the emitter resistor R_4 decoupled to prevent negative feedback being applied to the circuit. Capacitors C_1 and C_2 act as input and output coupling capacitors whose purpose is to prevent the d.c. conditions of the stage being upset by the source and load resistances.

At medium, or mid-band, frequencies, the reactances of all three capacitors are negligibly small and can be assumed to be zero. The emitter resistor R_4 is effectively short-circuited and so it does not appear in the a.c. equivalent circuit of the amplifier. Since the collector power supply has negligible internal resistance, the bias resistors R_1 and R_2 are effectively connected in parallel with one another across the base/emitter terminals of the transistor. Also, since C_2 has negligible reactance the external load

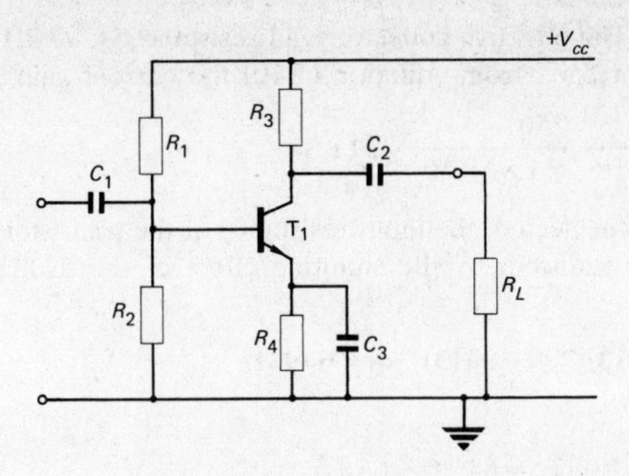

Fig. 3.1 Single-stage a.f. amplifier.

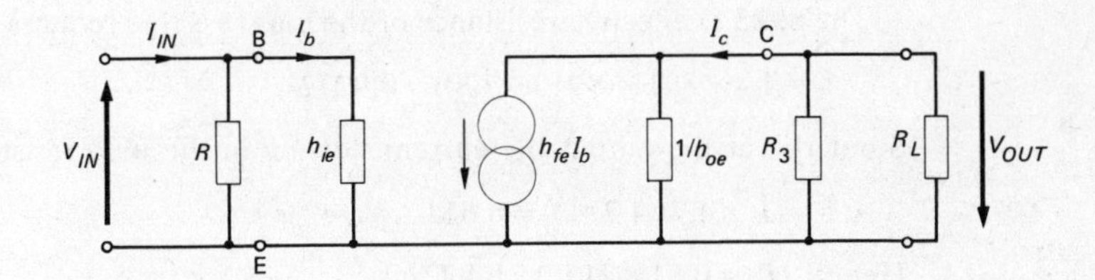

Fig. 3.2 h-parameter equivalent circuit of Fig. 3.1.

resistor appears in parallel with the collector resistor R_3.

The h-parameter equivalent circuit of the amplifier is given by Fig. 3.2 in which $R = R_1R_2/(R_1 + R_2)$. In the figure, h_{re} has been neglected and is not shown. Very often h_{oe} can also be omitted without the introduction of undue error.

Inspection of the equivalent circuit makes it clear that the current gain of the amplifier will always be less than the current gain of the transistor itself. This is because some of the input current I_{IN} is diverted away from the base of the transistor by the bias resistors R_1 and R_2 and because not all of the collector current flows into the load resistance R_L.

Example 3.1

Calculate the input resistance, the current gain and the voltage gain of the circuit shown in Fig. 3.3. The h-parameters of the transistor are $h_{ie} = 700\ \Omega$, $h_{oe} = 10^{-4}$ S and $h_{fe} = 250$ ($h_{re} = 0$). Determine also the percentage error in the calculation if h_{oe} is also neglected

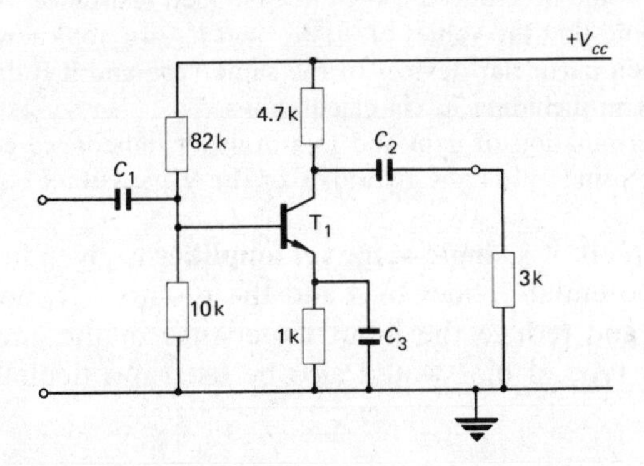

Fig. 3.3

Solution The effective collector load resistance is 4.7 kΩ in parallel with 3 kΩ or 1.83 kΩ. Therefore, from equation (1.10) the current gain A_i of the transistor is

$$A_i = \frac{250}{1+10^{-4}\times 1.83\times 10^3} = 211.3$$

Since h_{re} is neglected, the input resistance of the transistor is equal to h_{ie} or 700 Ω. This value is reduced, by the shunting effect of the 82 kΩ and 10 kΩ resistors in parallel, to

$$700\times 8913/(700+8913) \quad \text{or} \quad 649\ \Omega.$$

Therefore,

Input resistance = 649 Ω (*Ans*)

The 8913 Ω effective resistance of the bias resistors reduces the base current I_b to

$$I_b = I_{IN}\times 8913/(8913+700) = 0.927 I_{IN}$$

Similarly, at the output the current flowing in the load resistance R_L is

$$I_L = I_c\times 4.7/(4.7+3) = 0.61 I_c$$

Hence $I_L = 0.61\times 211.3\times 0.927 I_{IN}$
and the overall current gain A_{io} is

$$A_{io} = I_L/I_{IN} = 0.61\times 211.3\times 0.927 = 119.6 \quad (Ans)$$

From equation (1.12) the voltage gain A_v of the circuit is

$$A_v = \frac{119.6\times 3000}{649} = 552.9 \quad (Ans)$$

If h_{oe} is neglected, the current gain of the transistor is equal to h_{fe}, i.e. 250. Then the overall current gain A_{io} is

$$A_{io} = I_L/I_{IN} = 0.61\times 250\times 0.927 = 141.4 \quad (Ans)$$

Hence,

$$\text{Voltage gain} = \frac{141.4\times 3000}{649} = 653.6 \quad (Ans)$$

The percentage error introduced by neglecting h_{oe} is

$$(653.6-552.9)/552.9\times 100\% = +18.21\% \quad (Ans)$$

This error would be reduced if the effective load resistance were smaller. It should be borne in mind that the values of h_{ie}, h_{oe} and h_{fe} are not known accurately, since they vary between particular devices of the same type and it is doubtful whether there is much point in including h_{oe} in calculations.

The determination of gain and input/output impedance could, of course, also be carried out using either the hybrid-π or the y-parameter equivalent circuits.

2 The circuit of a **single-stage fet amplifier** is given in Fig. 3.4*a*. The circuit employs potential-divider bias and the resistors R_1 and R_2 shunt the input terminals and reduce the input impedance of the amplifier to $R_1R_2/(R_1+R_2)$. (This type of bias would also be used if a depletion-type mosfet were used.)

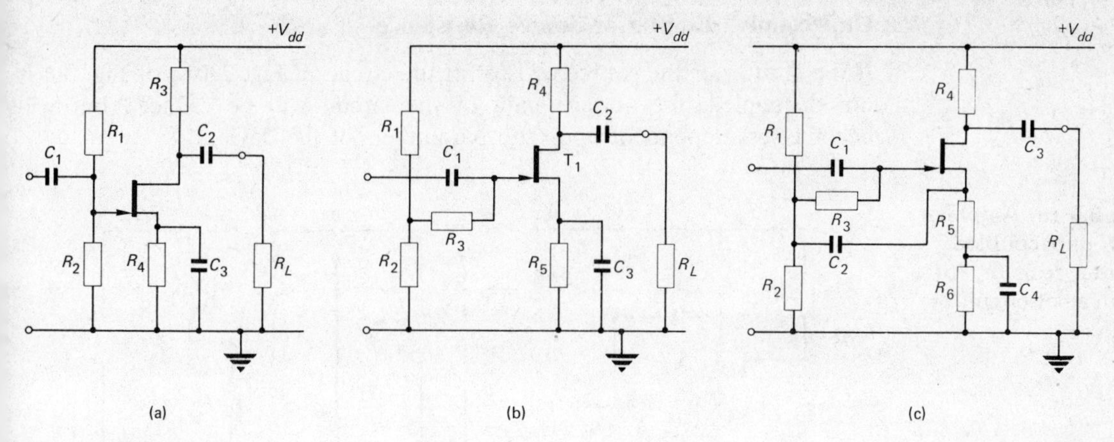

Fig. 3.4 Single-stage fet amplifier.

An increase in the input impedance can be achieved if the bias circuit is modified as shown in Fig. 3.4*b*. The input terminal is connected via the coupling capacitor C_1 to the gate of T_1 and the d.c. bias voltage is applied by connecting R_3 between the junction of R_1 and R_2 and the gate terminal. R_3 is selected to be 1 MΩ or more and so the input resistance of the amplifier is now $R_3+R_1R_2/(R_1+R_2)$ or very nearly just R_3 ohms.

A still further increase in the input resistance of the circuit can be achieved by the connection of a capacitor, C_2 in Fig. 3.4*c*, between the source terminal of T_1 and the junction of R_1 and R_2. The value of C_2 is such that it has negligible reactance at all the frequencies to be amplified. The signal voltage appearing at the source is very nearly equal to the signal voltage at the gate and thus both ends of R_3 are at almost the same potential. The effective impedance of R_3 is thereby increased to a very high value and the input impedance of the fet is very little affected by the bias circuit.

Example 3.2

The circuit of Fig. 3.4*b* has $R_1=56$ kΩ, $R_2=4.7$ kΩ, $R_3=1$ MΩ, $R_4=2$ kΩ and $R_5=1$ kΩ. The parameters of the fet are $g_m=4$ mS and $r_{ds}=80$ kΩ. Calculate the voltage gain and the input resistance of the circuit when a 3 kΩ load is connected across the output terminals. Assume all capacitors have negligible reactance.

Solution The effective drain load resistance is equal to 2 kΩ in parallel with 3 kΩ or 1.2 kΩ. This resistance is much smaller than the drain-source resistance r_{ds} of the fet and this permits the approximate expression for the voltage gain, i.e. g_mR_L, to be used. Hence,

$$A_v=4\times10^{-3}\times1.2\times10^3=4.8 \quad (Ans)$$

This value is very much smaller than the value calculated for the bipolar transistor circuit in the previous example and is an indication that generally the bipolar transistor will provide a much greater voltage gain than a fet.

The input resistance of the circuit is approximately equal to R_3 or 1M Ω (*Ans*)

Undecoupled Emitter or Source Resistance

If the emitter or the source resistor of the circuit of Fig. 3.1 or of Fig. 3.4 is not decoupled, the voltage gain of the circuit will be reduced but will become less dependent upon the parameters of the device.

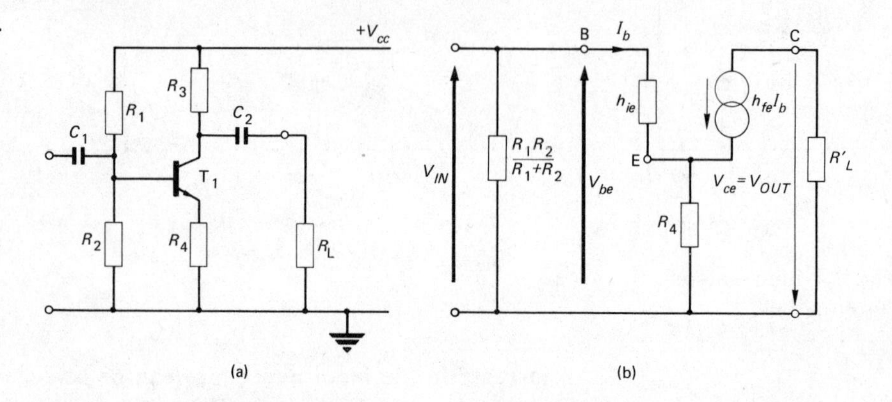

Fig. 3.5 (*a*) Amplifier with undecoupled emitter resistor, (*b*) equivalent circuit of (*a*).

Fig. 3.5*a* shows a common-emitter amplifier with its emitter resistor undecoupled and Fig. 3.5*b* shows its *h*-parameter equivalent circuit. Both h_{re} and h_{oe} have been neglected and

$$R'_L = R_3R_L/(R_3+R_L)$$

From Fig. 3.5*b* the current gain A_i of the transistor is

$$A_i = I_c/I_b = h_{fe}I_b/I_b = h_{fe}$$

Thus the current gain is not affected by the undecoupled emitter resistor.

The input resistance R_{IN} of the amplifier is equal to the input resistance of the transistor $R_{IN(T)}$ in parallel with the bias components R_1 and R_2. From Fig. 3.5*b*,

$$V_{be} = I_bh_{ie} + (1+h_{fe})I_bR_4$$

Hence,

$$R_{IN(T)} = V_{be}/I_b = h_{ie} + (1+h_{fe})R_4 \simeq h_{ie} + h_{fe}R_4 \tag{3.1}$$

Usually, $h_{fe}R_4 \gg h_{ie}$ and then

$$R_{IN(T)} = h_{fe}R_4 \tag{3.2}$$

Clearly, one effect of the undecoupled emitter resistor is to increase the input resistance of the circuit although this is reduced by the bias resistors.

The voltage gain A_v of the circuit is

$$A_v = V_{OUT}/V_{IN} = V_{ce}/V_{be} = \frac{I_cR'_L}{I_bR_{IN(T)}} = \frac{h_{fe}I_bR'_L}{I_b(h_{ie}+h_{fe}R_4)}$$

$$= \frac{h_{fe}R'_L}{h_{ie}+h_{fe}R_4} \tag{3.3a}$$

Very often $h_{fe}R_4 \gg h_{ie}$ and then

$$A_v = \frac{h_{fe}R'_L}{h_{fe}R_4} = \frac{R'_L}{R_4} \qquad (3.3b)$$

The voltage gain given by equations (3.3a) and (3.3b) is considerably less than the value obtained with the emitter decoupled but it is much more stable. The use of an undecoupled emitter resistor applies **voltage-current negative feedback** to the circuit and these results will be derived in an alternative manner in the next chapter.

Fig. 3.6 (*a*) Fet amplifier with undecoupled source resistor, (*b*) equivalent circuit of (*a*).

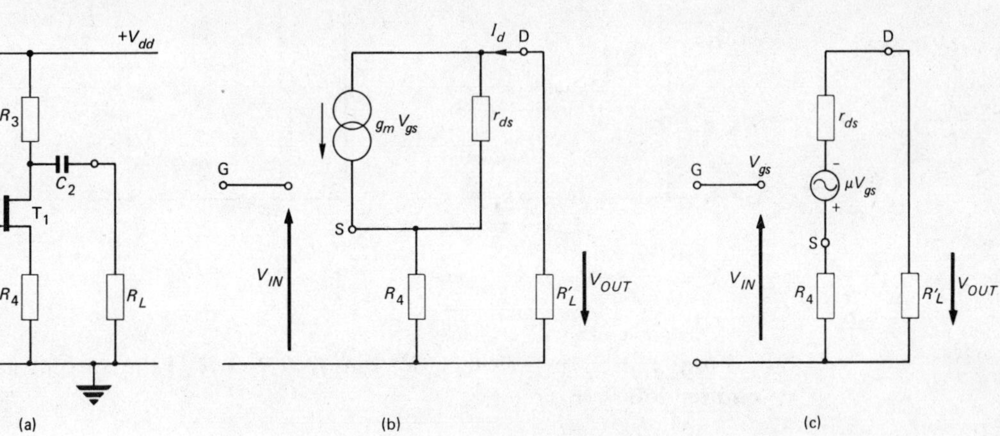

The circuit of a common source jfet amplifier with its source resistor undecoupled is shown in Fig. 3.6*a* and its equivalent circuit is in Fig. 3.6*b*.

Applying Thevenin's theorem to the drain-source terminals in Fig. 3.6*b* allows the equivalent circuit to be redrawn as shown by Fig. 3.6*c*. The voltage generator has an e.m.f. of μV_{gs} where

$$\mu = g_m r_{ds}$$

and is the **amplification factor** of the fet.

From Fig. 3.6*c* the drain current I_d is

$$I_d = \mu V_{gs}/(R_4 + r_{ds} + R'_L)$$

The gate-source voltage V_{gs} is equal to the difference between the input voltage V_{IN} and the voltage dropped across the source resistor R_4. Hence,

$$I_d = \mu(V_{IN} - I_d R_4)/(R_4 + r_{ds} + R'_L)$$

$$I_d = \frac{\mu V_{IN}}{R_4(1+\mu) + r_{ds} + R'_L} \qquad (3.4)$$

The output voltage V_{OUT} of the circuit is the voltage developed across the effective load resistance R'_L and so the voltage gain A_v is

$$A_v = V_{OUT}/V_{IN} = \frac{\mu R'_L}{R_4(1+\mu) + r_{ds} + R_L} \qquad (3.5)$$

Emitter and Source Followers

Figs 3.7a and b show, respectively, the circuit of an emitter follower and it a.c. equivalent circuit. When the input signal voltage goes positive, the bas current is increased and the current gain of the transistor gives an increase i the emitter current. This current passes through the emitter load an develops a positive-going voltage, i.e. the output voltage *follows* the inpu voltage.

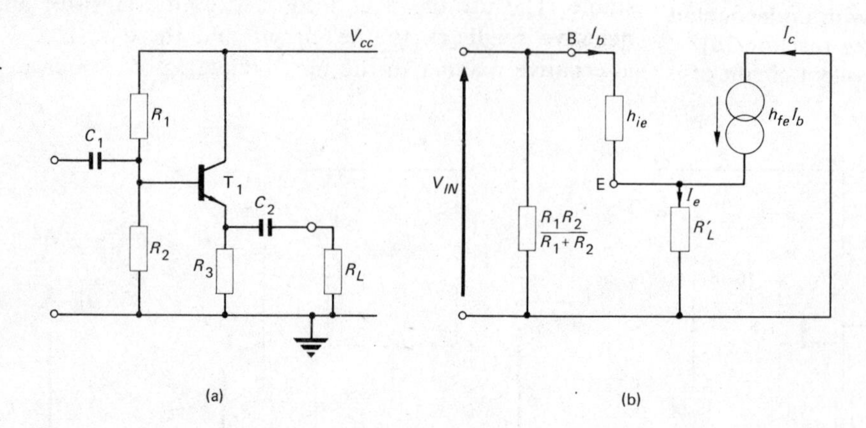

Fig. 3.7 (a) Emitter follower, (b) equivalent circuit of (a).

From Fig. 3.7b [in which $R'_L = R_LR_3/(R_L+R_3)$] the current gain of th **emitter follower** circuit is

$$A_i = I_e/I_b = I_b(1+h_{fe})/I_b = 1+h_{fe}$$

Also $V_{IN} = I_bh_{ie} + (1+h_{fe})I_bR'_L$

The input resistance $R_{IN(T)}$ of the transistor is the ratio V_{IN}/I_b and hence

$$R_{IN(T)} = h_{ie} + (1+h_{fe})R'_L \simeq h_{ie} + h_{fe}R'_L \qquad (3.$$

Usually, $h_{fe}R'_L \gg h_{ie}$ and then

$$R_{IN(T)} \simeq h_{fe}R'_L \qquad (3.$$

The input resistance of a transistor in the common-collector configuratio is high but this value is reduced by the shunting effect of the bias resisto R_1 and R_2.

The voltage gain A_v of the circuit is the ratio V_{OUT}/V_{IN} and so

$$A_v = \frac{I_b(1+h_{fe})R'_L}{I_b(h_{ie}+h_{fe}R'_L)} \simeq \frac{h_{fe}R'_L}{h_{ie}+h_{fe}R'_L} \qquad (3.$$

It should be clear from equation (3.8) that the voltage gain of an emitt follower must always be less than unity. The gain will more nearly approac unity as R_3 is increased in value but the maximum value that can b employed is determined by the collector supply voltage V_{cc} and the ma imum output signal voltage that is required. The emitter resistance can b still further increased in value if a second power supply voltage is availab (Fig. 3.8).

Fig. 3.8 Emitter follower with dual power supplies.

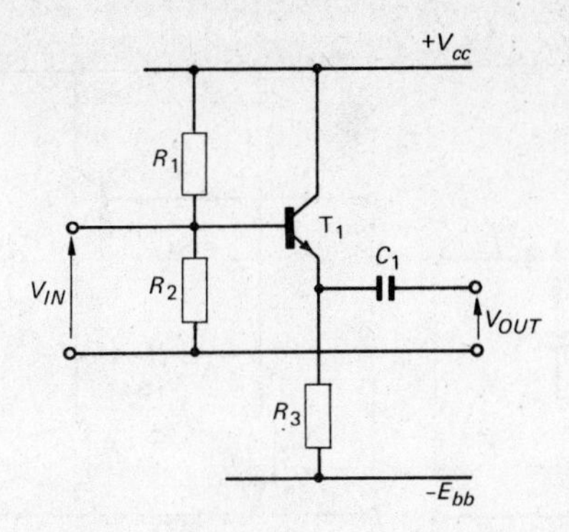

The output resistance of an emitter follower can best be determined by assuming that the input terminals are closed in an impedance equal to the source impedance and that a generator of e.m.f. E volts and zero internal resistance is connected across the output terminals (see Fig. 3.9). The output resistance of the transistor is $R_{OUT(T)} = E/I_e$ ohms and the output resistance R_{OUT} of the emitter follower is $R_{OUT(T)}$ in parallel with the emitter resistor R_3 (Fig. 3.9*b*).

Fig. 3.9 Determination of the output resistance of an emitter follower.

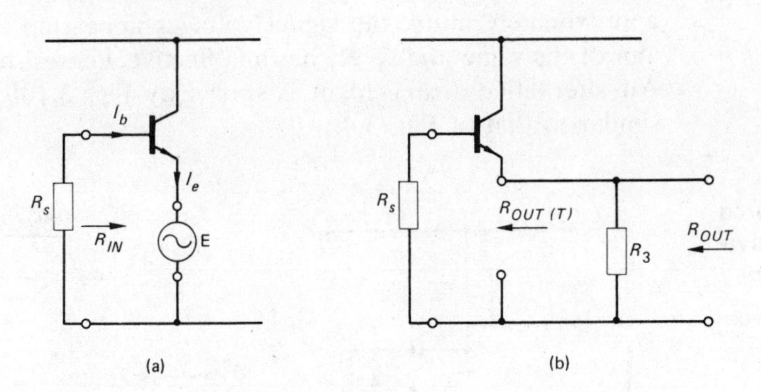

From Fig. 3.9*a*,

$$I_e = I_b(1+h_{fe}) = E(1+h_{fe})/(R_S+R_{IN})$$

Hence $\quad R_{OUT(T)} = E/I_e = (R_S+R_{IN})/(1+h_{fe})$

The emitter follower is commonly employed to connect a high impedance source to a low impedance load with little loss of signal voltage and so usually $R_S \gg R_{IN}$. Then,

$$R_{OUT(T)} \simeq R_S/h_{fe} \tag{3.9}$$

and hence the output resistance of an emitter follower is

$$R_{OUT} = \frac{\dfrac{R_S}{h_{fe}} \times R_3}{(R_S/h_{fe})+R_3} = \frac{R_S R_3}{R_S + h_{fe} R_3} \tag{3.10}$$

The high input impedance of an emitter follower is shunted by the bias

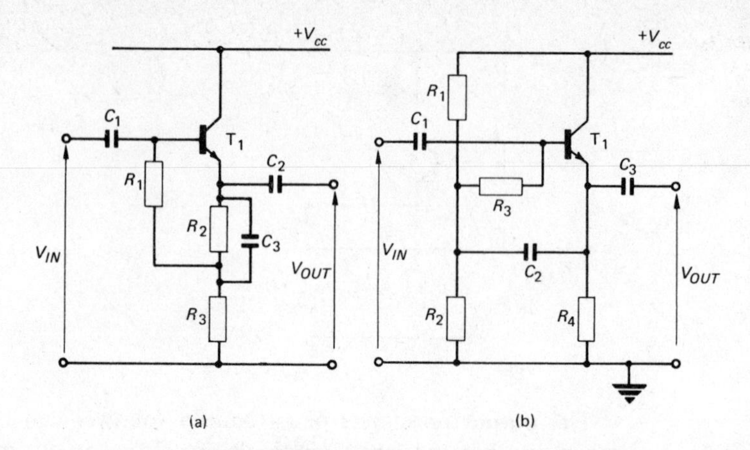

Fig. 3.10 Boot-strapped emitter follower.

resistors R_1 and R_2. When a very high input impedance is needed, th circuit must be **bootstrapped** (see Fig. 3.10). In Fig. 3.10*a* the emitte resistance has been divided into two parts so that the required bias voltag appears at their junction. The junction of R_2 and R_3 is connected to th base of T_1 by resistor R_1. Since the voltage gain of an emitter follower i approximately unity, the signal voltages appearing at each end of R_1 ar almost the same and so R_1 has an effective a.c. resistance that is very large An alternative arrangement is shown by Fig. 3.10*b*; its operation is ver similar to that of Fig. 3.4*c*.

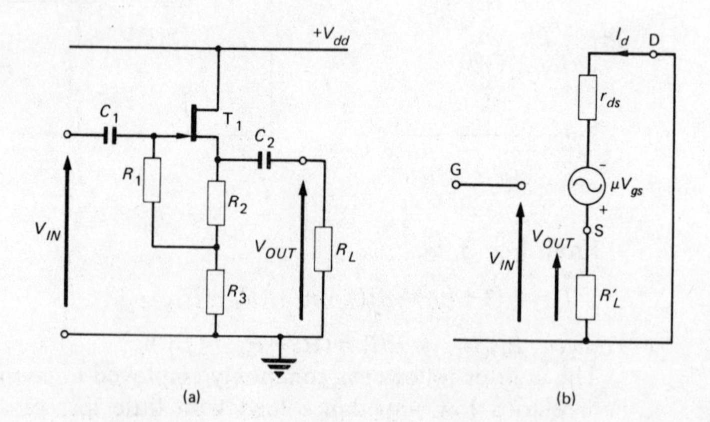

Fig. 3.11 (*a*) Source follower, (*b*) equivalent circuit of (*a*).

The circuit of a **source follower** is shown in Fig. 3.11*a* and its a.c equivalent circuit is given in Fig. 3.11*b*. The gate-source voltage V_{gs} is th difference between the input voltage V_{IN} and the voltage developed acros the effective source resistance R'_L. Hence,

$$I_d = \mu[V_{IN} - I_d R'_L]/(r_{ds} + R'_L)$$

$$I_d[r_{ds} + R'_L(1+\mu)] = \mu V_{IN}$$

$$I_d = \mu V_{IN}/[r_{ds} + R'_L(1+\mu)]$$

The output voltage $V_{OUT} = I_d R'_L$ and so the voltage gain A_v of the circuit is

$$A_v = V_{OUT}/V_{IN} = \frac{\mu R'_L}{r_{ds} + R'_L(1+\mu)} \tag{3.11}$$

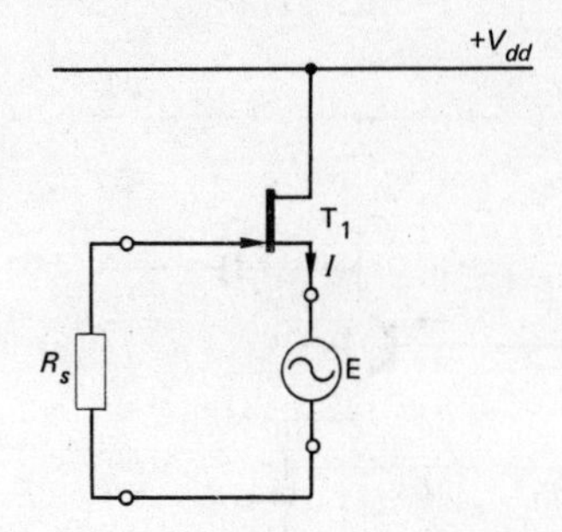

Fig. 3.12 Determination of the output resistance of a source follower.

The output resistance of the circuit can be determined in a similar manner to that employed with the emitter follower. Referring to Fig. 3.12,

$$R_{OUT(T)} = E/I = E/g_m E = 1/g_m \,\Omega$$

This resistance appears in parallel with the source resistance $(R_2 + R_3)$ and

$$R_{OUT} = \frac{R_2 + R_3}{1 + g_m[R_2 + R_3]} \tag{3.12}$$

Example 3.3

A mosfet is connected as a source follower to interface between a high-resistance source and a 75 Ω load. If the mosfet has a g_m of 2 mS and $r_{ds} = 20$ kΩ calculate i) the required source resistor to match the follower to its load and ii) the voltage gain then obtained.

Solution The output resistance of the mosfet is $1/g_m$ or 500 Ω and this value in parallel with the source resistance R_S is required to be equal to 75 Ω. Therefore,

$$75 = R_S \times 500/(R_S + 500) \quad \text{or} \quad R_S = 88\,\Omega \quad (Ans)$$

Hence $R'_L = 75 \times 88/(75 + 88) = 40.5\,\Omega$

From equation (3.11)

$$A_v = \frac{2 \times 10^{-3} \times 20 \times 10^3 \times 40.5}{20 \times 10^3 + 40.5(1 + 2 \times 10^{-3} \times 20 \times 10^3)} = 0.07 \quad (Ans)$$

Multiple Stages

Very often the voltage gain wanted from an amplifier is larger than can be provided by a single stage. Then two or more stages must be connected in cascade. The overall voltage gain of a **multi-stage amplifier** is the product of the individual stage gains. For example, a two-stage amplifier has an overall voltage gain $A_v = A_1 A_2$ and the gain of an n-stage amplifier is $A_v = A_1 A_2 \cdots A_n$. The signal voltage is applied to the input terminals of the first stage, amplified, and then the amplified signal is applied to the input terminals of the next stage and so on for each stage in the amplifier. The output signal voltage of one stage is the input signal of the next stage and sc some means of **coupling** the stages together is required. The overall current gain is not equal to the product of the current gains of the individual stages because of losses in the coupling circuits.

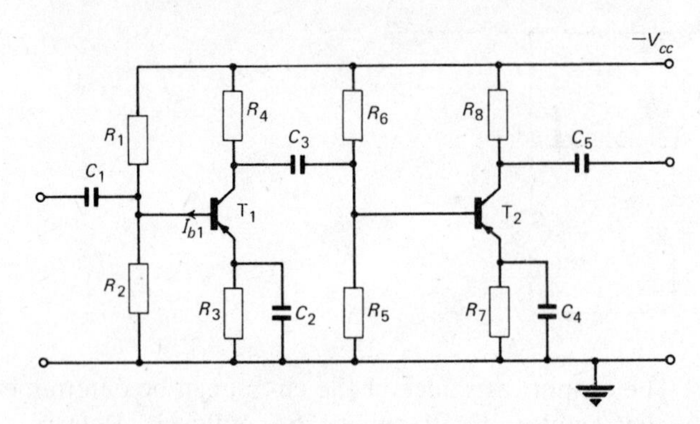

Fig. 3.13 *RC* coupled amplifier.

With discrete circuitry, amplifier stages are often coupled together by means of a capacitor and Fig. 3.13 shows the circuit of an *RC* couplec bipolar transistor amplifier. At most signal frequencies, the reactances of the coupling capacitors C_1, C_3 and C_5, and also of the decoupling capacitors C and C_4 are negligibly small and can be ignored. Then the overall gain of the amplifier has its maximum value. At both low and high frequencies the voltage gain is reduced because of capacitive effects, either resulting from the reactances of C_1, etc. no longer being negligibly small or from stray and transistor capacitances. Integrated circuit stages and many discrete design use direct coupling, examples of which are shown in Figs. 3.14 and 3.15.

If two n-p-n transistors are directly connected together, the collector o the first will be at the same potential as the base of the second. The emitte potential of the second transistor must then be set at a value slightly les than the base potential in order to obtain the required base/emitter bia voltage. This means that the collector potential of T_2 must be more positiv than the collector potential of T_1. This effect is repeated if more than tw stages are cascaded. The collector potentials of successive stages will becom more and more positive and the peak maximum output voltage will b reduced. This effect can be avoided if a combination of n-p-n and p-n- transistors is used as shown by the circuit of Fig. 3.15.

Example 3.4

Calculate the overall voltage gain of the two-stage amplifier shown in Fig. 3.16 whe its output terminals are connected to a 5 kΩ load resistor. The *h*-parameters of th transistors are $h_{ie} = 1000\ \Omega$ and $h_{fe} = 150$. Assume the reactances of the capacito C_1, C_2, C_3 and C_4 are all negligibly small.

Solution The effective collector load resistance of T_2 is 2.7 kΩ in parallel wi 5 kΩ or 1.75 kΩ, and hence the voltage gain of the second stage is

$$150 \times 1.75/1 = 262.5$$

Fig. 3.14 n-p-n/n-p-n d.c. amplifier.

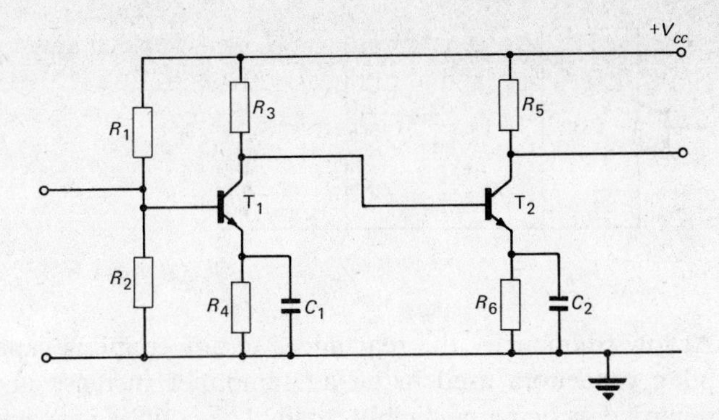

Fig. 3.15 n-p-n/p-n-p d.c. amplifier.

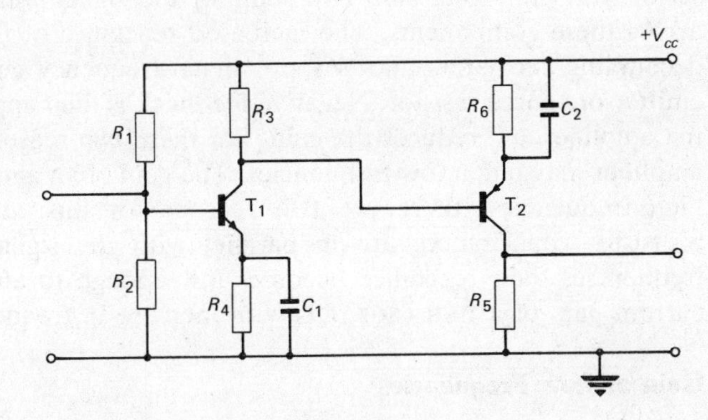

Fig. 3.16

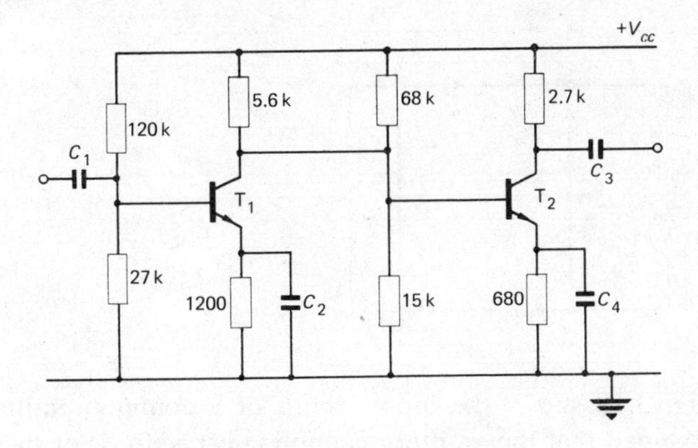

Transistor T_1 works into a load which consists of four resistances in parallel (see Fig. 3.17). Thus, its effective collector load is 794 Ω. Hence, the voltage gain of the first stage is

$$150 \times 794/1000 \quad \text{or} \quad 119.1$$

The overall voltage gain of the amplifier is

$$A_v = 119.1 \times 262.5 = 31\,263.8 \quad (Ans)$$

Fig. 3.17

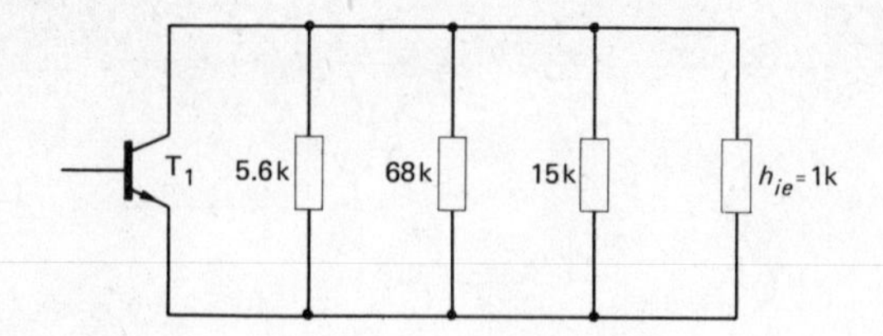

Variation of Gain with Frequency

At low frequencies the reactances of any coupling capacitors and/or decoupling capacitors used in an a.f. amplifier increase and may no longer be regarded as being negligibly small. The coupling capacitors are connected in series with the signal path and some of the signal voltage will be dropped across these components. The increased reactance of the emitter or source decoupling capacitors allows some signal-frequency current to flow in the emitter or source resistor. Negative feedback is then applied to each stage in the amplifier and reduces the gain. For these two reasons the gain of an a.f. amplifier may fall at low frequencies. The gain of an amplifier will also fall at high frequencies; there are two reasons for this: *a*) various stray and transistor capacitances are in parallel with the signal path and at high frequencies their reactance becomes low enough to affect the gain; *b*) the current gain of a transistor falls with increase in frequency.

Gain at Low Frequencies

The voltage gain of an audio-frequency amplifier will fall at **low frequencies** because of the time constants of the coupling and the decoupling circuits.

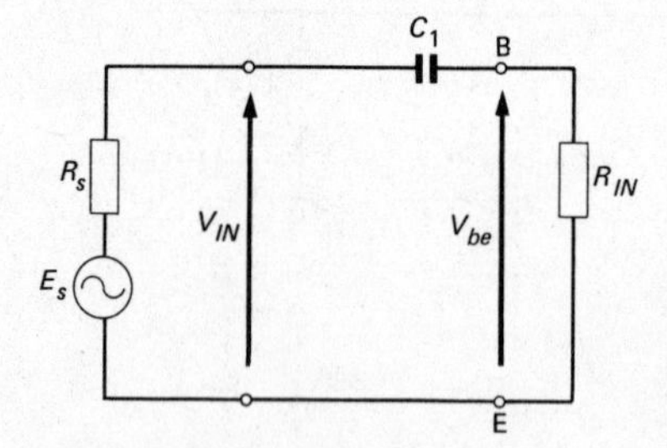

Fig. 3.18 Input circuit of a common-emitter stage.

Fig. 3.18 shows the input circuit of a common-emitter stage. It has been assumed that the emitter decoupling capacitor is of such a large value that it has negligible reactance at all the signal frequencies of interest. The input resistance R_{IN} of the stage is equal to the h_{ie} of the transistor in parallel with the bias components R_1 and R_2. At middle frequencies when C_1 has negligible reactance

$$V_{IN} = V_{be} = E_S R_{IN}/(R_S + R_{IN})$$

and the output voltage of the amplifier is

$$V_{OUT} = g_m V_{be} R_L$$

Hence the medium-frequency voltage gain V_{OUT}/V_{IN} is

$$A_{v(MF)} = g_m R_L$$

At low frequencies

$$V_{be} = \frac{E_S R_{IN}}{R_S + R_{IN} + 1/j\omega C_1}$$

and

$$V_{IN} = \frac{E_S(R_{IN} + 1/j\omega C_1)}{R_S + R_{IN} + 1/j\omega C_1}$$

Since $V_{OUT} = g_m V_{be} R_L$, the voltage gain $A_{v(LF)}$ at low frequencies is

$$A_{v(LF)} = V_{OUT}/V_{IN} = \frac{g_m R_L R_{IN}}{R_{IN} + 1/j\omega C_1} = \frac{A_{v(MF)}}{1 + 1/j\omega C_1 R_{IN}} \tag{3.13}$$

When the low-frequency gain has fallen by 3 dB on its mid-frequency value,

$$\left|\frac{A_{v(LF)}}{A_{v(MF)}}\right| = \frac{1}{\sqrt{2}} = \frac{1}{\sqrt{[1 + 1/\omega_1^2 C_1^2 R_{IN}^2]}} \quad \text{and} \quad \omega_1 = 1/C_1 R_{IN} \tag{3.14}$$

The expression for the low-frequency voltage gain can be written as

$$A_{v(LF)} = \frac{A_{v(MF)}}{1 - j\omega_1/\omega} \tag{3.15}$$

where $f_1 = \omega_1/2\pi$ is the lower 3 dB frequency of the amplifier.

If the voltage gain is taken as V_{OUT}/E_S then

$$A_{v(HF)} = \frac{g_m R_{IN} R_L}{R_S + R_{IN} + 1/j\omega C_1} = \frac{A_{v(MF)}}{1 + 1/j\omega C_1(R_S + R_{IN})}$$

and the lower 3 dB frequency is $\omega_1 = 1/C_1(R_S + R_{IN})$

For a fet amplifier the same analysis can be applied except that $R_{IN} \simeq h_{ie}$ is replaced by the *extremely* high input impedance of the device and V_{be} is replaced by V_{gs}.

Example 3.5

A bipolar transistor amplifier has a collector load resistance of 4.7 kΩ and is coupled by a capacitor to a 1 kΩ resistive load. Calculate the required value of the coupling capacitor to make the lower 3 dB frequency of the circuit equal to 50 Hz. Find also the loss introduced by the coupling capacitor at 25 Hz.

Solution Fig. 3.19 shows the coupling circuit. At middle frequencies the voltage across the 1 kΩ load is

$$g_m V_{be}[4700 \times 1000/(4700 + 1000)] = 825\ g_m V_{be}$$

At low frequencies the current flowing in the load is

$$\frac{g_m V_{be} \times 4700}{5700 + 1/j\omega C_1}$$

Fig. 3.19

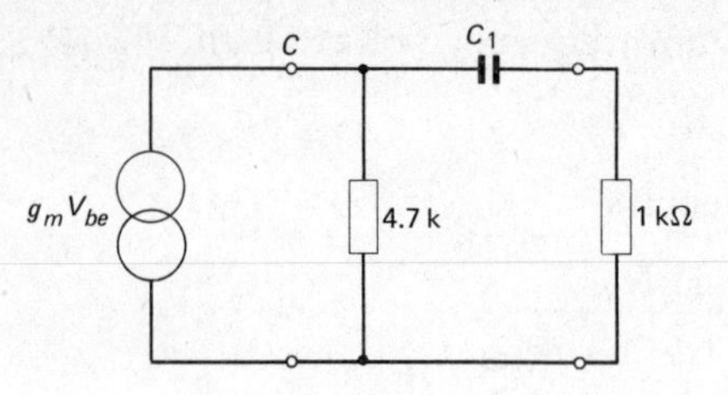

and so the load voltage is

$$\frac{g_m V_{be} \times 4700 \times 1000}{5700 + 1/j\omega C_1}$$

$$= \frac{g_m V_{be} \times 4.7 \times 10^6}{5700} \times \frac{1}{1 + \dfrac{1}{j\omega 5700 C_1}} = \frac{V_{L(MF)}}{1 + \dfrac{1}{j\omega 5700 C_1}}$$

The load voltage will be 3 dB down on its mid-frequency value when

$$C_1 = \frac{1}{5700\omega_1} = \frac{1}{5700 \times 2\pi \times 50} = 0.558\ \mu F \quad (Ans)$$

With this value of coupling capacitor, the load voltage at 25 Hz relative to the mid-frequency value is

$$\left|\frac{V_{L(MF)}}{V_{L(HF)}}\right| = \frac{1}{\sqrt{\left[1 + \dfrac{1}{(2\pi \times 25 \times 0.558 \times 10^{-6} \times 5700)^2}\right]}}$$

$$= \frac{1}{\sqrt{\left(1 + \dfrac{1}{0.25}\right)}} = 0.447 = -7\ \text{dB} \quad (Ans)$$

Decoupling of an emitter or source resistor becomes increasingly ineffective as the frequency is reduced and n.f.b. is applied to reduce the voltage gain. The analysis of this effect is considerably more complicated than that for the coupling capacitor and is beyond the scope of this book. The overall low-frequency response of a stage will depend upon the time constants of both the coupling and the decoupling capacitors. Many integrated circuit amplifiers do not employ decoupling capacitors because of the large values needed, and their low-frequency response is determined solely by any input and/or output coupling capacitors fitted.

Gain at High Frequencies

The voltage gain of an amplifier will fall at **high frequencies** because of the unavoidable transistor and stray capacitances. It is more convenient to use the hybrid-π equivalent circuit for analysis since its capacitances $C_{b'e}$ and $C_{b'c}$ account for the variation in the current gain of the transistor.

The hybrid-π equivalent circuit of a single-stage transistor amplifier is shown in Fig. 3.20, where $r_{b'c}$ has been omitted and R'_L is the effective

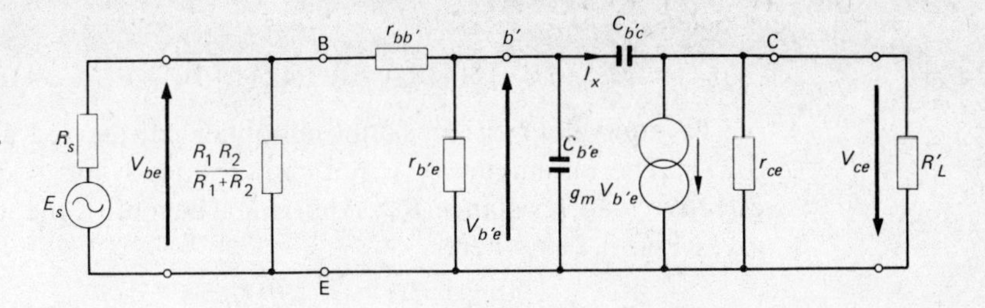

Fig. 3.20 High-frequency hybrid-π equivalent circuit of a single-stage amplifier.

collector load resistance. To simplify the analysis the bias resistors R_1 and R_2 will be assumed to have little effect and will be omitted from the circuit.

At medium frequencies the capacitances $C_{b'c}$ and $C_{b'e}$ have negligible effect on the performance of the circuit and then

$$\text{Voltage gain } A_v = V_{ce}/E_S = g_m R'_L r_{b'e}/(R_S + r_{b'e} + r_{bb'}) \quad \text{(see p. 25)}$$

$$\text{Voltage gain } V_{ce}/V_{b'e} = g_m R'_L$$

At high frequencies the current I_x flowing through $C_{b'c}$ is

$$I_x = (V_{b'e} - V_{ce}) j\omega C_{b'c} = (V_{b'e} + g_m R'_L V_{b'e}) j\omega C_{b'c}$$

and $I_x/V_{b'e} = (1 + g_m R'_L) j\omega C_{b'c}$

This equation shows that the input admittance of the circuit to the right of $r_{b'e}$ is

$$j\omega C_{b'e} + j\omega C_{b'c}(1 + g_m R'_L)$$

This represents an enhanced capacitance which is effectively connected between the b' and e terminals of the transistor. This means that $C_{b'e}$ and $C_{b'c}$ can be replaced in the equivalent circuit by a single capacitor, labelled C_{IN} in Fig. 3.21. The increase in the input capacitance can be quite large and is generally known as the *Miller Effect.*

Fig. 3.21 Simplified version of Fig. 3.20.

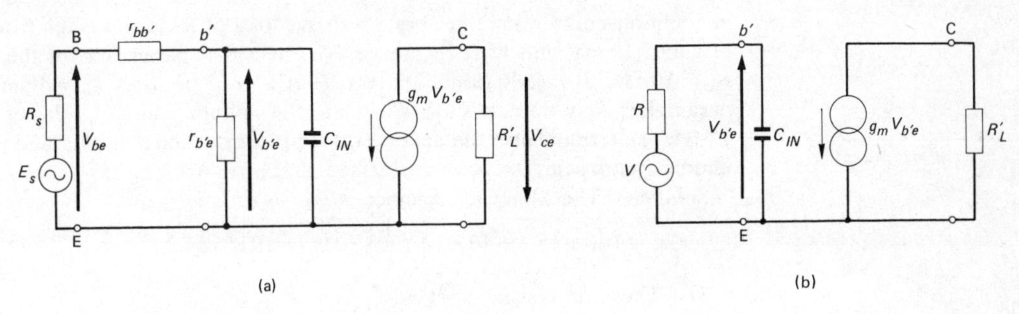

Example 3.6

The hybrid-π parameters of a bipolar transistor are $g_m = 30$ mS, $C_{b'e} = 40$ pF and $C_{b'c} = 4.5$ pF. If the effective collector load resistance is 2 kΩ calculate the input capacitance of the transistor.

Solution

$$C_{IN} = C_{b'e} + C_{b'c}(1 + g_m R'_L) = 40 + 4.5(1 + 30 \times 10^{-3} \times 2 \times 10^3) = 315 \text{ pF} \quad (Ans)$$

The equivalent circuit of the amplifier can be redrawn as shown by Fig. 3.21*a*. The parameter r_{ce} is not shown since it has been combined with the effective load resistance R'_L. Applying Thevenin's theorem to the left of $C_{b'e}$ gives Fig. 3.21*b* where

$$R = \frac{r_{b'e}(R_S + r_{bb'})}{r_{b'e} + r_{bb'} + R_S} \quad \text{and} \quad V = \frac{E_S r_{b'e}}{R_S + r_{bb'} + r_{b'e}} = V_{b'e}$$

at middle frequencies. At high frequencies, from Fig. 3.21*b*,

$$V_{b'e(HF)} = \frac{V \times 1/j\omega C_{IN}}{R + 1/j\omega C_{IN}} = V/(1 + j\omega C_{IN} R)$$

or $\quad V_{b'e(HF)} = V_{b'e(MF)}/(1 + j\omega C_{IN} R)$

The output voltage of the stage is $V_{OUT} = g_m V_{b'e} R'_L$ and hence

$$A_{v(HF)} = A_{v(MF)}/(1 + j\omega C_{IN} R) \tag{3.16}$$

The high-frequency voltage gain falls by 3 dB on its mid-frequency value at the *upper 3 dB frequency* $\omega_2/2\pi$ when

$$\left|\frac{A_{v(HF)}}{A_{v(MF)}}\right| = \frac{1}{\sqrt{2}} = \frac{1}{\sqrt{[1 + \omega_2^2 C_{IN}^2 R^2]}} \quad \text{or} \quad \omega_2 = 1/C_{IN} R \tag{3.17}$$

Equation (3.16) can now be rewritten as

$$A_{v(HF)} = A_{v(MF)}/(1 + j\omega/\omega_2) \tag{3.18}$$

Example 3.7

A common-emitter amplifier has a resistive load of 3 kΩ and is fed from a generator of e.m.f. 10 mV and 1 kΩ resistance. The hybrid-π parameters of the transistor are $r_{bb'} = 100\ \Omega$, $r_{b'e} = 800\ \Omega$, $C_{b'e} = 100$ pF, $C_{b'c} = 15$ pF and $g_m = 40$ mS; the other parameters may be neglected. Calculate the voltage gain V_{OUT}/E_S of the circuit at 1 MHz. Determine also the upper 3 dB frequency of the amplifier and the gain at the middle frequencies.

Solution The Miller capacitance is

$$C_{IN} = 100 + 15(1 + 40 \times 10^{-3} \times 3 \times 10^3) = 1915 \text{ pF}$$

The Thevenin resistance R is

$$800(100 + 1000)/(800 + 100 + 1000) = 463.2\ \Omega$$

and

$$V = \frac{10 \times 10^{-3} \times 800}{1000 + 100 + 800} = 4.21 \text{ mV}$$

Fig. 3.22

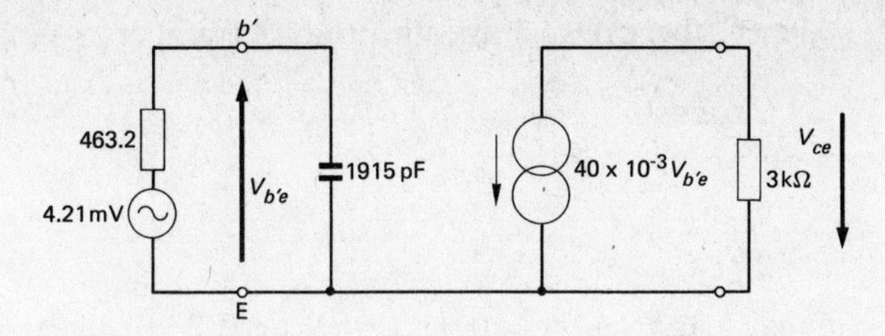

From Fig. 3.22

$$V_{b'e} = \frac{4.21 \times 10^{-3}}{1 + j2\pi \times 10^6 \times 1915 \times 10^{-12} \times 463.2} \quad \text{or} \quad |V_{b'e}| = 7.44 \times 10^{-4}\ \text{V}$$

Hence $|V_{OUT}| = 40 \times 10^{-3} \times 7.44 \times 10^{-4} \times 3 \times 10^3 = 0.09$ V

and the voltage gain is $0.09/10 \times 10^{-3} = 9$ (*Ans*)

From equation (3.17),

$$f_2 = \frac{1}{2\pi \times 1915 \times 10^{-12} \times 463.2} = 179.42\ \text{kHz} \quad (Ans)$$

At middle frequencies the transistor capacitances can be ignored. Then

$$V_{b'e} = \frac{10 \times 10^{-3} \times 800}{1000 + 100 + 800} = 4.21\ \text{mV}$$

and $V_{OUT} = 40 \times 10^{-3} \times 4.21 \times 10^{-3} \times 3 \times 10^3 = 0.51$ V.

Therefore Voltage gain $= 0.51/10 \times 10^{-3} = 51$ (*Ans*)

Gain-Bandwidth Product

When the output terminals of the transistor are short-circuited to alternating currents, its gain-bandwidth product is the parameter f_t. In similar fashion with a collector load R_L the gain-bandwidth product is the product of the voltage gain at middle frequencies and the upper 3 dB frequency.

The voltage gain at middle frequencies V_{OUT}/E_S is

$$g_m R_L r_{b'e}/(R_S + r_{bb'} + r_{b'e})$$

Also, the gain $V_{OUT}/V_{be} = g_m R_L r_{b'e}/(r_{bb'} + r_{b'e})$

and $f_2 = 1/2\pi C_{IN} R$

where $R = \dfrac{r_{b'e}(r_{bb'} + R_S)}{r_{b'e} + r_{bb'} + R_S}$ when generator resistance R_S is included

and $R = r_{b'e} r_{bb'}/(r_{b'e} + r_{bb'})$ when R_S is omitted.

Hence, the gain-bandwidth product is either

$$\frac{g_m R_L}{2\pi C_{IN}(r_{bb'} + R_S)} \tag{3.19}$$

or
$$\frac{g_m R_L}{2\pi C_{IN} r_{bb'}} \tag{3.20}$$

It is clear from equations (3.19) and (3.20) that increasing the collector load resistance will increase the gain of the amplifier but, since it will also increase C_{IN}, the bandwidth will be reduced. For a high gain-bandwidth product, the transistor must have a high f_t and a low value of $C_{b'c}$.

The FET at High Frequencies

At high frequencies the internal capacitances of a fet become important and its a.c. equivalent circuit is shown in Fig. 3.23*a*. R'_L is the effective drain load resistance and includes the drain/source resistance r_{ds}. Clearly, the circuit is very similar to the hybrid-π equivalent circuit of the bipolar transistor. Following the same steps as before the effective Miller input capacitance of the fet is

$$C_{IN} = C_{gs} + C_{gd}(1 + g_m R'_L)$$

and so the equivalent circuit can be redrawn as shown by Fig. 3.23*b*.

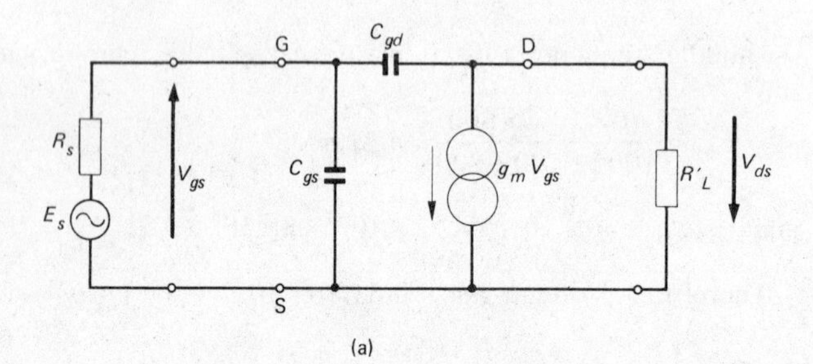

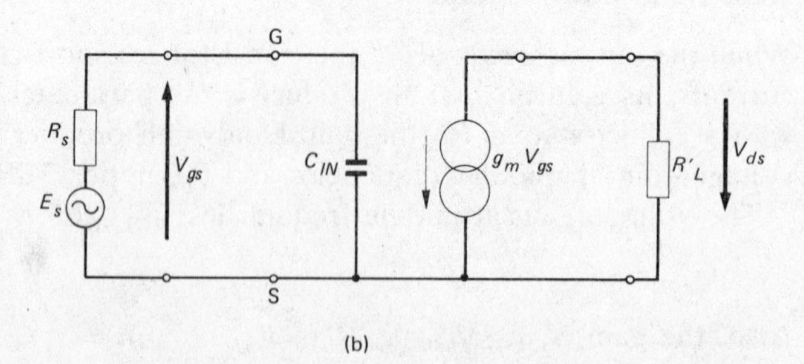

Fig. 3.23 (*a*) High-frequency equivalent circuit of a fet, (*b*) simplified equivalent circuit.

At middle frequencies $V_{gs} = E_S$ since the input impedance of the device is very high. At high frequencies

$$V_{gs} = \frac{E_S \cdot 1/j\omega C_{IN}}{R_S + 1/j\omega C_{IN}} = E_S/(1 + j\omega C_{IN} R_S)$$

Hence,

$$A_{v(HF)} = A_{v(MF)}/(1 + j\omega C_{IN} R_S) \tag{3.21}$$

The reader should note the similarity between this equation and equation (3.16).

Example 3.8

A junction fet has the following parameters: $C_{gd} = 10$ pF, $C_{gs} = 18$ pF, $g_m = 2$ mS, $r_{ds} = 50$ kΩ. Calculate *a*) the voltage gain at middle frequencies, *b*) the upper 3 dB frequency, and *c*) the gain-bandwidth product. The drain load resistance is 25 kΩ and the resistance of the generator is 10 kΩ.

Solution *a*) The mid-frequency voltage gain is

$$A_{v(MF)} = g_m R'_L = 2 \times 10^{-3} \times \frac{50 \times 10^3 \times 25 \times 10^3}{75 \times 10^3} = 33.33 \quad (Ans)$$

b) The Miller input capacitance is

$$C_{IN} = 18 \times 10^{-12} + 10 \times 10^{-12}(1 + 33.33) = 361.33 \text{ pF}$$

Hence $f_2 = \dfrac{1}{2\pi \times 361.33 \times 10^{-12} \times 10 \times 10^3} = 44.05$ kHz *(Ans)*

c) The gain-bandwidth product $= 33.33 \times 44.05 \times 10^3 = 1.468 \times 10^6$ *(Ans)*

Integrated Circuit Amplifiers

The majority of integrated circuit small-signal amplifiers utilize one kind or another of *operational amplifier* and these will be considered in Chapter 5. Relatively few low-power audio-frequency amplifiers are available and those that are are mainly either low-noise devices or stereo circuits. Some of the pre-amplifiers that are readily available are LM 381, LM 382, LM 387, and the CA 3 007.

The pin connections of the LM 381 are shown in Fig. 3.24*a* and Figs. 3.24*b* and *c* show two possible circuits using this device. It will be noticed that the LM 381 is a dual circuit and so the circuits *b* and *c* can be achieved using the same i.c.

Typical parameters for the LM 381 are as follows:
input impedance 150 kΩ.
open-loop voltage gain 3×10^5
supply voltage 10–40 V
output impedance 150 Ω
maximum output voltage $V_{cc} = -2$ V
small-signal bandwidth 15 MHz
large-signal bandwidth 75 kHz
maximum input voltage 300 mV
noise factor (see page 227) 1.25 dB.

The input signal is applied via the coupling capacitor C_1 to terminal 1. Sometimes a resistor is connected between pin 1 and earth. Negative

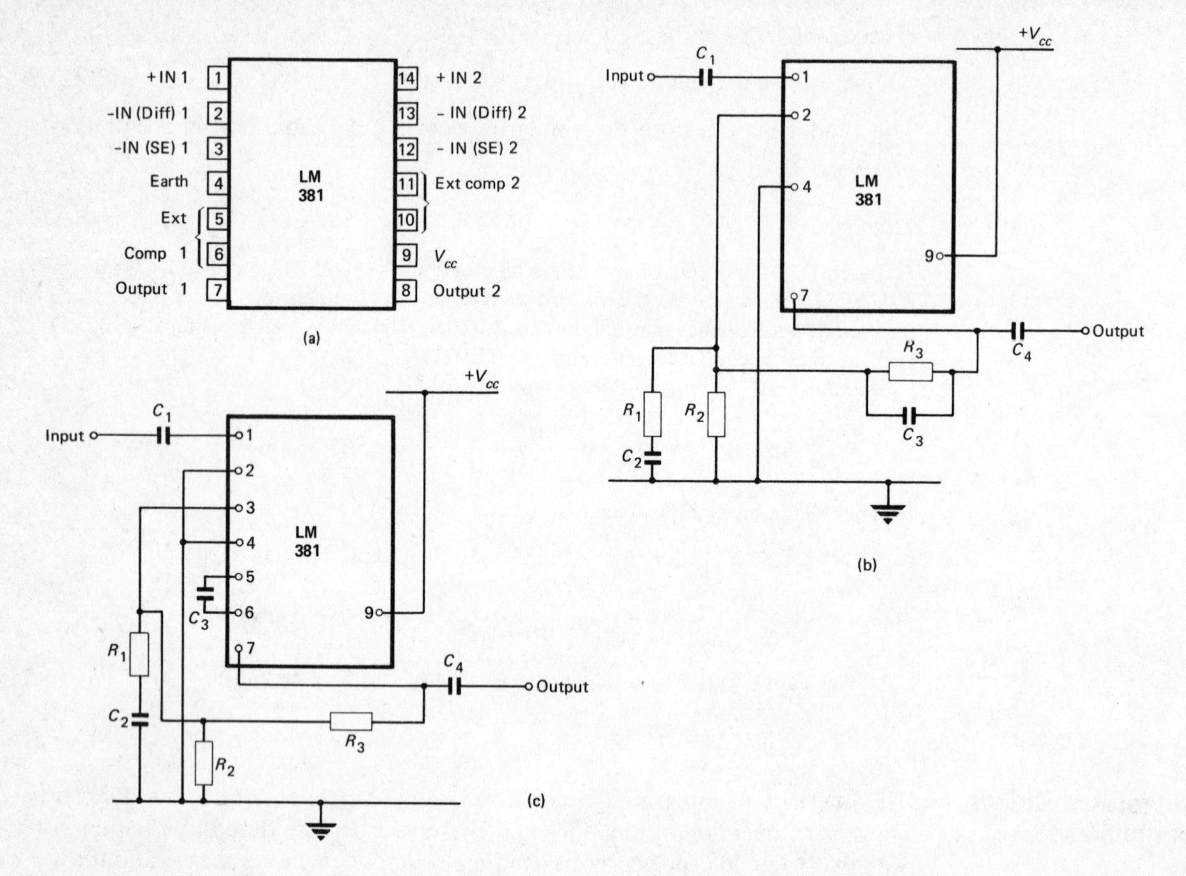

Fig. 3.24 (*a*) Pin connections of the LM 381, (*b*) and (*c*) circuits using the LM 381.

feedback is applied from the output terminal 7 to the – input terminal (2) via R_3.

The voltage gain of the circuit is $(R_1+R_3)/R_3$. (It will be seen in Chapter 5 that this means that the circuit is really a form of op-amp.)

Frequency compensation (Chapter 4) is provided in circuit *c*) by C_3 and is necessary to prevent high-frequency instability. Provided C_1 is of high value, the lower 3 dB frequency of the amplifier is determined by R_1 and C_2.

The d.c. bias voltage appearing at pin 2 is obtained by connecting a potential divider R_2+R_3 between the output pin 7 and earth.

The d.c. voltage at pin 7 is equal to $V_{cc}/2$ and hence the bias voltage is

$$\frac{V_{cc}}{2}\times\frac{R_2}{R_2+R_3}$$

Exercises 3

3.1 A common-emitter transistor has a collector load of 5 kΩ and is coupled by a large capacitor to another stage having an input impedance of 2.5 kΩ. Draw the a.c. equivalent circuit of the first stage at high frequencies and calculate *a*) the gain bandwidth product, *b*) the frequency at which the gain has fallen by 10 dB on its mid-frequency value, if the mid-frequency gain is 120 and the total shunt capacitance across the load is 100 pF.

3.2 A common-emitter transistor has a collector load of 5 kΩ and is coupled by a large capacitor to an external load of 2.5 kΩ. If the total shunt capacitance across the load is 200 pF calculate the relative voltage gain of the circuit at 1.2 MHz. Derive any expression used.

Fig. 3.25

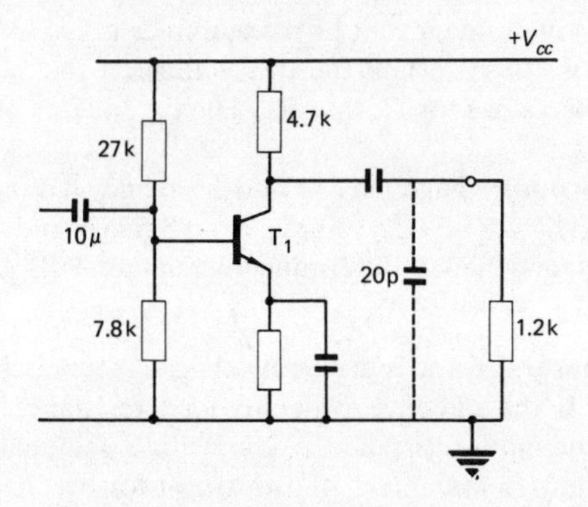

3.3 Draw the y-parameter equivalent circuit of the amplifier shown in Fig. 3.25. Use the circuit to calculate *a*) the input impedance, *b*) the mid-frequency voltage gain, and *c*) the upper 3 dB frequency of the amplifier, given $y_{ie} = 3 \times 10^{-3}$ S, $y_{fe} = 200 \times 10^{-3}$ S, $y_{re} = y_{oe} = 0$

3.4 Discuss the reasons for the fall-off in the voltage gain of an audio-amplifier at low frequencies.

An amplifier has a collector load resistor of 4.7 kΩ and is coupled to an external load resistance of 2 kΩ by a capacitor. Calculate the value of this coupling capacitor if the lower 3 dB frequency is to be 50 Hz. Calculate also the gain, relative to mid-frequencies, at 25 Hz.

3.5 Define each of the four y-parameters and draw the y-parameter equivalent circuit.

A common-emitter transistor has the following y-parameters: $y_{ie} = 500 \times 10^{-6}$ S, $y_{oe} = 40 \times 10^{-6}$ S, $y_{fe} = 0.03$ S, and $y_{re} \simeq 0$. If the collector load resistance is 8.2 kΩ, calculate *a*) the input resistance, *b*) the output resistance, *c*) the mid-frequency voltage gain, and *d*) the gain-bandwidth product of the circuit. The total capacitance in shunt with the collector load is 80 pF.

3.6 What is meant by the gain-bandwidth product of a fet amplifier that is coupled by a capacitor to its load?

A mosfet has $r_{ds} = 48$ kΩ, $g_m = 1$ mS and the drain load resistance is 20 kΩ. The mosfet is coupled to an external load by a large-valued capacitor. If the external load resistance is 15 kΩ and the total shunt capacitance is 100 pF, calculate *a*) the mid-frequency gain, *b*) the upper 3 dB frequency, *c*) the gain, relative to mid-frequencies, at three times the upper 3 dB frequency.

Fig. 3.26

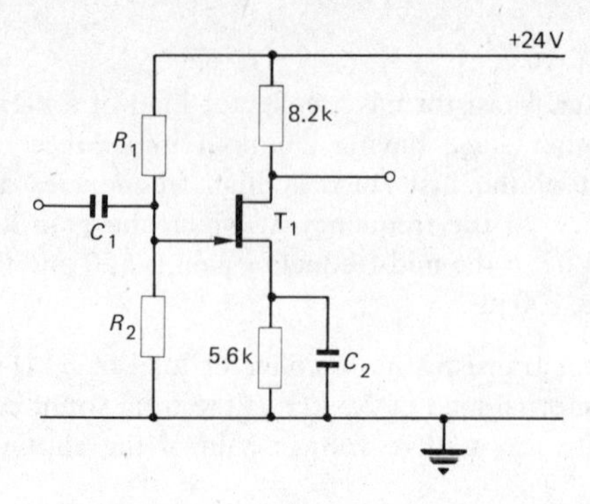

3.7 Calculate the steady drain current of the amplifier in Fig. 3.26 if, at mid-frequencies, the voltage gain is 10. Also find the d.c. voltage at the junction of R_1 and R_2 and determine possible values for R_1 and R_2. The parameters of the fet are $I_{dss} = 3.5$ mA and $V_p = -3$ V.

3.8 A jfet common-source stage has a mid-frequency voltage gain of 20. The jfet parameters are $C_{gs} = 4.8$ pF, $C_{gd} = 3$ pF, $I_{dss} = 3.5$ mA and $V_p = -3$ V. Draw the a.c. equivalent circuit and hence determine the upper 3 dB frequency of the stage if $R_S = 10$ kΩ.

3.9 A common-emitter transistor is driven by a signal source of e.m.f. 10 μV and output resistance 2 kΩ. If the effective collector load resistance is 4 kΩ calculate *a*) the current gain, *b*) the input resistance, *c*) the voltage gain, and *d*) the output resistance of the circuit. The *h*-parameters of the transistor are $h_{ie} = 4800$ Ω, $h_{fe} = 300$ and $h_{oe} = 22$ μS.

Fig. 3.27

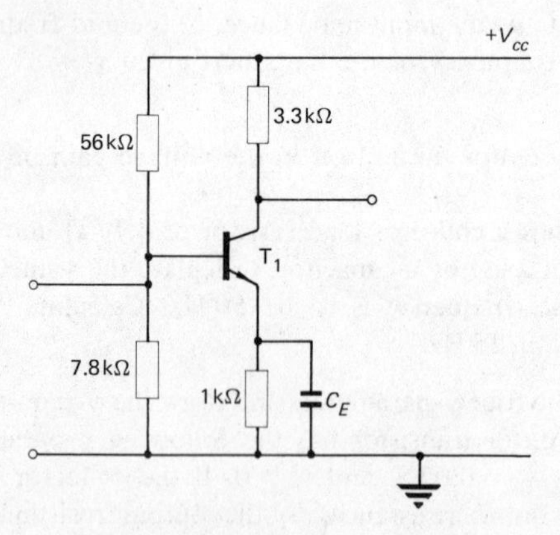

3.10 Simplify the circuit given in Fig. 3.27 and then calculate the value of emitter capacitance required to make the lower 3 dB frequency equal to *a*) 50 Hz, *b*) 25 Hz. The *h*-parameters of the transistor are $h_{fe} = 100$ and $h_{ie} = 1000$ Ω.

3.11 A single-stage transistor amplifier has an input resistance of 1000 Ω, an upper 3 dB frequency of 157 kHz, and a mid-frequency voltage gain of 200 when the collector load resistance is 2 kΩ and the impedance of the signal source is 1000 Ω. When the collector load resistance is reduced to 200 Ω the upper 3 dB frequency is 1 MHz and the mid-frequency gain is 20. Calculate the transistor capacitances $C_{b'e}$ and $C_{b'c}$

Fig. 3.28

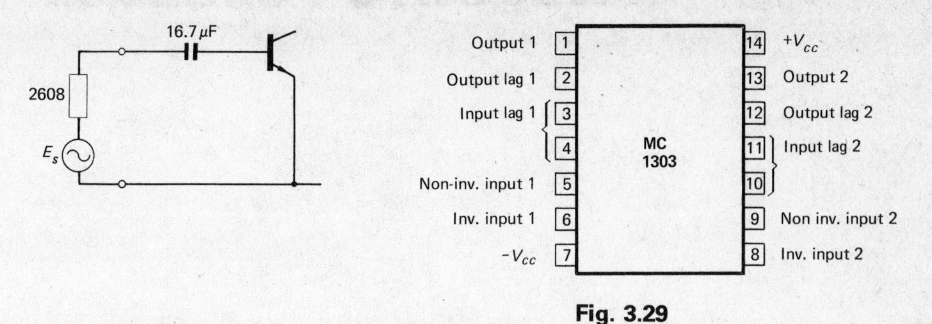

Fig. 3.29

Fig. 3.30

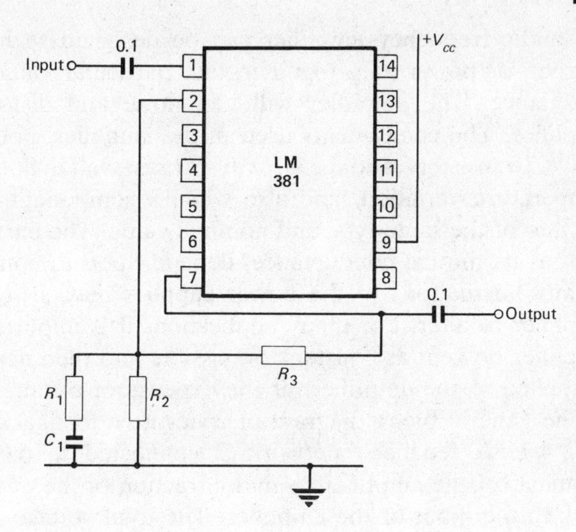

Short Exercises

3.12 An enhancement mode mosfet has $g_m = 3$ mS and a voltage gain of 12. Calculate the value of the drain load resistance, assuming it to be equal to r_{ds}.

3.13 If the transistor shown in Fig. 3.28 has an input resistance of 1000 Ω, calculate its lower 3 dB frequency.

3.14 Fig. 3.29 shows the pin connections of a MC 1303 low-noise pre-amplifier. Draw a circuit using this i.c. Note that terminals 3 and 4 (or 10 and 11) should be connected together by a 1 nF capacitor.

3.15 The LM 381 circuit of Fig. 3.30 is to have a voltage gain of 500 and a lower 3 dB frequency of 25 Hz. Calculate suitable values for R_1, R_2, R_3, and C_1.

3.16 Why is the input to an a.f. small-signal amplifier nearly always R-C coupled? Why does this affect the low-frequency gain of the circuit?

3.17 Write down the definition of each of the four h-parameters and determine the relationship between h_{fe} and g_m.

3.18 A transistor has $h_{ie} = 1200\ \Omega$, $h_{fe} = 150$, $h_{oe} = 20\ \mu$S; calculate the corresponding hybrid-π parameters.

4 Negative Feedback

Introduction

An audio-frequency amplifier can be designed to have a certain voltage, current, or power gain together with particular values of input and output impedance. The amplifier will add noise and distortion to any signal it amplifies. The components used in the amplifier, both passive (R & C) and active (transistors and i.c.s), will vary in value both with time and with temperature variation, and also when a component has to be replaced by another of the same type and nominal value. The parameters of a transistor, such as its mutual conductance, depend upon the operating conditions and so any fluctuations in the power supplies may also cause the gain of the amplifier to alter. For many applications it is important that the gain of an amplifier be kept as constant as possible and then **negative feedback** (n.f.b.) is applied to the amplifier, at the expense of a reduction in the gain.

The general block diagram of a negative feedback amplifier is shown in Fig. 4.1. A feedback network is connected in parallel with the output terminals of the amplifier so that a fraction of the output voltage can be fed back to the input of the amplifier. The total voltage applied to the input of the amplifier proper is the phasor sum of the input voltage and the fed-back voltage.

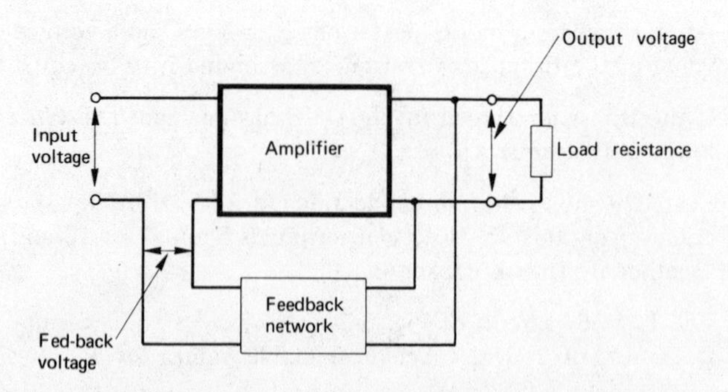

Fig. 4.1 Block diagram of a feedback amplifier.

If the fed-back voltage is in antiphase with the input voltage, the feedback is *negative* and the overall gain of the amplifier will be reduced. Conversely, if the input and fed-back voltage are in phase with one another, **positive feedback** is applied to the amplifier and its gain will be increased. Positive feedback will lead to the circuit becoming unstable and prone to oscillate at some particular frequency, and it is the basis of an oscillator circuit.

At middle frequencies where the reactances of the circuit capacitances can be neglected, it is easy to ensure that the feedback is negative. At low and high frequencies however, phase-shifts within the amplifier caused by capacitive reactances will ensure that the fed-back voltage is no longer in antiphase with the input voltage. It may often not be immediately evident whether the feedback is now negative or positive. If the internal phase shifts are large enough, negative feedback may be turned into positive feedback at certain low and high frequencies and there is then a danger that the circuit may oscillate.

Differences exist in the methods used to derive the fed-back signal and to introduce it into the input circuit. These differences lead to the classification of n.f.b. amplifiers into one of four types:

1 Voltage-series or voltage-voltage
2 Voltage-shunt or current-voltage
3 Current-series or voltage-current
4 Current-shunt or current-current.

In general these various types of n.f.b. have similar effects upon the performance of an amplifier other than on its input and output impedances. Each impedance may either be increased or decreased according to the type of n.f.b. (see Table 4.1).

Table 4.1 Types of negative feedback and effect on impedances

Type of n.f.b.	*Input Impedance*	*Output Impedance*
Voltage-voltage	Increased	Decreased
Voltage-current	Increased	Increased
Current-current	Decreased	Increased
Current-voltage	Decreased	Decreased

Voltage-Voltage Feedback

With **voltage-voltage feedback,** often known as **voltage-series** feedback, a *voltage*, that is proportional to the output *voltage* of the amplifier, is fed back and applied in series with the input signal voltage (Fig. 4.2). The amplifier has a voltage gain A_v before feedback is applied and β is the fraction of the output voltage fed back into the input circuit. The output

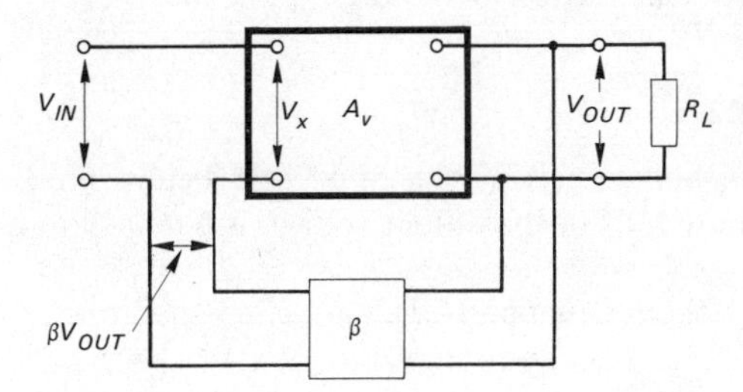

Fig. 4.2 Voltage-voltage feedback.

voltage of the amplifier is V_{OUT} and so the fed-back voltage is βV_{OUT}. The total voltage V_x applied to the amplifier is the phasor sum of βV_{OUT} and V_{IN} and hence

$$V_x = V_{IN} + \beta V_{OUT}$$
$$V_{OUT} = A_v V_x = A_v(V_{IN} + \beta V_{OUT})$$
$$V_{OUT}(1-\beta A_v) = A_v V_{IN}$$
$$\text{Voltage gain } A_{v(F)} = \frac{V_{OUT}}{V_{IN}} = \frac{A_v}{1-\beta A_v} \tag{4.1}$$

If the feedback is *definitely* negative, then *either* β or A_v must be negative so that the loop gain βA is negative and then

$$A_{v(F)} = \frac{A_v}{1-(-\beta A_v)} = A_v/(1+\beta A_v) \tag{4.2}$$

The requirement for the loop gain to be negative means that, if the amplifier has an even number of phase-inverting stages, its overall gain is $A_v\underline{/0°}$ and the β network must introduce the required 180° phase shift. If the amplifier has an odd number of phase-inverting stages so that its gain is $A_v\underline{/180°}$, then zero phase shift must be introduced by the feedback network. Very often the amount of feedback applied to a circuit is quoted in terms of the ratio (gain without n.f.b. to gain with n.f.b.) expressed in dB, i.e. feedback applied is

$$20 \log_{10} |A_v/A_{v(F)}| = 20 \log_{10} |1+\beta A| \tag{4.3}$$

Example 4.1

A voltage amplifier has a gain of 120 before n.f.b. is applied. Calculate its voltage gain if 1/10 of the output voltage is fed back to the input in antiphase with the input signal.

Solution From equation (4.2)

$$A_{v(F)} = 120/(1+120/10) = 9.23 \quad (Ans)$$

This example makes it clear that the application of n.f.b. to an amplifier results in a considerable reduction in its gain. At low-frequencies the presence of coupling and/or decoupling capacitors will reduce the voltage gain of the amplifier and introduce a phase shift. The loop gain βA of the circuit does not then have a phase angle of 180° and equation (4.1) must be used in any calculation of voltage gain.

Example 4.2

A voltage amplifier has a gain of $85\underline{/45°}$ before n.f.b. is applied. Calculate the voltage gain if 1/10 of the output voltage is fed back to the input in antiphase with the input signal.

Solution From equation (4.1),

$$A_{v(F)} = \frac{85\underline{/45°}}{1-\frac{85}{10}\underline{/45°}}$$
$$= \frac{85\underline{/45°}}{1-8.5\cos 45° - j8.5\sin 45°}$$

$$= \frac{85\angle 45^\circ}{-5.01 - j6.01}$$

$$= \frac{85\angle 45^\circ}{7.82\angle 230^\circ}$$

$$A_{v(F)} = 10.87\angle -185^\circ \quad (Ans)$$

The **overall voltage gain** of an amplifier is influenced by the relative values of the source and amplifier input impedances. The overall voltage gain $A_{vo(F)}$ is the ratio V_{OUT}/E_S.

Referring to Fig. 4.3,

$$E_S = I_{IN}(R_S + R_{IN}) - \beta V_{OUT}$$

$$= \frac{V_{IN}}{R_{IN}}(R_S + R_{IN}) - \beta V_{OUT}$$

$$A_{vo(F)} = V_{OUT}/E_S$$

$$= \frac{V_{OUT}}{\frac{V_{IN}}{R_{IN}}(R_S + R_{IN}) - \beta V_{OUT}} = \frac{V_{OUT}}{V_{IN}\left(\frac{R_S + R_{IN}}{R_{IN}} - \beta A_v\right)}$$

$$= \frac{A_v}{\frac{R_S + R_{IN}}{R_{IN}} - \beta A_v} \tag{4.4}$$

Clearly if $R_S \ll R_{IN}$, this expression reduces to equation (4.1).

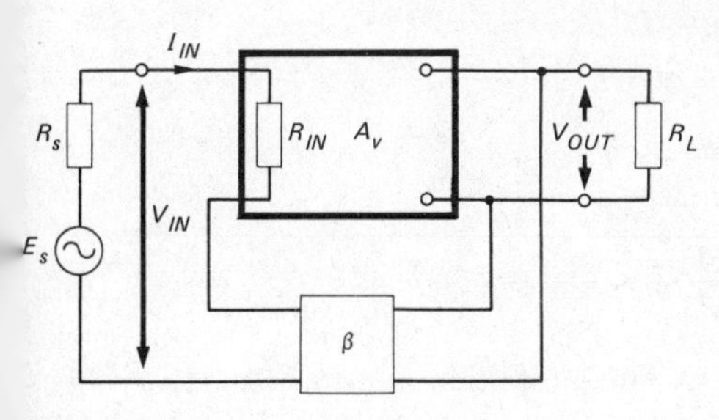

Fig. 4.3

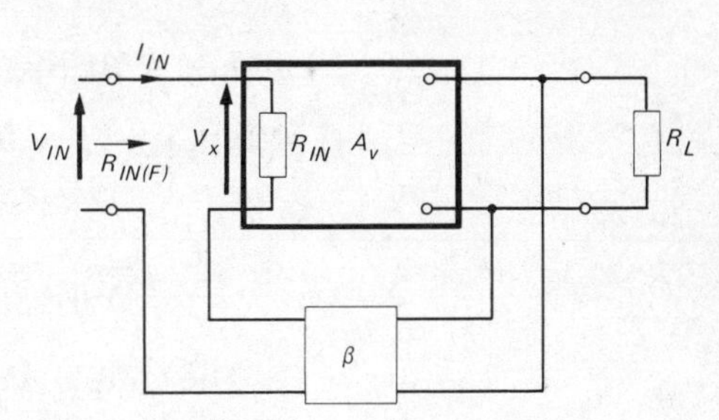

Fig. 4.4

In the derivation of the expressions for the **input and output impedances** of a feedback amplifier, it will be assumed that the feedback is negative and that the impedances are resistive.

The input impedance of an amplifier is the ratio V_{IN}/I_{IN}. From Fig. 4.4,

$$I_{IN} = \frac{V_{IN}}{R_{IN(F)}} = \frac{V_x}{R_{IN}}$$

But $V_x = V_{IN} - \beta V_x A_v$ or $V_{IN} = V_x(1 + \beta A_v)$ Hence

$$R_{IN(F)} = \frac{V_{IN}}{I_{IN}} = \frac{V_x}{I_{IN}}(1 + \beta A_v)$$

$$R_{IN(F)} = R_{IN}(1+\beta A_v) \tag{4.5}$$

Equation (4.5) shows that the input impedance of a voltage-voltage feedback amplifier is *increased.*

Fig. 4.5

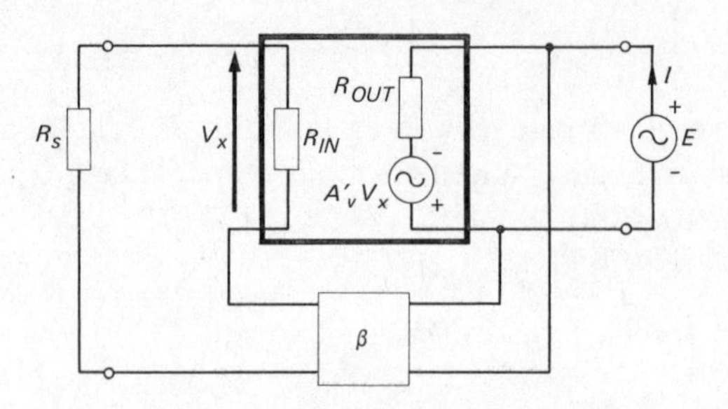

The output resistance of a n.f.b. amplifier is best determined by replacing the input signal source by an impedance equal to its internal resistance R_S, and connecting a voltage generator of e.m.f. E and zero internal impedance across the output terminals of the amplifier (see Fig. 4.5). The output impedance $R'_{OUT(F)}$ of the amplifier with the collector (or drain) load disconnected is then given by the ratio E/I.

R_{OUT} is the output impedance of the amplifier before negative feedback is applied and A'_v is the *open-circuit* voltage gain of the amplifier, i.e. the voltage gain with $R_L = \infty$.

From Fig. 4.5

$$E + A'_v V_x = IR_{OUT}$$

but $V_x = \beta E R_{IN}/(R_S + R_{IN})$ so that

$$E\left[1 + \frac{\beta A'_v R_{IN}}{R_S + R_{IN}}\right] = IR_{OUT}$$

$$\frac{E}{I} = R'_{OUT(F)} = \frac{R_{OUT}(R_S + R_{IN})}{R_S + R_{IN}(1+\beta A'_v)} \tag{4.6}$$

The output impedance $R_{OUT(F)}$ of the amplifier is equal to $R'_{OUT(F)}$ in parallel with the collector or drain load resistance. Note that A'_v is V_{OUT}/V_x, *not* V_{OUT}/E_S.

If $R_S = 0$ equation (4.6) reduces to

$$R'_{OUT(F)} = \frac{R_{OUT}}{1+\beta A'_v} \tag{4.7}$$

Equation (4.7) can also be used to determine $R_{OUT(F)}$ when $R_S \neq 0$ if A'_v is defined as the open-circuit gain V_{OUT}/V_S instead of V_{OUT}/V_x (see Example 4.3).

An example of series-voltage n.f.b. is the **emitter follower** (page 66 and Fig. 3.7*a*). The output voltage is in series with the input signal voltage but with the opposite polarity, thus *all* of the output voltage is applied as negative feedback ($\beta = 1$). From equation (4.2),

$$A_{v(F)} = \frac{A_i R_{L(eff)}/h_{ie}}{1 + A_i R_{L(eff)}/h_{ie}} = \frac{A_i R_{L(eff)}}{h_{ie} + A_i R_{L(eff)}} \tag{4.8}$$

Equation (4.8) should be compared with equation (3.8).

From equation (4.3)

$$R_{IN(F)} = h_{ie}\left(1 + \frac{A_i R_{L(eff)}}{h_{ie}}\right)$$

$$= h_{ie} + A_i R_{L(eff)} \simeq A_i R_{L(eff)} \tag{4.9}$$

[See equation (3.7).]

Finally, the output resistance of an emitter follower is (equation 4.6),

$$R_{OUT(F)} = \frac{(1/h_{oe})(h_{ie} + R_S)}{R_S + h_{ie}(1 + A_v')} \simeq \frac{(1/h_{oe})(h_{ie} + R_S)}{R_S + A_v' h_{ie}}$$

Now usually

$$h_{ie} \ll R_S \quad \text{and} \quad A_v' = \frac{(1/h_{oe}) + R_{L(eff)}}{R_{L(eff)}} \times \frac{h_{fe} R_{L(eff)}}{h_{ie}}$$

Therefore

$$R_{OUT(F)} = \frac{R_S/h_{oe}}{R_S + [(1/h_{oe}) + R_{L(eff)}]h_{fe}}$$

$$\simeq \frac{R_S/h_{oe}}{R_S + h_{fe}/h_{oe}} \simeq \frac{R_S/h_{oe}}{h_{fe}/h_{oe}} = R_S/h_{fe} \tag{4.10}$$

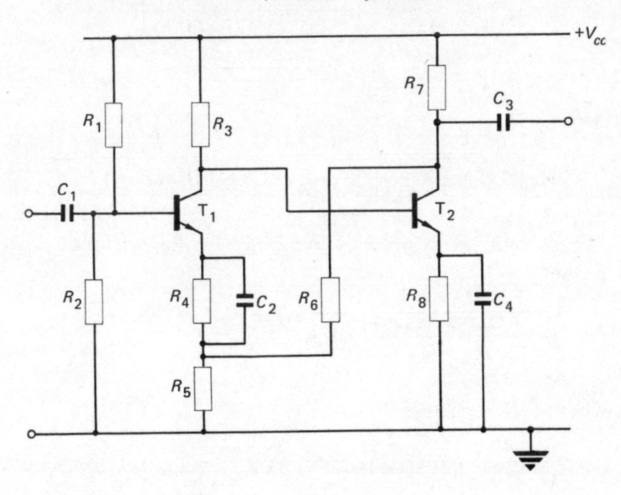

Fig. 4.6 A voltage-voltage feedback amplifier.

Another example of voltage-voltage (series) negative feedback is shown by Fig. 4.6. The overall gain of this amplifier is very nearly

$$A_{v(F)} = \frac{1}{\beta} = \frac{R_3 + R_4}{R_3}$$

The input impedance of the circuit is increased but its output impedance is decreased.

Example 4.3

A single-stage amplifier has $h_{fe}=150$, $h_{ie}=2\ \text{k}\Omega$ and $1/h_{oe}=6\ \text{k}\Omega$ and a collectc load resistance of 2000 Ω. The amplifier has negative feedback with $\beta=0.01$ applie and is used to amplify the signal provided by a source of e.m.f. 10 mV an impedance 1500 Ω. Calculate

a) the output resistance of the amplifier

b) the voltage developed across the load resistance.

Solution Fig. 4.7 shows the circuit.

Fig. 4.7

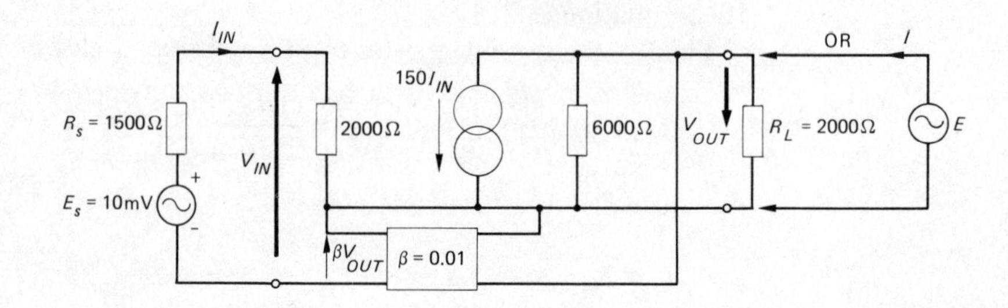

METHOD 1

a) With the load resistance R_L removed and replaced by the generator of E vol e.m.f.,

$$R_{OUT(F)}=E/I \quad\text{and}\quad I_{IN}=\beta E/(R_S+R_{IN})$$

$$I=h_{fe}I_{IN}+\frac{E}{R_{OUT}}=\left(\frac{150\times0.01}{1500+2000}+\frac{1}{6000}\right)E$$

$$=0.595\times10^{-3}\ E$$

Therefore,

$$R_{OUT(F)}=\frac{E}{I}=\frac{1}{0.595\times10^{-3}}=1680\ \Omega \quad (Ans)$$

This value is reduced by the 2000 Ω collector load resistance which is effectively parallel to 913 Ω (*Ans*)

b) $I_{IN}=\dfrac{E_S-\beta V_{OUT}}{R_S+R_{IN}}$ $\qquad V_{OUT}=h_{fe}I_{IN}R$

where $R=\dfrac{R_L}{1+h_{oe}R_L}$, so

$$V_{OUT}=150\left(\frac{10\times10^{-3}-0.01V_{OUT}}{3500}\right)\left(\frac{6000\times2000}{6000+2000}\right)$$

$$=64.286(10\times10^{-3}-0.01V_{OUT})=0.64286(1-V_{OUT})$$

$$V_{OUT}(1+0.64286)=0.64286$$

$$V_{OUT}=0.3913\ \text{V} \quad (Ans)$$

METHOD 2

a) With R_L disconnected,

$$V_{OUT(R_L=\infty)} = \frac{E_S}{R_{IN}} \cdot h_{fe} R_{OUT}$$

$$A_v' = V_{OUT}/V_x = \frac{h_{fe} R_{OUT}}{R_{IN}} = \frac{150 \times 6000}{2000} = 450$$

Hence, from equation (4.6)

$$R_{OUT(F)} = \frac{6000 \times 3500}{1500 + 2000(1 + 0.01 \times 450)} = 1680\ \Omega \quad (Ans)$$

Alternatively, defining A_v' as

$$V_{OUT}/E_S = \frac{h_{fe} R_{OUT}}{R_S + R_{IN}} = \frac{150 \times 6000}{3500} = 257.14$$

and substituting into equation (4.7)

$$R_{OUT(F)} = \frac{6000}{1 + 0.01 \times 257.14} = 1680\ \Omega \quad (Ans)$$

b) $$A_i = \frac{h_{fe}}{1 + h_{oe} R_L} = \frac{150}{1 + 2000/6000} = 112.5$$

$$A_v = \frac{A_i R_L}{R_{IN}} = \frac{112.5 \times 2000}{2000} = 112.5$$

From equation (4.5)

$$R_{IN(F)} = R_{IN}(1 + \beta A_v) = 2000(1 + 0.01 \times 112.5) = 4250\ \Omega$$

From equation (4.2)

$$A_{v(F)} = \frac{112.5}{1 + 0.01 \times 112.5} = 52.94$$

$$V_{IN} = \frac{E_S R_{IN(F)}}{R_S + R_{IN(F)}} = \frac{10 \times 10^{-3} \times 4250}{1500 + 4250} = 7.39\ \text{mV}$$

Therefore,

$$V_{OUT} = 7.39 \times 10^{-3} \times 52.94 = 391.3\ \text{mV} \quad (Ans)$$

Voltage-Current Feedback

The block diagram of a **voltage-current feedback** or **current-series** feedback is shown by Fig. 4.8. The fed-back voltage is proportional to the current flowing in the load and is applied in series with the input signal. Since the fed-back voltage is applied in series with the input circuit, the analysis

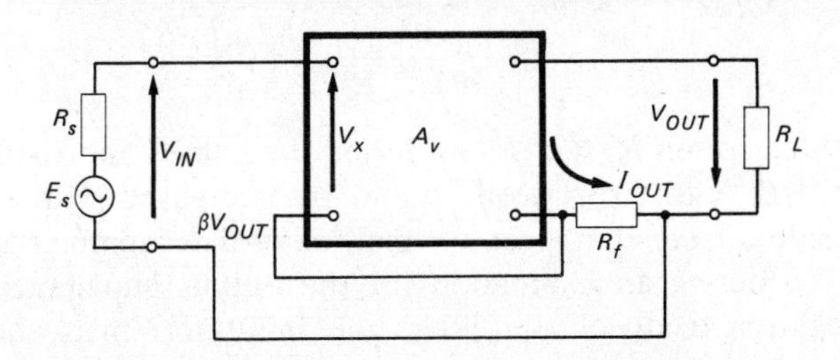

Fig. 4.8 Voltage-current feedback.

leading to equations (4.2) and (4.5) is again applicable. Hence, the voltage gain $A_v = V_{OUT}/V_{IN}$ is again

$$A_{v(F)} = A_v/(1+\beta A_v) \tag{4.2}$$

and the input impedance

$$R_{IN(F)} = R_{IN}(1+\beta A_v) \tag{4.5}$$

The feedback voltage βV_{OUT} is developed by the output current flowing in the feedback resistor R_f. Hence,

$$\beta V_{OUT} = I_{OUT}R_f = \frac{V_{OUT}R_f}{R_L}$$

and so

$$\beta = R_f/R_L$$

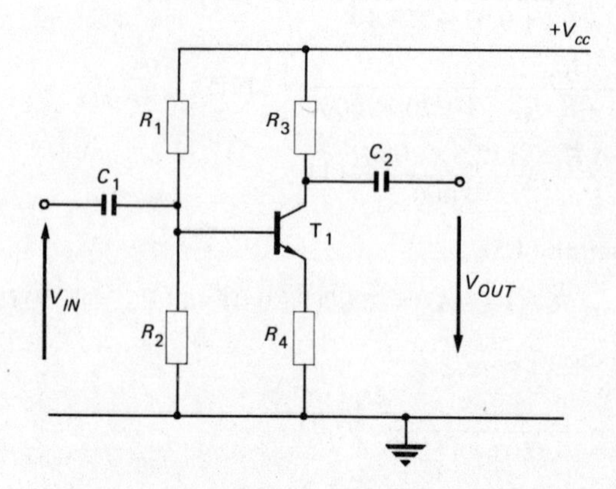

Fig. 4.9 A voltage-current feedback amplifier.

The simplest way of applying voltage-current negative feedback to a circuit is merely to leave the emitter, or source, resistor undecoupled as shown in Fig. 4.9 for a bipolar transistor circuit. The voltage gain before n.f.b. of this circuit is

$$A_v = h_{fe}R_3/h_{ie}$$

assuming the bias resistors are large enough to have negligible shunting effect upon h_{ie}. The feedback factor β is R_4/R_3 and hence the **voltage gain** with n.f.b. is

$$A_{v(F)} = \frac{V_{OUT}}{V_{IN}} = \frac{h_{fe}R_3/h_{ie}}{1+\dfrac{h_{fe}R_3}{h_{ie}}\cdot\dfrac{R_4}{R_3}} = \frac{h_{fe}R_3}{h_{ie}+h_{fe}R_4}$$

Very often $h_{ie} \ll h_{fe}R_3$ and then the gain is approximately given by R_3/R_4 or $1/\beta$. This expression should be compared with equation (3.3a) which resulted from an alternative method of determining the gain of the circuit.

To derive an expression for the **output impedance** of a voltage-current negative feedback amplifier, the input terminals should be closed in an

Fig. 4.10

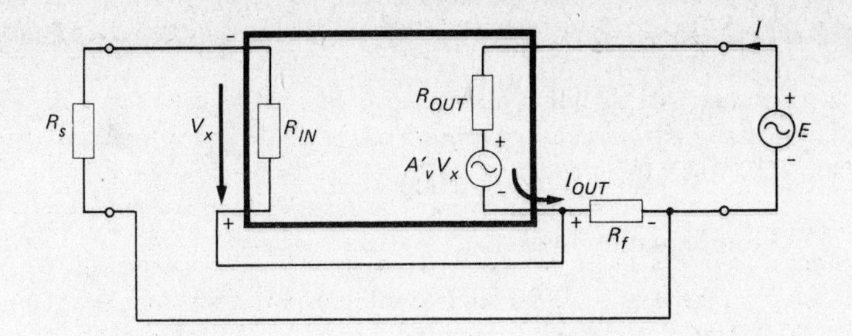

impedance equal to the source impedance, and a generator of e.m.f. E volts and zero impedance should be connected across the output terminals (Fig. 4.10). The output impedance is then E/I ohms.

Neglecting R_L for the moment, the voltage equation for the output circuit is

$$E - A'_v V_x = I_{OUT}(R_{OUT} + R_f)$$

(As before, A'_v = open-circuit voltage gain.)

But $$V_x = \frac{\beta V_{OUT} R_{IN}}{R_S + R_{IN}} = \frac{I_{OUT} R_f R_{IN}}{R_S + R_{IN}}$$

Hence

$$E - \frac{A'_v I_{OUT} R_f R_{IN}}{R_S + R_{IN}} = I_{OUT}(R_{OUT} + R_f)$$

$$E = I_{OUT}\left(R_{OUT} + R_f + \frac{A'_v R_f R_{IN}}{R_S + R_{IN}}\right)$$

Therefore,

$$R_{OUT(F)} = E/I = R_{OUT} + R_f\left(1 + \frac{A'_v R_{IN}}{R_S + R_{IN}}\right) \tag{4.11}$$

If the source resistance can be neglected as small compared with the input impedance R_{IN} of the amplifier, this equation reduces to

$$R_{OUT(F)} = R_{OUT} + R_f(1 + A'_v) \tag{4.12}$$

The feedback can lead to a large increase in the output impedance of the active device but this will be reduced by the shunting effect of R_L.

Example 4.4

An amplifier has an input impedance of 1 kΩ, an output impedance of 10 kΩ and a voltage gain of 600 before the application of n.f.b. A collector load resistor of 4.7 kΩ is used. Negative feedback is applied to the amplifier by removing the decoupling capacitor from its 330 Ω emitter resistor.

Calculate i) the input impedance, ii) the output impedance, and iii) the voltage gains V_{OUT}/V_{IN} and V_{OUT}/E_S of the amplifier after n.f.b. has been applied. Assume a source resistance of 1000 Ω.

Solution

$\beta = 330/4700 = 0.07.$

$$A_v = 600 = \frac{A'_v \times 4700}{4700 + 10\,000} \quad \text{or} \quad A'_v = 1877$$

From equation (4.5)

$$R_{IN(F)} = 1000(1 + 0.07 \times 600) = 43\ \text{k}\Omega \quad (Ans)$$

From equation (4.11)

$$R_{OUT(F)} = 10\,000 + 330\left(1 + \frac{1877}{2}\right) = 320.04\ \text{k}\Omega \quad (Ans)$$

This impedance is shunted by the collector load resistor to give an overall outpu impedance of

$$320.04 \times 4.7/(320.04 + 4.7) = 4.63\ \text{k}\Omega \quad (Ans)$$

Thus, the overall output impedance of the amplifier is almost unaltered by the applica tion of the n.f.b.

From equation (4.2)

$$A_{v(F)} = \frac{600}{1 + 600 \times 0.07} = 13.95 \quad (Ans)$$

and from equation (4.4)

$$A_{vo(F)} = \frac{600}{\dfrac{1000 + 1000}{1000} + 600 \times 0.07} = 13.63 \quad (Ans)$$

This last answer can be obtained in an alternative manner. Since the input impe dance, with n.f.b. applied, of the amplifier is 43 kΩ, the input voltage V_{IN} of th amplifier is

$$V_{IN} = E_S \times 43/(43 + 1) = 0.9773E_S$$

and hence the overall voltage gain is

$$0.9773 \times 13.95 = 13.63 \quad (Ans)$$

Current-Current Feedback

With **current-current feedback**, or **current-shunt** feedback, a fraction of th output current is fed back in parallel with the input circuit. The bloc diagram of the arrangement is shown by Fig. 4.11 in which the directions c the currents shown assume that the feedback is negative. Since a current fed back it is better to work in terms of the **current gain** of the amplifier

From the Fig. 4.11,

$$\beta I_{OUT} = \frac{I_{OUT} R_{f1}}{R_{f1} + R_{f2} + R}$$

where R is the parallel combination of R_S and R_{IN}, i.e.

$$R = R_S R_{IN}/(R_S + R_{IN}).$$

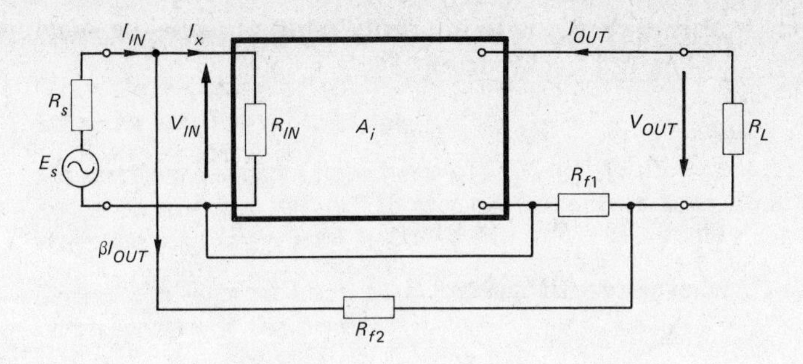

Fig. 4.11 Current-current feedback.

Hence $\beta = R_{f1}/(R_{f1}+R_{f2}+R)$ (4.13)

Often $R_{f2} \gg R$, then

$$\beta = R_{f1}/(R_{f1}+R_{f2}) \simeq R_{f1}/R_{f2} \tag{4.14}$$

Also $I_S = I_{IN} - \beta I_{OUT}$

$$I_{OUT} = A_i I_S = A_i(I_{IN} - \beta I_{OUT})$$

$$I_{OUT}(1+\beta A_i) = A_i I_{IN}$$

$$A_{i(F)} = \frac{I_{OUT}}{I_{IN}} = \frac{A_i}{1+\beta A_i} \tag{4.15}$$

The simplest method of deriving an equation for the **input impedance** of a current-shunt n.f.b. amplifier is to work in terms of admittances. Hence, from Fig. 4.11

$$I_{IN} = I_S + \beta I_{OUT}$$

$$= I_S(1+\beta A_i)$$

$$\frac{I_{IN}}{V_{IN}} = Y_{IN(F)} = \frac{I_S}{V_{IN}}(1+\beta A_i) = Y_{IN}(1+\beta A_i)$$

$$R_{IN(F)} = 1/Y_{IN(F)} = R_{IN}/(1+\beta A_i) \tag{4.16}$$

Thus, the input impedance of an amplifier is reduced by the application of current-shunt n.f.b.

For the calculation of the **output impedance** with n.f.b., the input terminals of the circuit should be closed with an impedance equal to the source impedance and a voltage generator of e.m.f. E volts and zero internal impedance connected across the output terminals (see Fig. 4.12). The load

Fig. 4.12

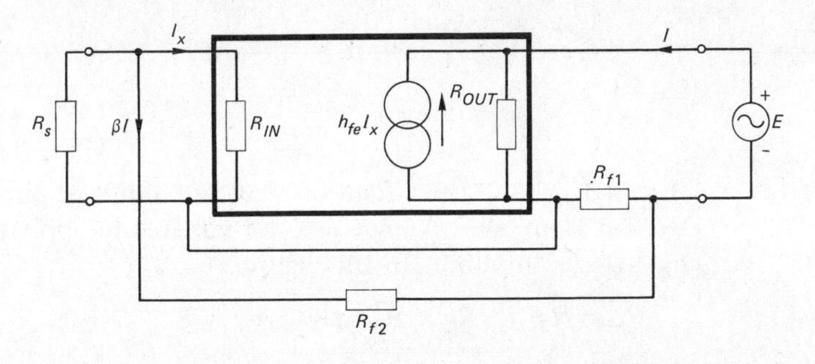

resistance R_L is temporarily omitted from the diagram.

The fed-back current is βI where

$$\beta = \frac{R_{f1}}{R_{f1}+R_{f2}+R} \quad \text{with} \quad R = R_S R_{IN}/(R_S + R_{IN})$$

Then $E = R_{OUT}(I + h_{fe}I_x) + IR_{f1}$

where $I_x = \beta I R_S/(R_S + R_{IN})$

$$R_{OUT(F)} = E/I = R_{OUT}\left(1 + \frac{h_{fe}R_S}{R_S + R_{IN}} \cdot \frac{R_{f1}}{R_{f1}+R_{f2}+R}\right) + R_{f1} \tag{4.17}$$

If $R_{f2} \gg R$, $R_{f2} \gg R_{f1}$ and $R_S \gg R_{IN}$, this expression reduces to

$$R_{OUT(F)} = R_{OUT}\left(1 + \frac{h_{fe}R_{f1}}{R_{f2}}\right) + R_{f1} \tag{4.18}$$

Thus, the output impedance of the amplifier is increased by the application of current-shunt n.f.b. However, $R_{OUT(F)}$ is shunted by the load resistance R_L so that the overall output impedance $R_{OUT(OF)}$ of the amplifier is

$$R_{OUT(OF)} = R_{OUT(F)}R_L/(R_{OUT(F)} + R_L) \tag{4.19}$$

and this is usually approximately equal to R_L.

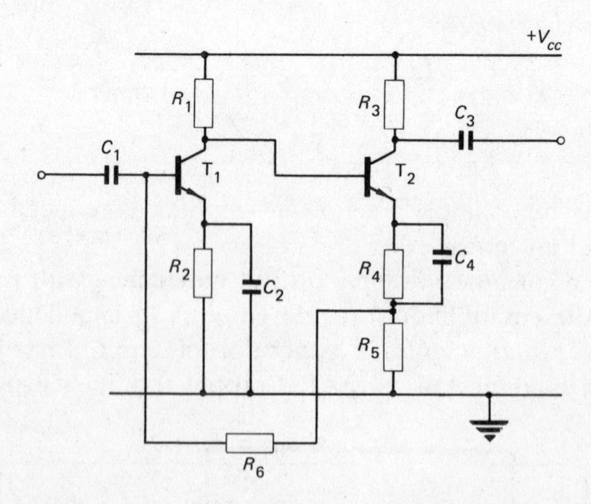

Fig. 4.13 A current-current feedback amplifier.

Fig. 4.13 shows the circuit of a current-shunt negative feedback amplifier. Two stages are shown since it is not possible to apply this form of n.f.b. to a single-stage amplifier. In this figure

$$R_{f1} = R_5 \qquad R_{f2} = R_6 \quad \text{and} \quad R_L = R_3.$$

Current-Voltage Feedback

The fourth type of feedback involves feeding back to the input circuit a current whose magnitude is proportional to the output voltage. The block diagram of **current-voltage feedback** or **voltage-shunt** is shown in Fig. 4.14.

$$\beta I_{OUT} = \frac{V_{OUT}}{R_f + R} \quad \text{where} \quad R = R_S R_{IN}/(R_S + R_{IN})$$

$$\frac{\beta I_{OUT}}{I_{OUT}} = \frac{V_{OUT}}{(R_f + R)V_{OUT}/R_L}$$

$$\beta = R_L/(R_f + R) \tag{4.20}$$

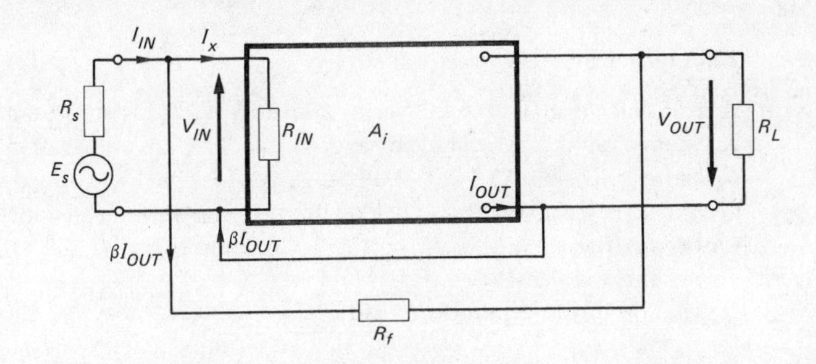

Fig. 4.14 Current-voltage feedback.

The **current gain** and the **input resistance** with feedback of the amplifier are obtained in the same way as for the current-shunt circuit and leads to the same equations, i.e. (4.15) and (4.16).

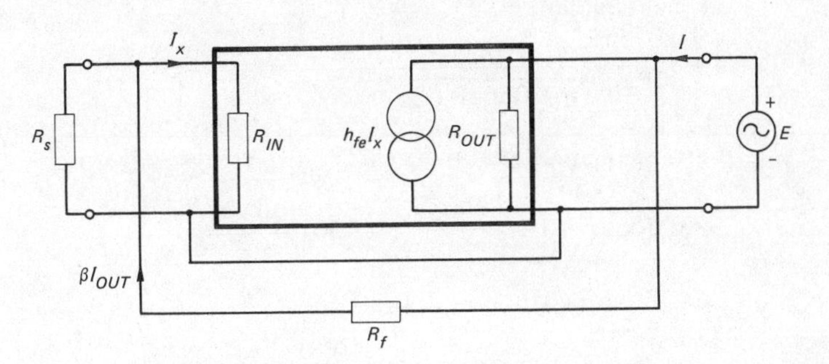

Fig. 4.15

To determine an expression for the **output impedance** of the amplifier, the circuit given in Fig. 4.15 must be employed. From Fig. 4.15,

$$I_x = \frac{E}{R_f + R} \cdot \frac{R_S}{R_S + R_{IN}}$$

and

$$I_{OUT} = \frac{h_{fe} E R_S}{(R_f + R)(R_S + R_{IN})} + \frac{E}{R_{OUT}} + \frac{E}{R_f + R}$$

Therefore,

$$Y_{OUT(F)} = \frac{I_{OUT}}{E} = \frac{h_{fe}R_S}{(R_f + R)(R_S + R_{IN})} + \frac{1}{R_{OUT}} + \frac{1}{R_f + R}$$

$$\simeq \frac{h_{fe}R_S}{(R_f + R)(R_S + R_{IN})} + \frac{1}{R_{OUT}} \qquad (4.21)$$

and the overall output admittance, taking account of R_L is

$$Y_{OUT(OF)} = \frac{h_{fe}R_S}{(R_f + R)(R_S + R_{IN})} + \frac{1}{R_{OUT}} + \frac{1}{R_L} \qquad (4.22)$$

Example 4.5

A transistor has $h_{ie} = 6\ \text{k}\Omega$, $h_{fe} = 200$ and $h_{oe} = 1 \times 10^{-4}$ S and has a collector load resistance of 2.2 kΩ. The voltage source has an e.m.f. of 10 mV and an internal resistance of 1 kΩ. Voltage-shunt negative feedback is applied to the amplifier using a feedback resistor of 100 kΩ. Calculate i) the input resistance, ii) the output resistance, iii) the voltage gain of the n.f.b. amplifier, iv) the signal voltage appearing across the load resistor.

Solution From equation (4.20)

$$\beta = \frac{2200}{100 \times 10^3 + \dfrac{6 \times 10^3 \times 10^3}{7 \times 10^3}} = 0.022$$

i) $$A_i = \frac{h_{fe}}{1 + h_{oe}R_L} = \frac{200}{1 + 10^{-4} \times 2200} = 164$$

From equation (4.15),

$$A_{i(F)} = \frac{164}{1 + 164 \times 0.022} = 35.6$$

From equation (4.16),

$$R_{IN(F)} = \frac{6000}{1 + 164 \times 0.022} = 1302\ \Omega \quad (Ans)$$

ii) From equation (4.22),

$$Y_{OUT(OF)} = \frac{200 \times 10^3}{(100 \times 10^3 + 857)(7 \times 10^3)} + 1 \times 10^{-4} + \frac{1}{2200} = 8.378 \times 10^{-4}$$

Therefore,

$$R_{OUT(OF)} = Y_{OUT(OF)} = 1194\ \Omega \quad (Ans)$$

iii) $$A_{v(F)} = \frac{A_{i(F)}R_L}{R_{IN(F)}} = \frac{35.6 \times 2200}{1302} = 60.15 \quad (Ans)$$

iv) $$V_{IN} = \frac{E_S R_{IN(F)}}{R_S + R_{IN(F)}} = \frac{10 \times 10^{-3} \times 1302}{1000 + 1302} = 5.66\ \text{mV}$$

Hence Output voltage = 5.66 × 60.15 = 340 mV (*Ans*)

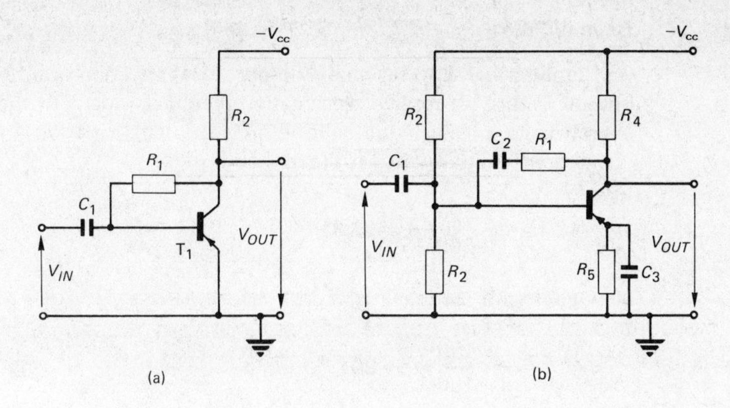

Fig. 4.16 Two current-voltage feedback circuits.

Note that the voltage gain of the amplifier before n.f.b. was applied was

$$A_v = A_i R_L / R_{IN} = 164 \times 2200/6000 = 60.15$$

and this shows that shunt feedback reduces the current gain of an amplifier but does not affect its voltage gain. The simplest voltage-shunt n.f.b. circuit is shown in Fig. 4.16*a*; the feedback current βI_{OUT} is applied to the input terminals via the feedback resistor R_1, connected between the base and collector terminals. R_1 also provides d.c. bias and stabilization for the circuit. Sometimes the value of R_1 required to provide a desired amount of feedback is not the same as that required for correct bias; then a capacitor C_2 can be connected in series with R_1 and the bias provided by some other means, such as that shown in Fig. 4.16*b*.

When n.f.b. is applied to a fet amplifier, the approach used for operational amplifiers in the next chapter is more appropriate because of the extremely high input impedance of the device [see equation (5.10)].

Advantages of Negative Feedback

The application of any of the four types of negative feedback to an amplifier has the effect of reducing either the voltage gain or the current gain of that amplifier. This is obviously a disadvantage, but on the credit side, a number of desirable changes in the amplifier performance also take place. These are

1 The gain stability is increased.
2 Distortion and noise produced within the feedback loop *may* be reduced.
3 The input and output impedances of the amplifier can be modified to almost any desired value.

Stability of Gain

For many applications it is important that an amplifier should have a more or less **constant gain** even though various parameters, such as power supply voltages and transistor parameters, may alter with time and/or with change in ambient temperature. The application of n.f.b. will considerably reduce the effect of such parameter variations on the overall gain of the amplifier.

Example 4.6

An amplifier has a voltage gain, before the application of n.f.b., of 500. If 1/25 of the output voltage is applied as negative feedback calculate the percentage change in overall voltage gain if the gain before n.f.b. falls to 100.

Solution Initial gain with feedback is

$$A'_{v(F)} = \frac{500}{1+500/25} = 23.81$$

New gain with feedback is

$$A''_{v(F)} = \frac{100}{1+100/25} = 20$$

Hence, the percentage change in overall voltage gain is

$$\frac{20-23.81}{23.81} \times 100\% = -16\% \quad (Ans)$$

This is to be compared with 80% change in the gain without feedback. Clearly a considerable increase in gain stability has been achieved.

If the loop gain, βA_v or βA_i, is much larger than unity the equations (4.2) and (4.15) reduce to

$$A_{v(F)} = A_{i(F)} = 1/\beta \tag{4.23}$$

The gain of the amplifier is now merely a function of the components making up the feedback circuit, generally one or two resistors. Any changes in the parameters of the transistor, etc. will now have very little effect on the overall gain of the amplifier.

Amplitude-Frequency Distortion

Amplitude-frequency distortion of a complex signal is caused by the different frequency components of the signal being amplified by different amounts. The variation of gain with frequency of an amplifier was considered in Chapter 3 where it is was shown to be caused by capacitive effects. The gain at low frequencies of an audio amplifier can be written as

$$A_{v(LF)} = \frac{A_{v(MF)}}{1 - j\omega_{3dB}/\omega} \tag{3.15}$$

and the gain at high frequencies as

$$A_{v(HF)} = \frac{A_{v(MF)}}{1 + j\omega/\omega_{3dB}} \tag{3.18}$$

The expression for the overall low-frequency gain with feedback is now

$$A_{v(F)(LF)} = \frac{\dfrac{A_{v(MF)}}{1 - j\omega_{3dB}/\omega}}{1 + \dfrac{\beta A_{v(MF)}}{1 - j\omega_{3dB}/\omega}} = \frac{A_{v(MF)}}{1 + \beta A_{v(MF)} - j\omega_{3dB}/\omega}$$

$$= \frac{\dfrac{A_{v(MF)}}{1 + \beta A_{v(MF)}}}{1 - \dfrac{j\omega_{3dB}/\omega}{1 + \beta A_{v(MF)}}} = \frac{A_{v(F)(MF)}}{1 - \dfrac{j\omega_{3dB}/\omega}{1 + \beta A_{v(MF)}}}$$

The lower 3 dB frequency ω'_{3dB} now occurs when

$$1 = \frac{\omega_{3dB}}{\omega'_{3dB}(1 + \beta A_{v(MF)})}$$

$$\omega'_{3dB} = \frac{\omega_{3dB}}{1 + \beta A_{v(MF)}} \qquad (4.24)$$

This is a lower frequency than that obtained without feedback.

Similarly, at high frequencies the overall gain with feedback is

$$A_{v(F)(HF)} = \frac{\dfrac{A_{v(MF)}}{1 + j\omega/\omega_{3dB}}}{1 + \dfrac{\beta A_{v(MF)}}{1 + j\omega/\omega_{3dB}}} = \frac{A_{v(F)(MF)}}{1 + \dfrac{j\omega/\omega_{3dB}}{1 + \beta A_{v(MF)}}}$$

and the upper 3 dB frequency ω'_{3dB} is

$$\omega'_{3dB} = \omega_{3dB}(1 + \beta A_{v(MF)}) \qquad (4.25)$$

This simple analysis *indicates* that the bandwidth of an amplifier is increased in the same ratio as the gain is decreased by the application of n.f.b. However the results would only be true if the amplifier contained just the one time constant at low-frequencies.

Fig. 4.17 shows the gain/frequency characteristic of a typical amplifier before and after the application of n.f.b. Clearly, the feedback has made the gain of the amplifier much "flatter" over most of the frequency band.

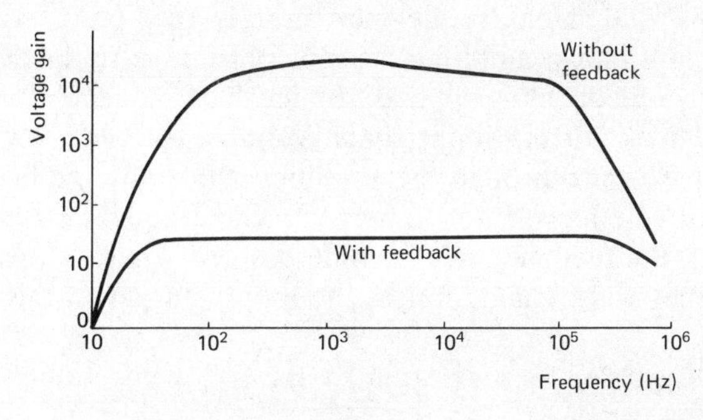

Fig. 4.17 Showing the effect of negative feedback on the gain/frequency characteristic of an amplifier.

Non-linearity Distortion

The dynamic transfer and mutual characteristics of bipolar and field effect transistors exhibit some non-linearity and as a result the output and input signal waveforms are not identical. **Non-linearity distortion** is said to have occurred. The distortion exists because the output waveform contains components at frequencies that were not present in the input signal. If these new, unwanted frequencies are harmonically related to the input signal frequencies, **harmonic distortion** is said to be present. If the new frequencies are equal to the sums and differences of frequencies combined in the input signal or harmonics of them, **intermodulation distortion** has taken place. Very often both harmonic and intermodulation distortion occur at the same time.

Suppose, for simplicity, that the input signal is sinusoidal and the distortion produced is due to a second-harmonic component only, say D.

The output of the amplifier is then

$$V_{OUT} = A_v V_x + D = A_v(V_{IN} - \beta V_{OUT}) + D$$

$$V_{OUT}(1+\beta A_v) = A_v V_{IN} + D$$

$$V_{OUT} = \frac{A_v V_{IN}}{1+\beta A_v} + \frac{D}{1+\beta A_v} \qquad (4.26)$$

According to equation (4.26) distortion is reduced by the same factor as the gain and so the percentage distortion is unchanged. Distortion is a function of the voltage level and is mainly produced in the final large-signal stage of an amplifier. For the same output level before and after the application of n.f.b., the input signal voltage with feedback must be increased $(1+\beta A_v)$ times. Now the percentage distortion is reduced by the factor $(1+\beta A_v)$. The extra signal must be provided by an earlier small-signal stage but this will introduce little, if any, distortion.

Equation (4.26) also assumes that the gain A_v of the amplifier is constant for all input signal levels; this of course is never true and, if it were, the gain characteristic would be linear and there would be no distortion in the first place.

Stability of Negative Feedback Amplifiers

At midband signal frequencies there is 180° phase shift through each stage of a multi-stage amplifier (except perhaps a follower stage). When overall negative feedback is applied, the fed-back voltage must be arranged to be in *antiphase* with the input signal voltage. At both low and high frequencies the phase shift through the amplifier will be altered because of the inevitable capacitive effects. If, at some particular frequency, the loop phase shift $\underline{/\beta A}$ is 360° the feedback will become *positive*. There is then a possibility that the amplifier will oscillate unless the loop gain βA at this frequency is less than unity.

The problem is illustrated by the following example.

Example 4.7

A three-stage amplifier has three equal non-interacting 0.1 sec time constants at low frequencies. If the midfrequency voltage gain is $100\underline{/180^\circ}$, determine the greatest value of feedback factor β that can be used if the circuit is not to oscillate. If a greater value of β is used, calculate the frequency of the resulting oscillations.

Solution The "gain" of each *RC* circuit in the amplifier can be written in the form

$$A_{(LF)} = \frac{A_{(MF)}}{1 - j/\omega\tau} = \frac{A_{(MF)}}{1 - j/0.1\omega}$$

For three such identical stages,

$$\frac{A_{(LF)}}{A_{(MF)}} = \frac{1}{(1 - j/0.1\omega)^3} = \left[\frac{1}{\sqrt{[1 + (1/0.1\omega)^2]}}\right]^3 3\underline{/\tan^{-1} 1/0.1\omega}$$

The amplifier introduces a 180° phase shift and for a loop phase shift of 360° the *RC* circuits must give a total phase shift of 180°.

Hence, the phase shift per circuit = 180°/3 = 60°, and

$$\tan^{-1} 1/0.1\omega = 60^\circ$$

$$\omega = 10/\tan 60^\circ = 10/1.732 = 5.786 \text{ rad/s}$$

Therefore $f = 5.786/2\pi = 0.92$ Hz (*Ans*)

At this frequency the magnitude of the gain, before feedback, of the amplifier is

$$\frac{100}{\{\sqrt{[1 + (1/0.1\omega)^2]}\}^3} = \frac{100}{\{\sqrt{[1 + (1.732)^2]}\}^3}$$

$$= 100/(\sqrt{4})^3 = 100/8$$

The amplifier will oscillate if $|\beta A| = 1$ or if $\beta = 8/100$ (*Ans*)

There are two main methods available by which the stability of a n.f.b. amplifier can be predicted. These two methods are the *Nyquist diagram* and the *Bode plot.*

1 The Nyquist Diagram The loop gain of a feedback amplifier is a complex quantity, in that it has both amplitude and phase. The loop gain can be plotted on a graph using polar coordinates to produce the Nyquist diagram of the amplifier.

Nyquist's criterion states that: if the locus of βA does *not* enclose the point $1,j0$ the circuit will be stable but, if the point $1,j0$ is enclosed, the circuit will be unstable.

A circle of unity radius and centred upon the point $1,j0$ can be drawn on the same axes, and then whenever the locus of βA is *inside* this circle the feedback is *positive*; conversely, whenever the locus of βA is *outside* the circle the feedback is *negative.* Consider the typical Nyquist diagram shown in Fig. 4.18.

It can be seen that the point $1,j0$ is not enclosed by the locus of βA and this means that the amplifier is stable, i.e. it will not oscillate. The locus does enter the unity circle at the points XX and so the feedback becomes positive

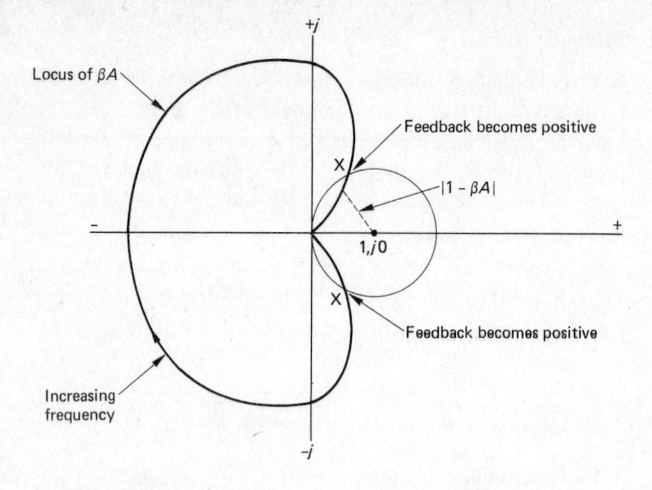

Fig. 4.18 Nyquist diagram of a feedback amplifier.

at both low and high frequencies. When the locus of βA is inside the circle, it is clear that the line $|1-\beta A|$ must have a length less than 1 and so

$$|A_{v(F)}| = \frac{A}{|1-\beta A|}$$

must be greater than A and the feedback is positive.

Example 4.8

An amplifier has a voltage gain before feedback given by

$$A_v = \frac{-1200}{(1+jf\times 10^{-5})^3}$$

and a feedback factor of $0.04\underline{/0^\circ}$. Draw the Nyquist diagram for the amplifier and use it to determine whether or not the amplifier is stable.

Solution

$$\beta A = \frac{-1200\times 0.04}{(1+jf\times 10^{-5})^3} = \frac{48\underline{/180^\circ}}{\{\sqrt{[1+(10^{-5}f)^2]}\}^3 3\tan^{-1}(10^{-5}f)}$$

Tabulating:

$\underline{/\beta A}$	−45°	0°	45°	90°	135°	180°
$3\tan^{-1}(10^{-5}f)$	225°	180°	135°	90°	45°	0°
$\tan^{-1}(10^{-5}f)$	75°	60°	45°	30°	15°	0°
f(kHz)	373.2	173.2	100	57.7	26.8	0
$\sqrt{[1+(10^{-5}f)^2]}$	3.864	2.0	1.414	1.155	1.035	1
$\{\sqrt{[1+(10^{-5}f)^2]}\}^3$	57.7	8.0	2.828	1.539	1.11	1
$\dfrac{48}{\{\sqrt{[1+(10^{-5}f)^2]}\}^3}$	0.83	6	16.97	31.19	43.26	48

Plotting the locus of βA from these figures gives the Nyquist plot of Fig. 4.19. The point $1,j0$ *is* enclosed and so the amplifier is unstable.

Fig. 4.19

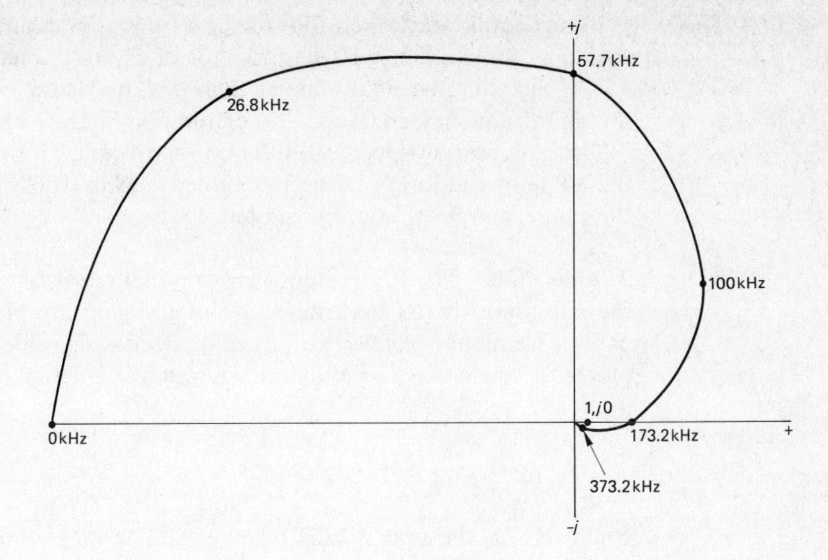

Gain and phase margin If an amplifier is shown by its Nyquist diagram to be stable it is of considerable importance to be able to determine how near the circuit is to instability. There are two ways in which the relative stability of an amplifier can be specified, namely the *gain margin* and the *phase margin.*

Fig. 4.20 (*a*) Gain margin, (*b*) phase margin.

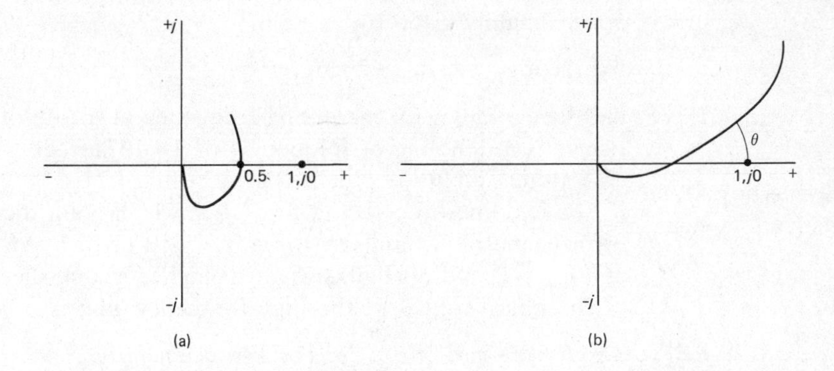

The **gain margin** of an amplifier is the amount by which the gain must be increased in order to produce instability. Generally, the gain margin is expressed in decibels. Thus, referring to Fig. 4.20*a*, the βA locus crosses the real axis at a frequency f_1 at the point $0.5,j0$. The increase in the magnitude of the loop gain $|\beta A|$ needed to produce instability, i.e. for the locus to reach the point $1,j0$, is $1/0.5$ or 2. Thus, the gain margin is $20 \log_{10} 2$ or 6 dB.

The **phase margin** is $\underline{/\beta A}$ at the frequency at which $|\beta A|$ is unity. It is the amount by which the phase of the loop gain must be altered to make the amplifier unstable. The phase margin is shown as in Fig. 4.20*b*.

It is generally reckoned that for good stability an amplifier should have a gain margin of at least 10 dB (a factor of 5) and a phase margin of at least 45°. If the gain margin is smaller than this, undesirable peaks will appear in the gain/frequency characteristic of the amplifier.

The gain margin specification is not appropriate for an amplifier in which the angle of the loop phase $\underline{/\beta A}$ never reaches 180°; for such an amplifier the phase margin should be quoted.

2 Bode Plots The Bode plot of an amplifier consists of two graphs in which the amplitude *in dB* and the phase of the gain are plotted separately to a base of frequency *plotted to a logarithmic scale.* Consider first of all the Bode plot of a single-stage amplifier whose gain/frequency characteristic is given by

$$A_{v(HF)} = \frac{A_{v(MF)}}{1 + jf/f_{3dB}}$$

where f_{3dB} is the upper 3 dB frequency. The magnitude in dB of $A_{v(HF)}$ is

$$20 \log_{10} |A_{v(HF)}| = 20 \log_{10} |A_{v(MF)}| - 20 \log_{10}\sqrt{[1 + f^2/f_{3dB}^2]}$$

If this equation is plotted, a response similar to that given in Fig. 4.17 will be obtained. For a Bode plot, however, it is more usual to proceed in the following way.

For frequencies up to the upper 3 dB frequency f_{3dB}, take $|A_{v(HF)}|$ dB as equal to $|A_{v(MF)}|$ dB.

For frequencies above the upper 3 dB frequency where $f > f_{3dB}$, the gain is approximately given by

$$20 \log_{10} |A_{v(MF)}| - 20 \log_{10} f^2/f_{3dB}^2$$

and hence falls with increasing frequency at a rate of 6 dB per octave (an octave is a doubling of frequency) or 10 dB/decade.

A Bode amplitude plot is drawn in Fig. 4.21*a*.

The two lines intersect at $f = f_{3dB}$ at which point there is obviously +3 dB error in the plot. Similarly there is +1 dB error at $f/f_{3dB} = 0.5$, +1 dB error at $f/f_{3dB} = 2$ and negligible error at higher frequencies.

The phase shift θ of the high-frequency gain is

$$\theta = 0° - \tan^{-1} f/f_{3dB} \qquad [\text{or } 180° - \tan^{-1} f/f_{3dB}]$$

For frequencies very much below f_{3dB}, $\theta \simeq 0°$ and when $f = f_{3dB}$, $\theta = -45°$. Also, for frequencies greater than about $10f_{3dB}$, θ tends towards $-90°$. The Bode phase plot is drawn with a slope of 45° per decade in Fig. 4.21*b*. It can be determined that this idealized phase characteristic is never in greater error than 6°.

The same principles apply when more than one time constant is involved. For an amplifier with two identical high-frequency time constants, the line decreases at the rate of 12 dB/octave and the phase plot changes at 90° per decade. If the two time constants differ from one another, the Bode amplitude plot is obtained using

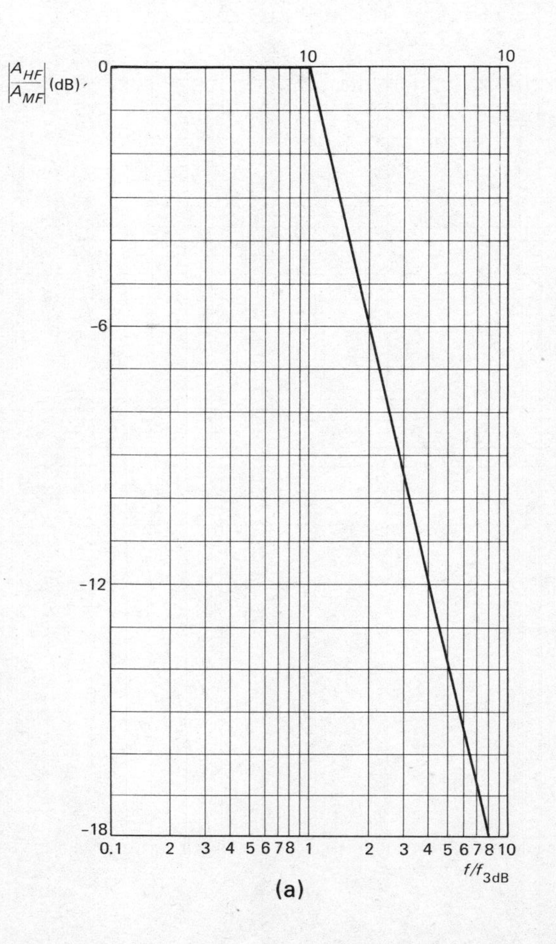

Fig. 4.21 Bode amplitude and phase diagrams.

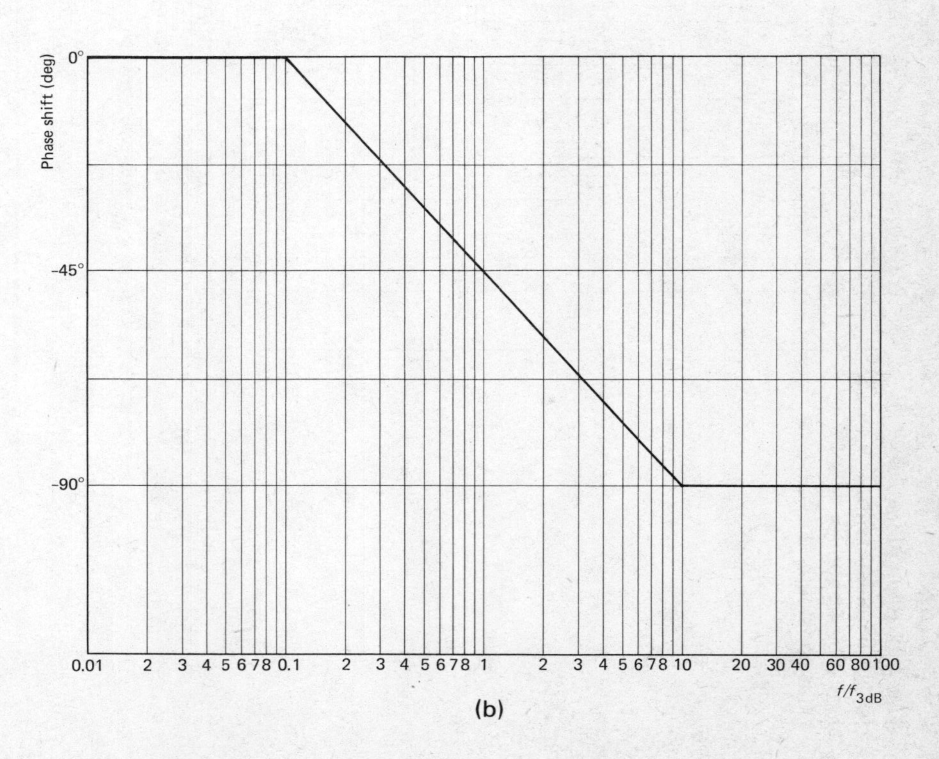

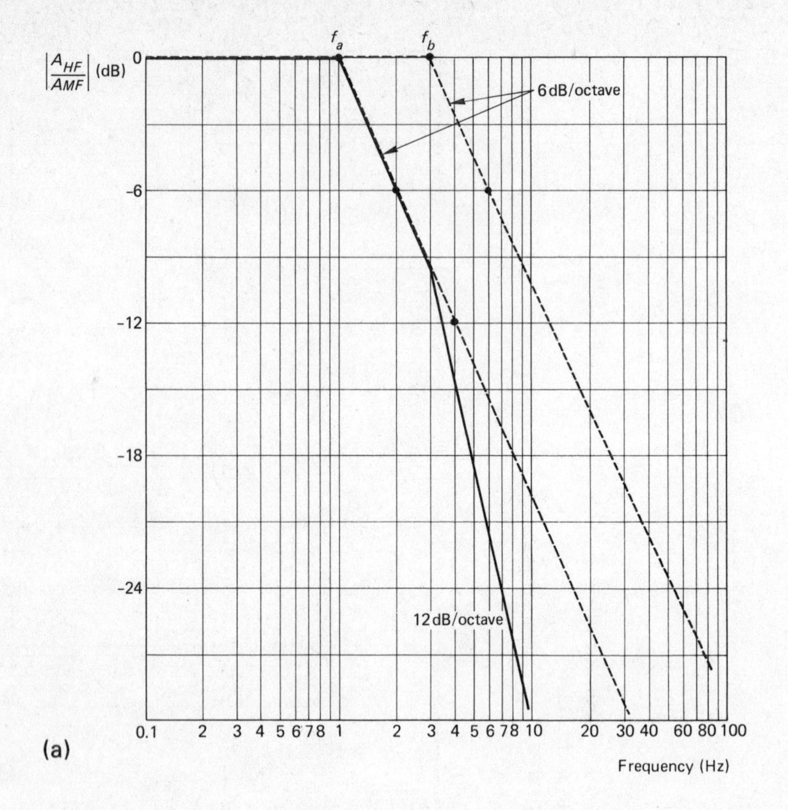

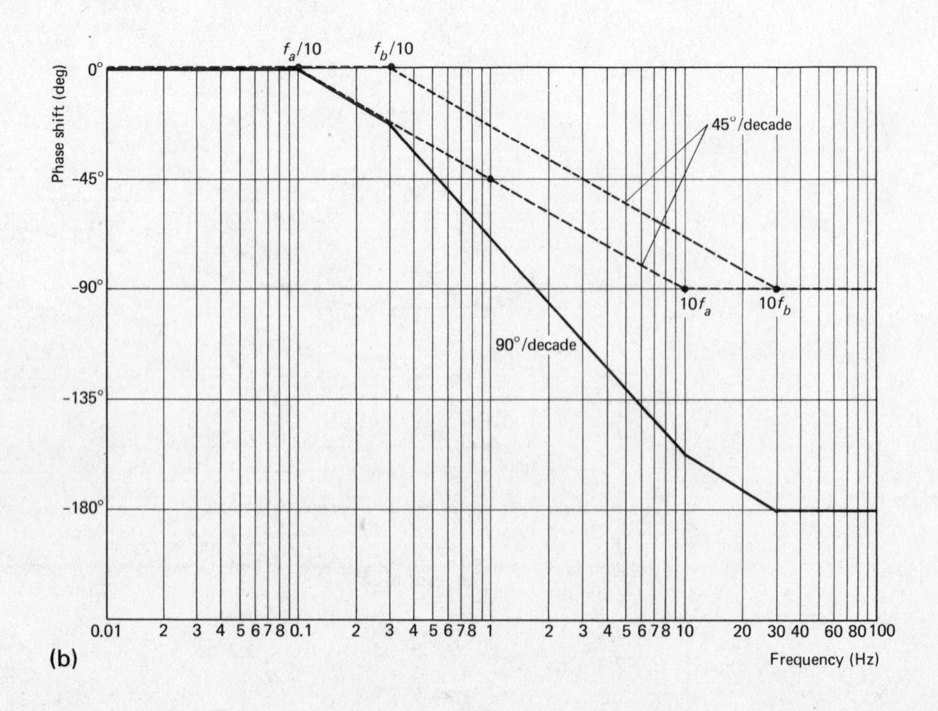

Fig. 4.22 Bode plots for an amplifier with two high-frequency time constants.

$$|A_{(HF)}|\,\text{dB} = 20\log_{10}|A_{(MF)}| - 20\log_{10}\sqrt{[1+(f/f_a)^2]} - 20\log_{10}\sqrt{[1+(f/f_b)^2]}$$

and this means the plot will consist of the sum of

i) a horizontal straight line up to $f = f_a$

ii) a straight line decreasing at 6 dB/octave starting from f_a

iii) a straight line with a slope of −6 dB/octave starting from f_b.

The three individual lines are shown dotted, and their resultant solid, in Fig. 4.22*a*.

Similarly the phase response can be written as

$$\theta = 0° - \tan^{-1} f/f_a - \tan^{-1} f/f_b$$

The phase plot is represented by the sum of

i) a horizontal straight line $\theta = 0°$ from $f = 0$ to $f = f_a/10$

ii) a straight line with −45°/decade slope from $f = f_a/10$ to $f = 10\,f_a$

iii) a third straight line also with −45°/decade slope from $f_b/10$ to $10f_b$

iv) horizontal straight lines from $10f_a$, and $10f_b$, upwards in frequency

These four lines are drawn dotted in Fig. 4.22*b* and their resultant is shown by the solid line. Now that the basic principles of the Bode method of representing the gain/frequency characteristic of an amplifier are known, its use for the prediction of the stability of a feedback amplifier can be considered. Consider, as an example, the amplifier which was the subject of Example 4.8, i.e.

$$A_v = \frac{-1200}{(1+jf10^{-5})^3} \quad \text{and} \quad \beta = 0.04\underline{/0°}$$

The amplifier will oscillate if $\beta A_v = 1\underline{/180°}$

$$|A_v|\,\text{dB} = 20\log_{10} 1200 - 60\log_{10}\sqrt{[1+f^2\times 10^{-10}]}$$
$$\simeq 61.6\,\text{dB} - 60\log_{10} 10^{-5}f$$

The 3 dB frequency is 100 kHz and so

$$f_{3dB}/10 = 10\,\text{kHz} \quad \text{and} \quad 10f_{3dB} = 1\,\text{MHz}$$

The Bode diagram for this amplifier is shown in Fig. 4.23. From the figure, $\theta = 360°$ when the frequency is approximately 205 kHz, $A_v = 42$ dB. For oscillations to occur $|\beta A_{v(HF)}| = 1$ or

$$20\log_{10}|A_{(HF)}| + 20\log_{10}|\beta| = 0$$

Hence $20\log_{10}|A_{(HF)}| = 20_{10}\log|1/\beta|$

So for oscillations $|1/\beta| = |A_v| = 42\,\text{dB} = 126$. This is greater than the actual value of β, i.e. $1/\beta = 1/0.04 = 25$ and so the amplifier is unstable.

For stability, the magnitude of the gain must be made to fall at a faster rate to ensure that the loop gain is less than unity at 205 kHz.

If a particular phase margin is required, say 30°, then $\underline{/\beta A} = 360° - 30° = 330°$ which occurs at a frequency of 130 kHz when $|A_v| = 55$ dB. The phase margin is defined for $|\beta A_v| = 1$ or 0 dB and hence $|1/\beta| = |A_v|$ to achieve the required margin. In this case the required β value is 1.78×10^{-3}. A similar

Fig. 4.23

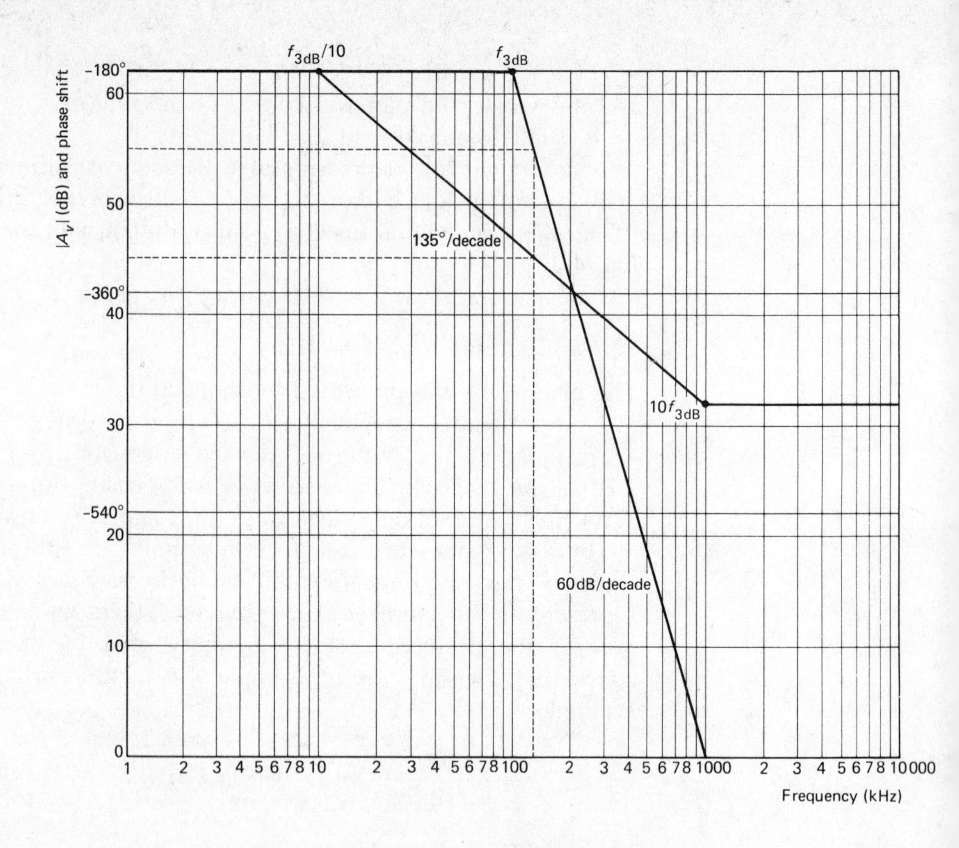

analysis can be applied to the low-frequency end of the gain characteristic.

An amplifier cannot become unstable with only two time constants and this means that low-frequency stability can probably be assured by the use of direct coupling between stages.

High-frequency stability can be achieved by the use of **phase-lead compensation**. This consists of the connection of a capacitor C_f in parallel with the feedback resistor R_f as shown in Fig. 4.24.

Fig. 4.24 Phase-lead compensation.

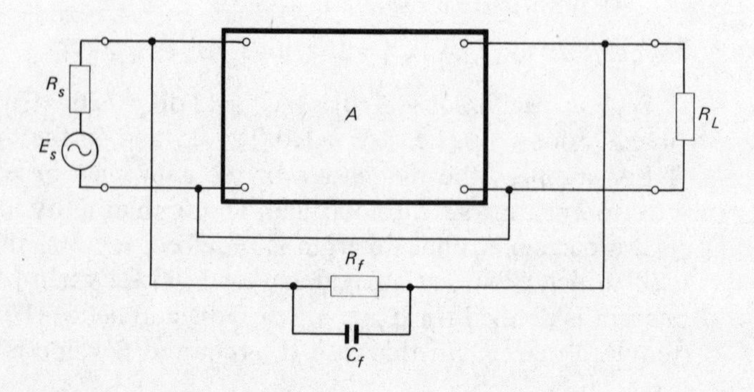

From equation (4.20), $\beta = R_L/(R_f + R) \simeq R_L/R_f$. At high frequencies where the reactance of C_f is low enough to affect the response of the amplifier $\beta_{(HF)} = R_L/Z_f$ where

$$Z_f = \frac{R_f}{1 + j\omega C_f R_f} = \frac{R_f}{1 + j\omega/\omega_{3dB}}$$

Therefore,

$$\beta_{(HF)} = \frac{R_L}{R_f}(1 + j\omega/\omega_{3dB}) = \beta(1 + j\omega/\omega_{3dB}) = \beta(1 + j\omega\tau)$$

The high-frequency gain of an amplifier with three h.f. time constants is of the form

$$A_{v(HF)} = \frac{A_{v(MF)}}{(1 + j\omega\tau_a)(1 + j\omega\tau_b)(1 + j\omega\tau_c)}$$

where τ_a, τ_b and τ_c are the three h.f. time constants. The gain with feedback is

$$A_{v(HF)} = \frac{A_{v(MF)}/(1 + j\omega\tau_a)(1 + j\omega\tau_b)(1 + j\omega\tau_c)}{1 + \dfrac{\beta(1 + j\omega\tau)A_{v(MF)}}{(1 + j\omega\tau_a)(1 + j\omega\tau_b)(1 + j\omega\tau_c)}}$$

If τ is chosen to be equal to one of the three amplifier time constants, one term in the denominator of $A_{v(HF)}$ will be cancelled out. The loop gain then has the high-frequency phase characteristic of a two-stage amplifier and will be stable.

Example 4.9

An amplifier has three high-frequency time constants having 3 dB frequencies of, respectively, 10^5 Hz, 2×10^5 Hz and 10^6 Hz. Phase lead compensation is to be used to cancel the 3 dB frequency at 2×10^5 Hz. Calculate the value of capacitance needed. The feedback resistor is 200 kΩ.

Solution

$$C_f = \frac{1}{2\pi f R_f} = \frac{1}{2\pi \times 2 \times 10^5 \times 2 \times 10^5} = 3.98 \text{ pF} \quad (Ans)$$

Log-Gain Plots

It is possible to draw a Bode plot of an amplifier's characteristics *after* negative feedback has been applied. Consider the amplifier whose Bode amplitude and phase diagrams are shown in Fig. 4.25. The amplifier has a low- and medium-frequency current gain of 3162 or 70 dB and upper 3 dB frequencies or **break points** at 10^4 Hz, 3×10^4 Hz and 10^5 Hz. For a phase margin of 45° the amplifier gain without feedback is (by projection) 61 dB or 1122. Thus, the required feedback factor is $1/1122 = 8.9 \times 10^{-4}$.

If the loop gain βA is large relative to unity, the gain with feedback is approximately equal to $1/\beta$ or 1122. This is the same value as the gain

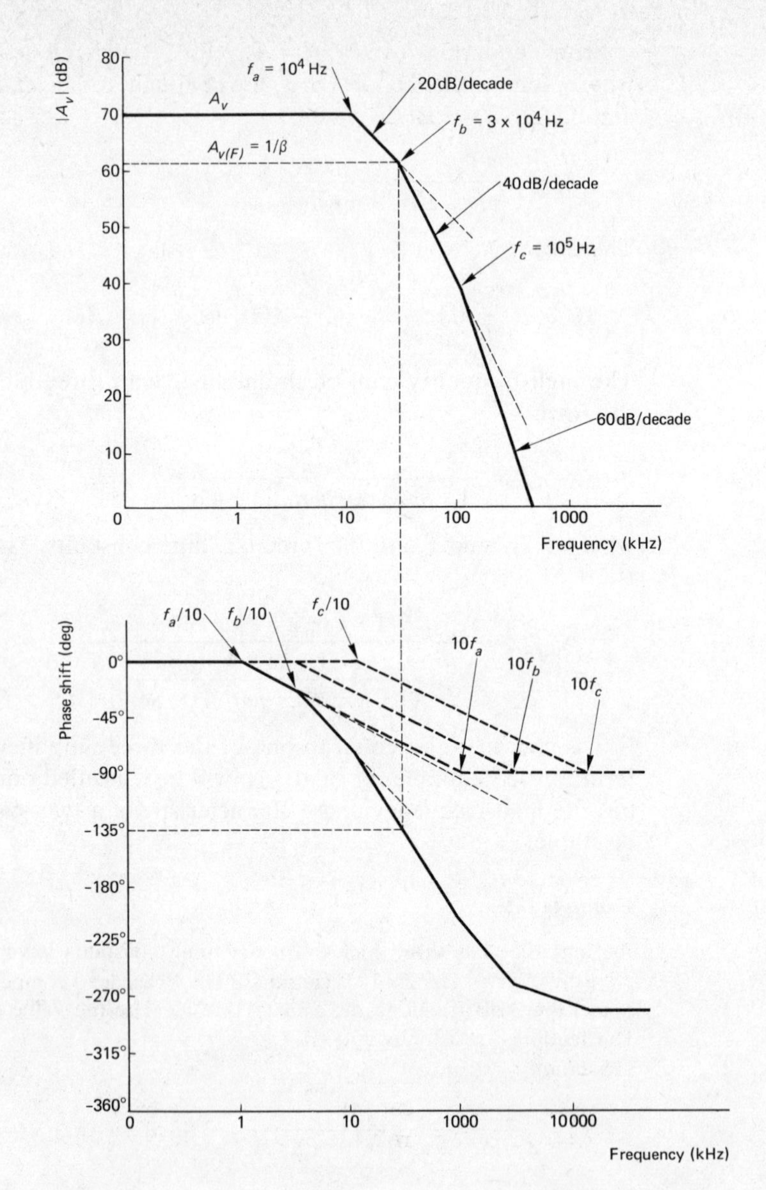

Fig. 4.25 Log-gain plot of an amplifier.

without feedback at the frequency at which the phase margin is 45° (or whatever phase margin is wanted). This means that the *closed loop* and the *open loop* gain lines intersect at the frequency at which the loop gain is unity. This is shown by the line labelled $A_{v(F)} = 1/\beta$. For stability with 45° phase margin, this intersection must take place at a frequency at or below the second break frequency. At higher frequencies the loop gain βA is less than unity and the open and closed loop gains are approximately equal—absolutely equal on the straight line approximation of the Bode diagram.

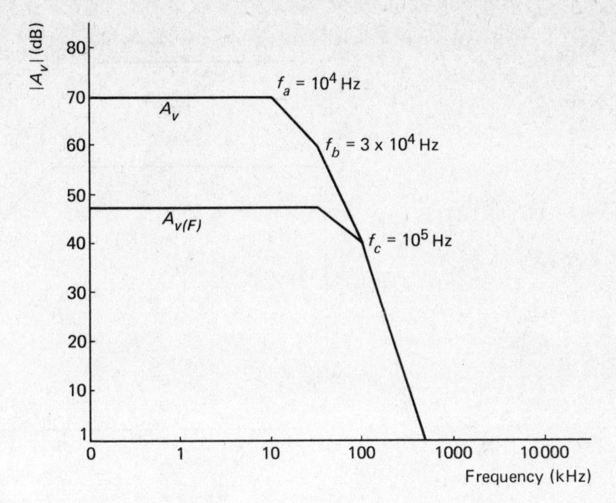

Fig. 4.26 Showing the effect of phase-lead compensation.

The upper 3 dB frequency of the amplifier with feedback is very nearly the first break frequency of the closed loop curve. This, of course, occurs at the frequency at which the closed loop and the open loop curves intersect.

The effect of phase-lead compensation is shown in Fig. 4.26. The gain with feedback is approximately

$$1/\beta_{(HF)} = 1/\beta(1 + j\omega\tau)$$

and it therefore has a break frequency at $f = 1/2\pi\tau$. The figure has been drawn assuming that $f = 3 \times 10^4$ Hz so that the middle break point of the open loop gain is cancelled. Now the A_v and $A_{v(F)}$ lines intersect at the highest break frequency of 10^5 Hz, and the upper 3 dB frequency of the compensated amplifier is 3×10^4 Hz. The difference in the slopes of A_v and $A_{v(F)}$ at the point of intersection must not be greater than 20 dB/decade and this means that $A_{v(F)}$ must be lower when phase-lead compensation is used than if the amplifier is uncompensated.

An alternative method of improving the stability of a feedback amplifier is to apply extra shunt capacitance to one of the stages and so reduce its upper 3 dB frequency. The magnitude of the loop gain becomes less than unity at a frequency where the amplifier introduces negligible phase shift. For stability, the line representing $A_{v(F)} = 1/\beta$ should intersect the A_v line at or below the second break frequency, and clearly this is made easier if the lowest break frequency f_a is shifted to the left (Fig. 4.25).

Example 4.10

An amplifier has a mid-frequency gain before feedback of 6000 and upper break frequencies of 3×10^4 Hz, 10^5 Hz and 3×10^5 Hz respectively. Calculate i) the maximum β possible for 45° phase margin if compensation is not applied, ii) the upper break frequency of the feedback amplifier, iii) the maximum β possible if phase-lead compensation is applied with a break frequency at 10^5 Hz, iv) the voltage gain and the upper break frequency with phase-lead compensation.

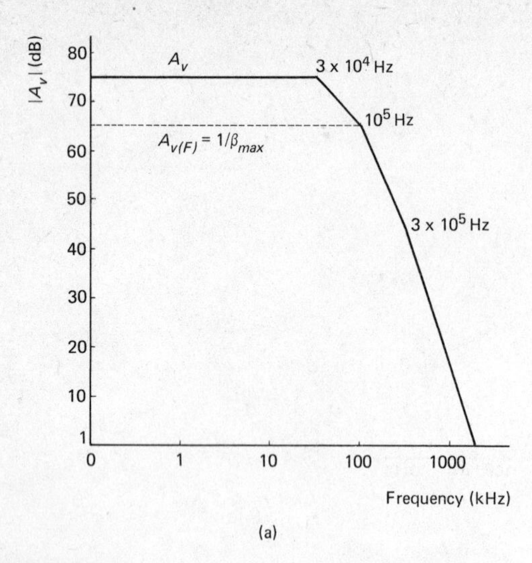

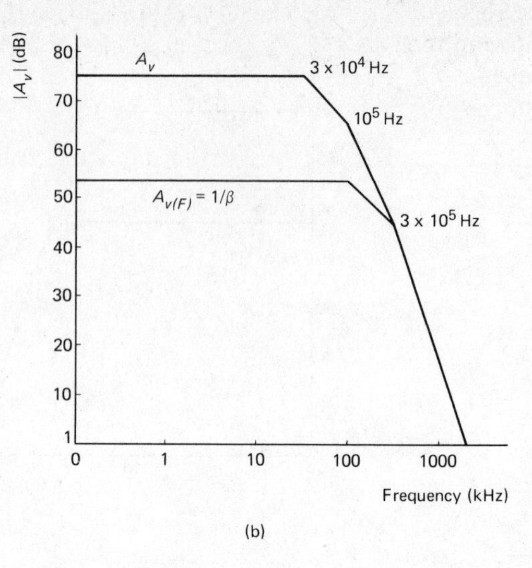

Fig. 4.27

Solution The Bode diagram for the uncompensated amplifier is shown by Fig 4.27*a*. (6000 = 75.6 dB.)

i) The maximum β is obtained when the intersection of the A_v and $A_{v(F)}$ lines is a the second break frequency. Hence, from the figure

$$A_{v(F)} = 1/\beta = 65 \text{ dB} = 1778 \quad \text{and so } \beta_{max} = 5.62 \times 10^{-4} \quad (Ans)$$

ii) The upper break frequency $= 10^5$ Hz (*Ans*)

iii) With phase-lead compensation, the $A_v = 1/\beta$ line has a break frequency at one o the three open-loop break frequencies but usually the middle one is chosen. (Not that a higher β is possible if the highest break point is cancelled.) The compensate Bode diagram is shown in Fig. 4.27*b*. From this

$$A_{v(F)} = 54 \text{ dB} = 501 \quad \text{and so } \beta = 1/501 = 2 \times 10^{-3} \quad (Ans)$$

iv) Also from Fig. 4.27*b*,

the upper break frequency is 10^5 Hz (*Ans*)

Exercises 4

4.1 The transistor used in the circuit of Fig. 4.28 has the following parameters $h_{ie} = 800\ \Omega$, $h_{fe} = 120$ and $h_{oe} = 90\ \mu$S. Calculate the current gain, input resistanc and output resistance of the circuit. Assume the source resistance to be 600 Ω.

4.2 For the transistor used in Fig. 4.29, $h_{ie} = 11$ kΩ, $h_{fe} = 150$, $h_{oe} = 100 \times 10^{-6}$ S. Calcu late the voltage gain, input resistance and output resistance of the circuit. Assum the source resistance is 2000 Ω.

Fig. 4.28

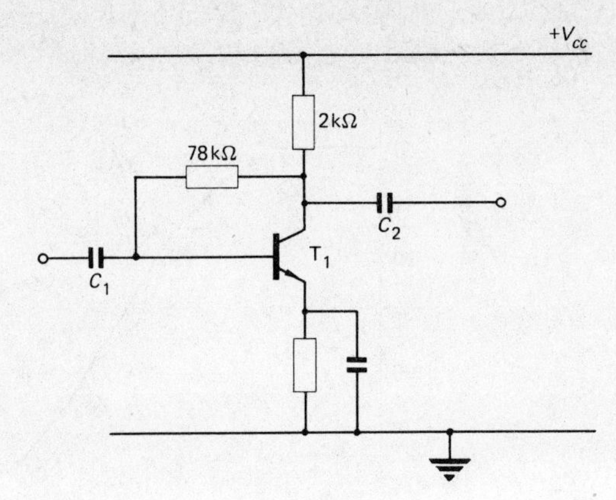

Fig. 4.29

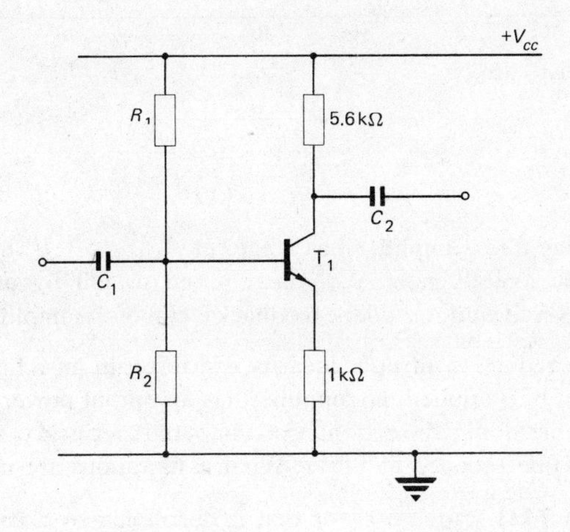

4.3 An amplifier has a voltage gain of

$$A_v = \frac{-1000}{(1+j10^{-5}f)^3}$$

before feedback is applied. The feedback factor is 1/250. Plot the Nyquist diagram and use it to confirm that the amplifier is stable. Determine the gain margin of the amplifier.

4.4 Fig. 4.30 shows the Bode amplitude and phase plots of an amplifier. Determine the feedback factor required for a phase margin of i) 30°, ii) 60°.

4.5 Draw the circuit of an emitter follower and explain why bootstrap biasing is sometimes used.

An emitter follower is connected between a 6000 Ω source and a 100 Ω load. If the follower is to be matched to the load, determine the possible values for h_{fe} and the emitter resistance R_E. What is then the input resistance of the circuit?

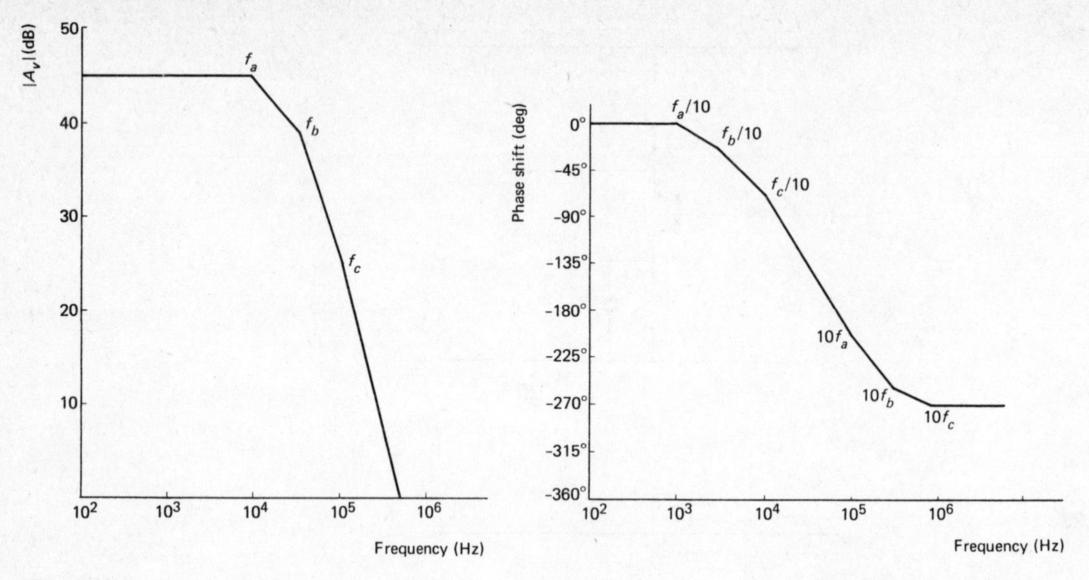

Fig. 4.30

4.6 A three-stage n.f.b. amplifier has a gain of $A_{v(F)} = 50$. If the gain of each stage fall by 10% the overall gain $A_{vo(F)}$ is required to fall by only 0.5%. Calculate th necessary overall gain A_v and the feedback factor of the amplifier before n.f.b. is applied

4.7 Discuss the reduction of non-linearity distortion in an n.f.b. amplifier.

Before n.f.b. is applied, an amplifier has an output power of 5 W in a load of 12 with 10% harmonic distortion. Calculate the necessary value of βA for the % distortion to be reduced to 0.5%. What assumptions are made?

4.8 A fet has a 2 kΩ source resistor that is decoupled by a capacitor and a drain loa resistor R_L of 22 kΩ. Calculate the voltage gain of the circuit i) with, ii) without th decoupling capacitor if $g_m = 3$ mS. Derive any expressions used.

4.9 An amplifier has series-voltage negative feedback applied to it. If the gain befor feedback is $460\underline{/0°}$ at 2 kHz and $230\underline{/-60°}$ at 20 kHz and $\beta = 0.01$ calculate the gai at i) 2 kHz, ii) 20 kHz.

4.10 The specification of an integrated circuit amplifier includes i) nominal gain 80 dB, i maximum gain variation 75–85 dB, iii) input impedance 120 kΩ. The i.c. is used wit series-voltage n.f.b. to reduce the gain to 60 dB. Calculate *a*) the maximum gai variation, *b*) the input impedance of the feedback amplifier.

4.11 Explain why an amplifier may become unstable at either a low or a high frequenc How many time constants are necessary for instability?

A particular amplifier has three identical stages each with a time constant 0.004 S. Calculate the frequency at which the amplifier may oscillate.

Short Exercises

4.12 For the feedback amplifier whose Bode plot is given in Fig. 4.31 calculate i) β, ii) β at low and medium frequencies.

Fig. 4.31

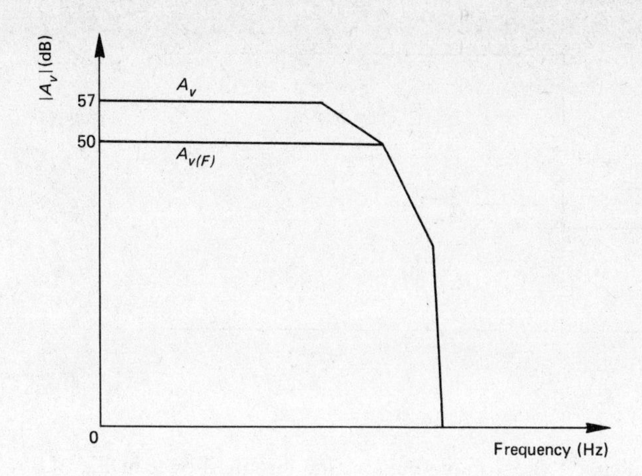

Fig. 4.32

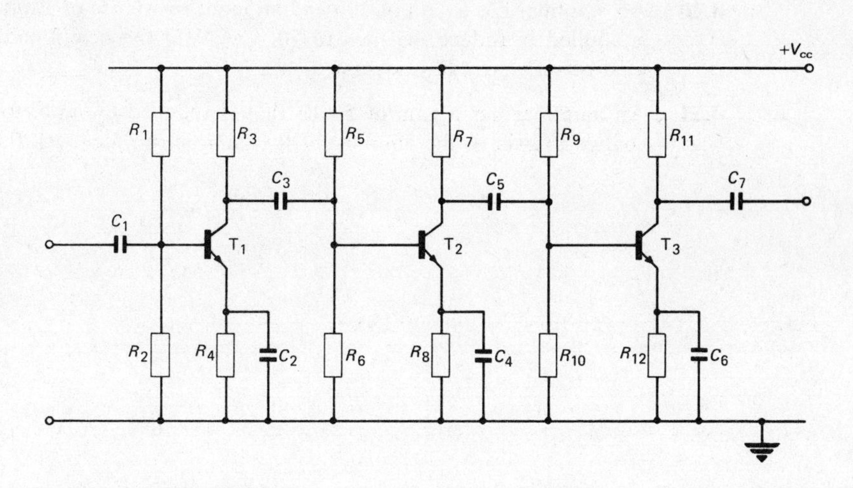

4.13 Draw the block diagram of a feedback amplifier and use it to clearly explain the terms negative and positive feedback.

4.14 Fig. 4.32 shows the circuit of a transistor amplifier. Draw diagrams to show how overall i) series-voltage, ii) shunt-voltage, iii) series-current n.f.b. could be applied to the circuit.

4.15 List the effects of applying negative feedback to an amplifier.

4.15 Show, by means of block diagrams, the four types of negative feedback.

4.16 Draw the Nyquist diagram of a n.f.b. amplifier that has i) a gain margin of 3 dB, ii) a phase margin of 40°. Also draw the Nyquist diagram of an *RC* oscillator.

4.17 Draw the Bode amplitude and phase diagrams of an unstable n.f.b. amplifier.

4.18 An amplifier has a gain before feedback of $160\angle 60°$. Feedback is applied with $\beta = 0.01\angle 50°$. Determine whether the feedback is positive or negative.

Fig. 4.33

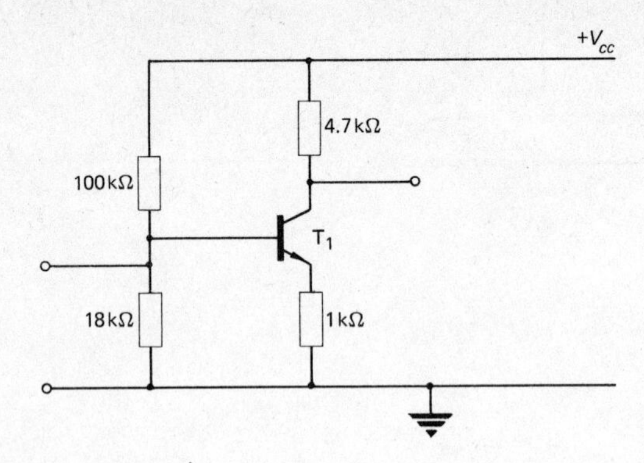

4.19 Calculate the voltage gain of the circuit shown in Fig. 4.33.

4.20 An amplifier has a gain of 200 and an input resistance of 1200 Ω. Negative feedback is applied to reduce the gain to 50. Calculate the new input resistance. Say which kind of n.f.b. you have assumed.

4.21 An amplifier has a gain of 50 dB that is reduced by n.f.b. to 30 dB. Calculate the change in gain of the amplifier if the gain before feedback falls to 44 dB.

5 Operational Amplifiers

Operation and Types

An operational amplifier, or **op-amp**, is a monolithic integrated circuit amplifier which has a very high voltage gain, a high input impedance, and (for most types) a low output impedance. These are, of course, the desirable parameters of a voltage amplifier and they make the op-amp suitable for use in a wide variety of applications. A few types are known as operational transconductance amplifiers (o.t.a.) and these have a high output impedance. A monolithic op-amp is essentially a d.c. differential amplifier of high gain, and the symbol for the device is shown in Fig. 5.1.

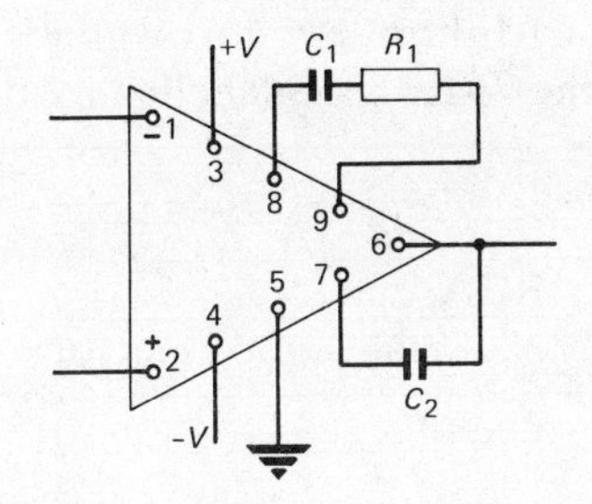

Fig. 5.1 The operational amplifier.

Two input terminals, 1 and 2, are provided. One of them, labelled −, is known as the inverting terminal since a voltage applied to this terminal appears at the output terminal with the opposite polarity, i.e. a sinusoidal input signal will experience a phase shift of 180°. The other input terminal is labelled + and this is the non-inverting terminal; a signal applied to it is amplified with zero phase shift. Only a very small difference in potential between the two inputs is needed to produce a large output voltage. When this difference is equal to zero, the output voltage should also be zero but, for reasons to be explained later, this is not always the case. The voltage gain is large and unpredictable and is *always* reduced by negative feedback, when the device is used as an amplifier. Two further terminals, 3 and 4, are provided for the connection of positive and negative power supply voltages. Two polarities are necessary so that the output voltage can vary either side of zero volts, although at least one type of op-amp, the LM 10, operates from a single positive supply voltage.

Most types of op-amp will operate satisfactorily from a wide range of supply voltages although it is generally advisable to decouple each power supply line with a suitable value of capacitance. Another terminal, 5, is provided to which the mid-point of the power supplies must be connected;

in most cases this point is earth potential as shown in the diagram. Further terminals, 7, 8 and 9 may be provided for the connection of external frequency compensation components.

Most op-amps are supplied in either 8-pin or 14-pin **dual-in-line** packages and, particularly for the second package, not all the pins are connected. The pin connections of some of the more popular op-amps are given later in the chapter.

The majority of op-amps in use today are of the **741 type** even though many alternatives offering improved characteristics are available. This situation persists because the well known 741 is cheap, readily available, and its characteristics are quite adequate for many applications.

Most op-amps, including the popular 741, employ bipolar transistors throughout their internal circuitry. An increased input resistance and reduced input bias current are possible if the input stage uses field effect transistors. Op-amps with a junction fet input stage are known as BiFET devices while another technology that uses pmosfets in the input stage is known as BiMOS. Also available are cmos op-amps.

The many types of op-amp are manufactured by several different companies with the standard devices being produced by most of them although, as is only to be expected, each firm uses its own labelling scheme.

Consider, for example, the 741; this is labelled by different makers in the following ways:

Signetics	A741 CV	Fairchild	A741 TC
Motorola	MC 1741 CP	RCA	CA 741 CS
Texas	SN 2741 P	National Semiconductor	LM 741

Parameters of Operational Amplifiers

The ideal op-amp would have the following parameters:

Infinite input impedance
Zero output impedance
Infinite voltage gain
Infinite bandwidth
Zero output voltage when equal voltages are applied to its two input terminals.

In practice, of course, the ideal op-amp does not exist and practical devices have various limitations.

1 Input Resistance The input resistance of an op-amp depends upon the type of transistors used in its input stage and whether or not they are Darlington-connected. Typically, the input resistance of a bipolar op-amp may be in the range 100 kΩ to 10 MΩ and perhaps 10^{12} Ω for a BiFET or BiMOS type.

2 Output Resistance Since the op-amp is essentially a voltage amplifier, its output resistance should be low as possible and is typically some 50 Ω to 4 kΩ.

3 Voltage Gain Commercially available op-amps have open-loop gains

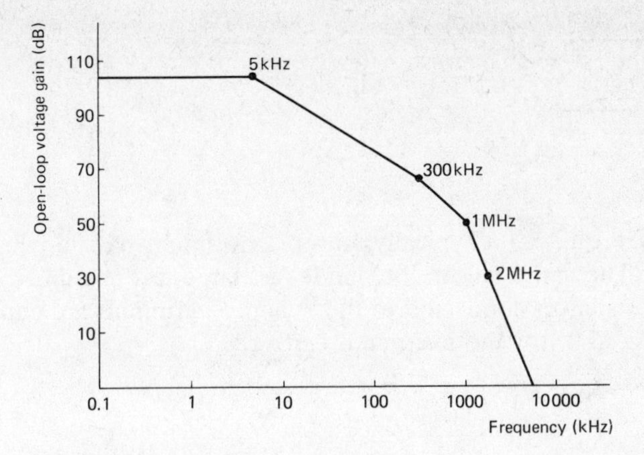

Fig. 5.2 Bode diagram of a typical op-amp gain/frequency characteristic.

that vary from one type to another but may be somewhere in the range 10 000 to 200 000.

4 Bandwidth The open-loop voltage gain of an op-amp is not constant at all frequencies but falls because of capacitive effects. A typical Bode diagram is shown in Fig. 5.2. The bandwidth of an op-amp can be specified in more than one way:

a) The first break point.

b) The frequency at which the gain has fallen to unity.

c) The full-power bandwidth; this is the maximum frequency at which a sinusoidal output voltage equal to the maximum peak output is obtained. This figure is always much less than the unity gain figure because of *slew-rate limiting.*

d) The gain-bandwidth product; thus a 741 has a gain-bandwidth product of approximately 1 MHz and this means that if a gain of 100 is wanted the maximum frequency is 10 kHz and so on.

5 Common-mode Rejection Ratio The output voltage of an op-amp is proportional to the *difference* between the voltages applied to its inverting and non-inverting terminals. When these two voltages are equal, the output should ideally be zero. A signal that is simultaneously applied to both input terminals is known as a *common-mode* signal and it is nearly always an unwanted noise voltage. Because of slightly different gains between the two input terminals, common-mode signals do *not* entirely cancel at the output. The ability of an op-amp to suppress common-mode signals is expressed in terms of its common-mode rejection ratio (c.m.r.r.).

The c.m.r.r. is defined by

$$\textbf{c.m.r.r.} = 20\log_{10}\left[\frac{\text{Voltage gain for signal applied to + or − terminal}}{\text{Voltage gain for common-mode signal}}\right]\text{dB} \qquad (5.1)$$

Alternatively, the c.m.r.r. is often expressed thus:

$$\text{c.m.r.r.} = 20\log_{10}\left[\frac{\text{Common-mode input voltage}}{\text{Differential input voltage for the same output}}\right]\text{dB} \qquad (5.2)$$

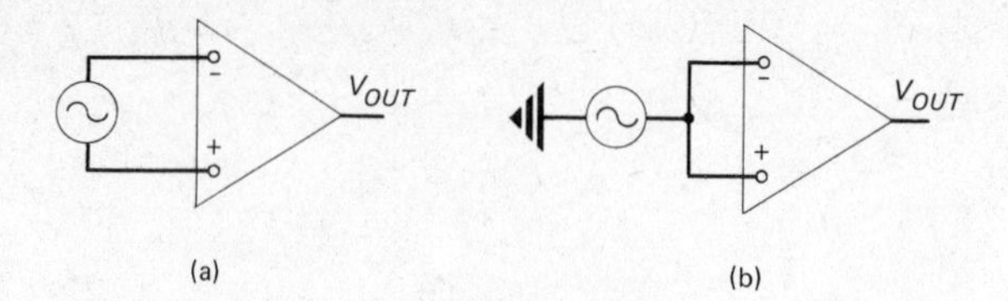

Fig. 5.3 Common-mode rejection ratio.

(see Fig. 5.3). Typically, an op-amp might have a c.m.r.r. of 90 dB.

The c.m.r.r. can be made as large as possible by ensuring that the resistances connected to the − and + terminals are equal. Thus, referring to Fig. 5.4, for the maximum c.m.r.r,

$$R_3 = R_1R_2/(R_1+R_2)$$

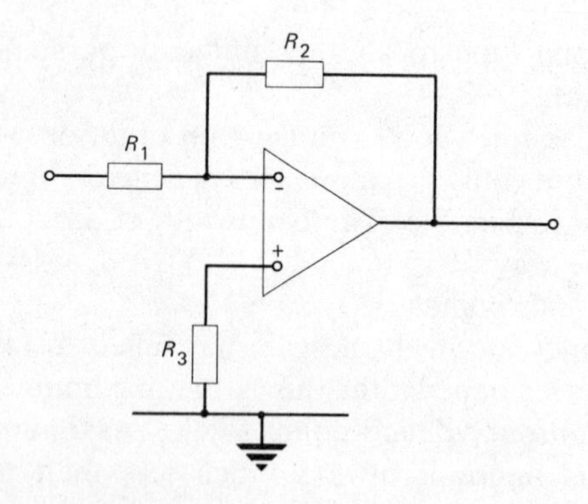

Fig. 5.4 Reduction of c.m.r.r.

Example 5.1

An op-amp has an inverting gain of 150 000 and a non-inverting gain of 149 960. Calculate its c.m.r.r.

Solution When a common-mode signal is applied

$$V_{OUT(cm)} = (-150\,000 + 149\,960)\,V_{IN(cm)} = -40\,V_{IN(cm)}$$

Therefore, the common-mode gain is 40.

The average gain is $(150\,000 + 149\,960)/2 = 149\,980$

Hence the c.m.r.r. $= 20 \log_{10}[149\,980/40] = 71.5$ dB (*Ans*)

6 Slew Rate The slew rate of an op-amp is the maximum rate, in V/μs, at which its output voltage is capable of changing when the maximum output voltage is being supplied, i.e.

$$\text{Slew rate} = \frac{dV_{OUT(max)}}{dt}\ \text{V}/\mu\text{s} \qquad (5.3)$$

When a signal at a given frequency is applied to an op-amp, the maximum permissible output voltage is determined by the slew rate; should a greater output voltage be developed, the signal waveform will be distorted. Typical slew rates are in the region of 1–30 V/μs.

The slew rate and the power bandwidth are related since

$$v_{OUT(max)} = V_{OUT(max)} \sin 2\pi f_m t \quad (5.4)$$

where f_m is the power bandwidth upper frequency limit in Hz. Therefore,

$$\frac{dv_{OUT(max)}}{dt} = 2\pi f_m V_{OUT(max)} \cos 2\pi f_m t \quad (5.5)$$

and since the maximum rate of change occurs as the waveform crosses the zero axis then

$$\text{Slew rate} = 2\pi f_m V_{OUT(max)} \quad (5.6)$$

Equation (5.6) illustrates that, if the maximum peak output voltage swing is reduced, the operating frequency can be increased, and vice versa.

Example 5.2

An op-amp has an output voltage for full-power response of 12 V and a slew rate of 5 V/μs. Calculate its power bandwidth.

Solution From equation (5.6).

$$f_m = \text{slew rate}/2\pi V_{OUT} = 5 \times 10^6 / 2\pi \times 12 = 66.3 \text{ kHz} \quad (Ans)$$

If the slew rate of an op-amp is inadequate for the signal applied to it, the output signal waveform will be distorted.

7 Input Offset Current An op-amp will have a small bias current flowing into its + and − terminals even when the input voltages are zero. The magnitude of the difference between these two bias currents is known as the *input offset current.*

The *input bias current* is one-half the sum of the two bias currents when the output voltage is zero. Typical values are 20 nA for the input offset current and 100 nA for the bias current.

The input bias current will result in the input voltage having a non-zero value even when there is no input signal voltage. Referring to Fig. 5.4 and supposing that R_3 is zero, then since the − terminal will be a virtual earth, no current flows in R_1 so that the output voltage is equal to R_2 times the bias current.

8 Input Offset Voltage Ideally, the output voltage of an op-amp should be zero when equal voltages are applied to its two input terminals but for any op-amp an output voltage will be developed. This voltage appears because of various imperfections within the amplifier but it is supposed to be caused by an *offset voltage.*

The input offset voltage of an op-amp is equal to the output voltage produced by equal input voltages divided by the open loop gain of the amplifier. Alternatively stated: the input offset voltage is the voltage that must be applied between the input terminals to obtain zero output voltage.

Typically the input offset voltage is about 1 mV. Many op-amps are provided with a pair of terminals between which a variable resistor can be connected that can be adjusted for the minimum offset voltage, e.g. a 10 kΩ variable resistor between terminals 1 and 5 in the case of a 741.

Inverting Amplifier

When an operational amplifier is used as an amplifier, a large amount of negative feedback is used to accurately specify the gain. There are two ways in which an op-amp can be connected to act as an amplifier; one method gives an inverting gain and the other gives a non-inverting gain.

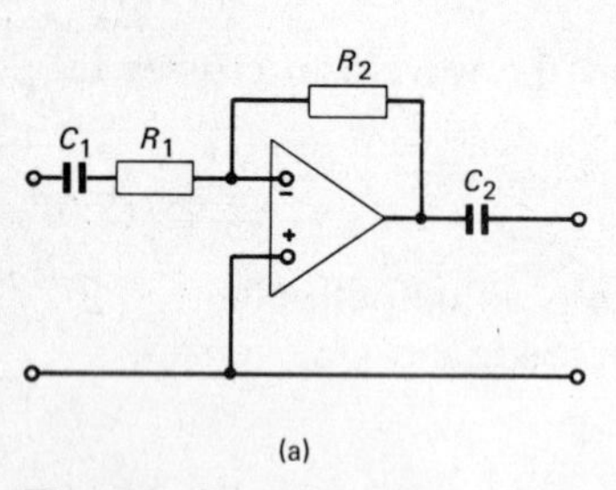

Fig. 5.5 Op-amp connected to provide an inverting gain.

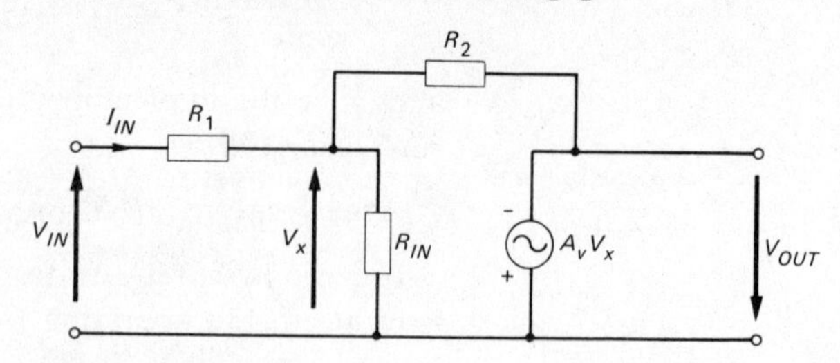

Fig. 5.6 Equivalent circuit of an inverting op-amp.

Fig. 5.5 shows how an **inverting gain** is obtained. A resistor R_1 is connected between the input terminal of the circuit and the $-$ terminal of the op-amp, and another resistor R_2 is connected between the $-$ terminal and the output terminal. The $+$ terminal is connected to earth either directly or via a third resistor whose value is chosen to optimize the c.m.r.r. of the circuit. The open-loop gain of an op-amp is very high and this means that the voltage at the $-$ terminal must be very small.

The $-$ terminal voltage will be at very nearly the same potential as the $+$ terminal and it is therefore said to be a **virtual earth**. This means that the input voltage is dropped across R_1 and so the input current is V_{IN}/R_1. The impedance of the op-amp is very high and so very little current flows into the op-amp itself. All the input current flows via R_2 and the voltage developed across R_2 is equal to the output voltage V_{OUT} of the circuit. Therefore,

$$V_{IN}/R_1 = -V_{OUT}/R_2 \quad \text{or} \quad A_{v(F)} = V_{OUT}/V_{IN} = -R_2/R_1 \tag{5.7}$$

$$\text{Input impedance } Z_{IN(F)} = V_{IN}/I_{IN} = R_1 \tag{5.8}$$

This derivation assumes that the input impedance and the open loop voltage gain of the op-amp are both very high and that the output impedance is very low.

If the input impedance and the voltage gain are both taken as being of finite value, but the output impedance is still assumed to be of negligible value, the equivalent circuit of the amplifier is as shown by Fig. 5.6. Now

$$\frac{V_{IN} - V_x}{R_1} + \frac{V_{OUT} - V_x}{R_2} = \frac{V_x}{R_{IN}}$$

But $V_x = -V_{OUT}/A_v$ so

$$\frac{V_{IN} + V_{OUT}/A_v}{R_1} + \frac{V_{OUT}(1 + 1/A_v)}{R_2} = \frac{-V_{OUT}}{A_v R_{IN}}$$

$$V_{IN}\left[\frac{1}{R_1}\right]=-V_{OUT}\left[\frac{1}{A_vR_{IN}}+\frac{1}{A_vR_1}+\frac{1+1/A_v}{R_2}\right]$$

$$=-V_{OUT}\left[\frac{1}{A_vR_{IN}}+\frac{1}{A_vR_1}+\frac{1+A_v}{A_vR_2}\right]$$

Therefore,

$$A_{v(F)}=\frac{V_{OUT}}{V_{IN}}=-\frac{1}{R_1}\times\frac{1}{\dfrac{1}{A_vR_{IN}}+\dfrac{1}{A_vR_1}+\dfrac{1+A_v}{A_vR_2}}$$

$$=-\frac{R_2}{R_1}\times\frac{1}{\dfrac{1}{A_v}\left[\dfrac{R_2}{R_{IN}}+\dfrac{R_2}{R_1}+1+A_v\right]}$$

$$A_{v(F)}=-\frac{R_2}{R_1}\times\frac{1}{1+\dfrac{1}{A_v}\left[1+\dfrac{R_2}{R_{IN}}+\dfrac{R_2}{R_1}\right]} \quad (5.9)$$

If A_v is large this equation reduces to equation (5.7).

If a resistor is connected between the + terminal and earth (Fig. 5.4), it is in series with the *much* larger R_{IN} and does not affect the gain expression.

The input resistance of the inverting amplifier can be similarly determined:

$$\text{Voltage across } R_2 = V_x(1+A_v)$$

$$I_{IN}=V_x(1+A_v)/R_2$$

$$V_{IN}=I_{IN}R_1+\frac{I_{IN}R_2}{1+A_v}$$

Hence Input impedance $R_{IN(F)}=V_{IN}/I_{IN}=R_1+\dfrac{R_2}{1+A_v}$ (5.10a)

If A_v is large, as it always is in practice, $R_{IN(F)}\simeq R_1$ as before. The effect of the output resistance R_{OUT} of the op-amp on the voltage gain will not be determined because it involves a rather lengthy analysis but its effect is so small that it can nearly always be neglected. To determine the expression for the output resistance R_{OUT}, short-circuit the input terminals of the op-amp and connect a generator of e.m.f. E volts and zero internal impedance across the output terminals. Then

$$V_x=ER_1/(R_1+R_2)$$

$$A_vER_1/(R_1+R_2)+E=IR_{OUT}$$

$$E[R_1(1+A_v)+R_2]/(R_1+R_2)=IR_{OUT}$$

$$R_{OUT(F)}=\frac{E}{I}=\frac{R_{OUT}(R_1+R_2)}{R_1(1+A_v)+R_2}\simeq\frac{R_{OUT}(R_1+R_2)}{A_vR_1} \quad (5.10b)$$

Since shunt-voltage n.f.b. is applied, the output resistance is reduced to a very low value, generally less than 1 Ω.

Non-inverting Amplifier

Fig. 5.7 shows an op-amp connected to form a **non-inverting amplifier**. Assuming initially that the input impedance of the op-amp is very high and the output impedance is very low, then

$$V_{OUT} = A_v(V_{IN} - V_y)$$

where V_y is the voltage that appears at the − terminal and A_v is the open loop voltage gain of the op-amp.

$$V_y = V_{OUT}R_1/(R_1+R_2) \quad (5.11)$$

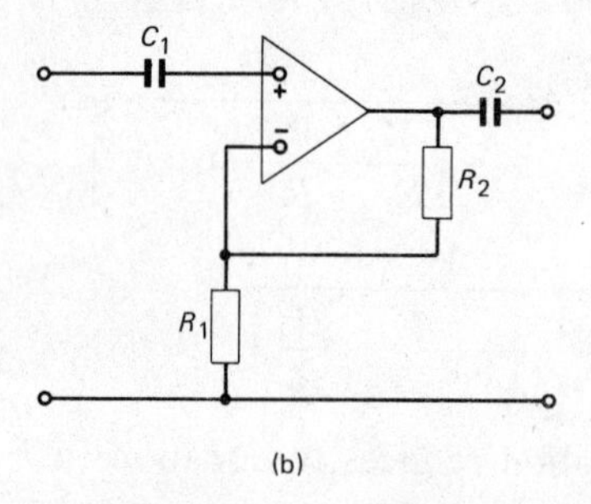

Fig. 5.7 Op-amp connected to provide a non-inverting gain.

Therefore,

$$V_{OUT} = A_v[V_{IN} - V_{OUT}R_1/(R_1+R_2)]$$

$$V_{OUT}[1+A_vR_1/(R_1+R_2)] = A_vV_{IN}$$

and

$$A_{v(F)} = V_{OUT}/V_{IN} = \frac{A_v}{1+A_vR_1/(R_1+R_2)} \quad (5.12)$$

The open loop voltage gain A_v of the op-amp is very large and so

$$A_vR_1/(R_1+R_2) \gg 1$$

$$A_{v(F)} = \frac{A_v}{A_vR_1/(R_1+R_2)} = \frac{R_1+R_2}{R_1} \quad (5.13)$$

If the input and output impedances of the op-amp are taken into account it is found that the voltage gain is still very nearly given by equation (5.13).

To determine an equation for the input impedance of the non-inverting op-amp consider Fig. 5.8.

$$R_{IN(F)} = V_{IN}/I_{IN} \quad \text{where} \quad I_{IN} = (V_{IN} - V_y)/R_{IN}$$

so $$R_{IN(F)} = \frac{V_{IN}R_{IN}}{V_{IN} - V_y} = \frac{R_{IN}}{1 - V_y/V_{IN}}$$

From equation (5.12)

$$V_{IN} = \frac{V_{OUT}[1+A_vR_1/(R_1+R_2)]}{A_v}$$

and from equation (5.11)

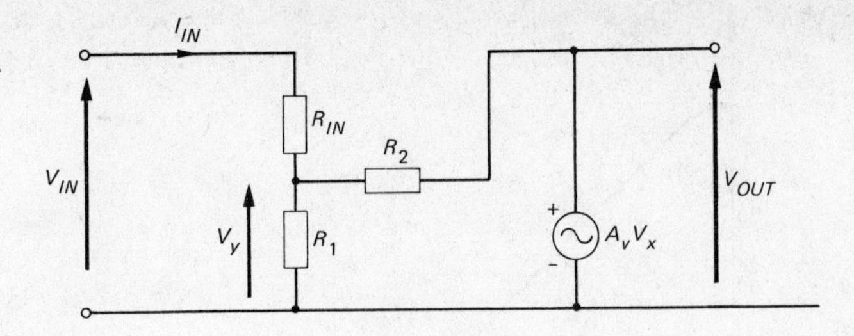

Fig. 5.8 Equivalent circuit of a non-inverting op-amp.

$$V_y = V_{OUT}R_1/(R_1+R_2)$$

Hence

$$V_y/V_{IN} = \frac{R_1A_v}{(R_1+R_2)\left[1+\dfrac{A_vR_1}{R_1+R_2}\right]} = \frac{A_vR_1}{R_2+R_1(1+A_v)}$$

and

$$R_{IN(F)} = \frac{R_{IN}}{1-\dfrac{A_vR_1}{R_2+R_1(1+A_v)}} \simeq \frac{R_{IN}R_2+R_{IN}R_1A_v}{R_1+R_2}$$

$$\simeq \frac{R_{IN}R_1A_v}{R_1+R_2} = \frac{A_vR_{IN}}{1+R_2/R_1} \tag{5.14}$$

Since $A_v \gg 1+R_2/R_1$, the input impedance of the non-inverting op-amp is considerably increased. This result should have been expected since it is clear from Fig. 5.8 that voltage-series n.f.b. has been applied to the circuit. Also, the output impedance of the circuit is reduced to a very low value.

Compensation of Operational Amplifiers

A high-gain op-amp is connected for operation as an inverting or a non-inverting amplifier by the application of n.f.b. Because its open-loop gain is very high, there is always the risk of gain instability unless the amplifier is **frequency compensated**. Many types of op-amp are internally compensated but others require the connection of external compensation components between the appropriate terminals of the device. The open-loop response of A 741 op-amp is shown by Fig. 5.9. The gain-bandwidth product is 1 MHz; this means that the gain at 1 MHz is unity, and the gain at 10 kHz is 100 and so on. This is a gain/frequency slope of −20 dB/decade and is produced by a single 30 pF compensating capacitor within the chip.

If n.f.b. is applied to the device to reduce its gain to 60 dB, then the response is flat (because $\beta A_v \gg 1$ and $A_{v(F)} \simeq 1/\beta$) up to the point at which the closed loop gain line intersects with the open loop gain curve. It can be seen that this will occur at approximately 1 kHz. Similarly, if the closed loop gain is set to 40 dB, the gain is flat up to approximately 10 kHz. Note that in each case the gain-bandwidth product is constant.

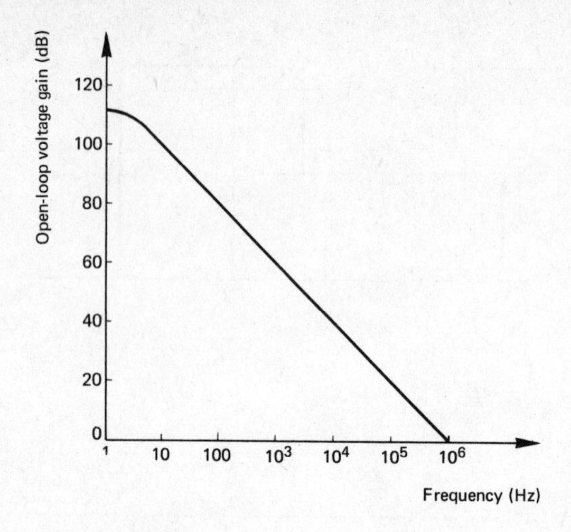

Fig. 5.9 Open loop response of a 741.

Example 5.3

A 741 op-amp has a closed loop gain of 10. What is the maximum frequency at which this gain is obtained? What is the maximum output voltage obtainable at this frequency if the slew rate is 0.5 V/μs?

Solution From Fig. 5.9 the maximum frequency for $A_{v(F)} = 10$ or

$$20\text{ dB} = 10^6/10 = 100\text{ kHz} \quad (Ans)$$

From equation (5.6).

$$V_{OUT} = \frac{0.5 \times 10^6}{2\pi \times 10^5} = 0.796\text{ V} \quad (Ans)$$

Frequency Compensation

The reasons why a high-gain n.f.b. amplifier may be unstable and the principles of frequency compensation have been discussed in the previous chapter. The ideas involved will now be briefly repeated.

Fig. 5.10 shows a Bode plot of the open loop response of an op-amp; clearly the amplifier has three break frequencies, labelled respectively as f_a, f_b and f_c. When n.f.b. is applied, the closed loop gain is approximately equal to $1/\beta$ as long as $\beta A_v \gg 1$ and this curve is also plotted. At the intersection of the two curves

$$20 \log_{10} |A_v| = 20 \log_{10} |1/\beta| \qquad 20 \log_{10} |\beta A_v| = 0$$

or $|\beta A_v| = 1$.

For stability the difference in slope between the two curves at the point of intersection must be *less than* 12 dB/octave or 40 dB/decade.

1 If only a small amount of feedback is applied, the point of intersection will occur where the open loop gain has a slope of only −6 dB/octave or 20 dB/decade. The circuit is then inherently stable and so no compensation is needed.

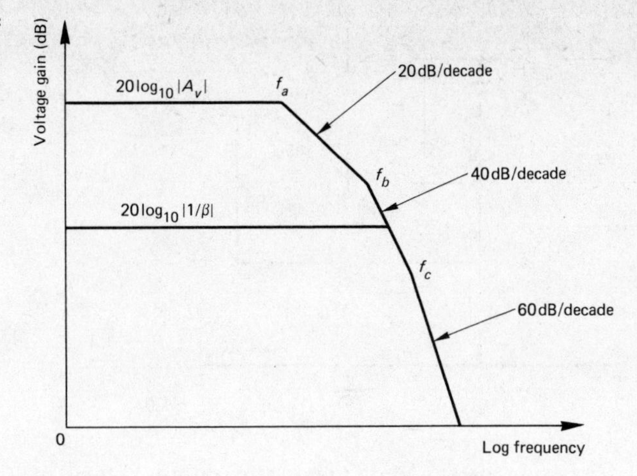

Fig. 5.10 Bode plot of an op-amp open-loop gain response.

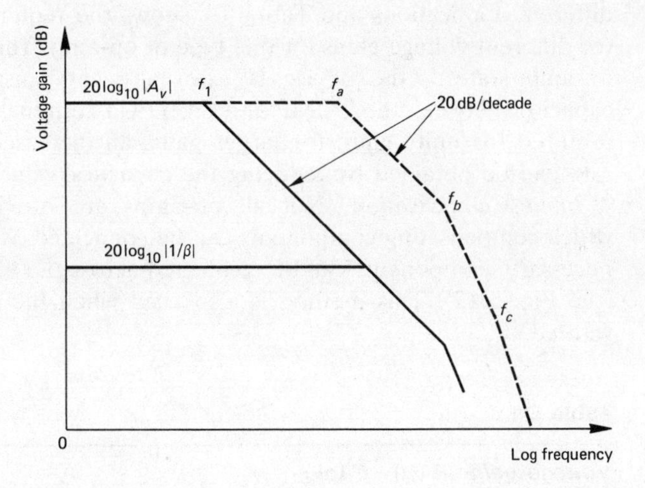

Fig. 5.11 Frequency compensation of an op-amp.

2 Suppose a capacitor (usually some 5–50 pF) is connected across the appropriate terminals to introduce a break frequency f_1 which is lower than the lowest uncompensated break frequency f_a. The effect of this, shown by Fig. 5.11, is to ensure that the point of intersection occurs where the open loop gain has a slope of only −6 dB/octave. The capacitance value must be chosen so that the new frequency of intersection is lower than f_b. This method of compensation reduces both the bandwidth and the slew rate of the amplifier.

3 Another method of compensation consists of shifting the second break frequency f_b to a new frequency f_1 that is lower than f_a by means of an R-C series circuit connected across the appropriate pins. The capacitor C causes the gain to fall off at 6 dB/octave at f_1 up to f_a; above f_a the resistance R predominates and nullifies the effect of C. This means that at frequencies higher than f_a the characteristic is the inherent response of the op-amp (see Fig. 5.12).

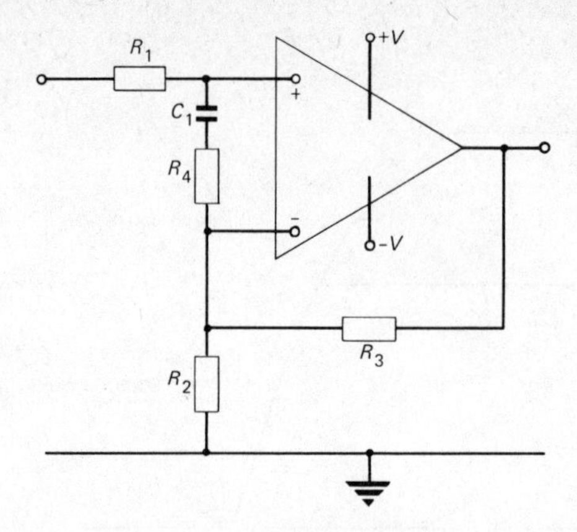

Fig. 5.12 Another method of frequency compensation.

Manufacturer's data generally quote values of *C* and *R* required for different applications and Table 5.1 shows the required values of *C* and *R* for different voltage gains for one type of op-amp. The data sheet of another op-amp states: "the device is externally compensated by connecting a capacitor between the COMP and OUTPUT terminals. A 33 pF capacitor is required for unity gain; for larger gains an increased bandwidth and slew rate can be obtained by reducing the capacitor value".

4 *Input Compensation* Not all op-amps are provided with pins across which compensating components can be connected. If some compensation is necessary, components can be connected across the input terminals + and − (see Fig. 5.12). This method is also used when the maximum slew rate is wanted.

Table 5.1

Voltage gain	*R* (Ω)	*C* (pF)
1000	30 k	1000
100	10 k	2200
10	1 k	2200
1	390	2200

Practical Operational Amplifiers

The most commonly used op-amp at the time of writing is the 741. Its characteristics are perfectly adequate for many applications, it is cheap and easily obtainable. The 741 has internal frequency compensation which allows 100% n.f.b. to be applied without instability. Some variations of the 741 are also available:

the μA 791 is a 741 with a high power output stage;

the 748 is a 741 without the internal compensation; this allows a greater gain-bandwidth product to be achieved;

the 747 is two 741s in the same package.

Most op-amps require dual voltage (±) power supplies but a few like the LM 10 do not. The main characteristics of a number of popular op-amps are given in Table 5.2.

Table 5.2 Characteristics of popular op-amps

Op-amp	*Open-loop voltage gain* (dB)	*Input resistance* (MΩ)	*Slew rate* (V/μs)	*Full power bandwidth* (kHz)	*Input offset current* (nA)	*Input offset voltage* (mV)	*c.m.r.r.* (dB)	*Unity gain bandwidth* (MHz)
LM 301	88	2	0.4	10	3	2	90	10
741	106	2	1.0	10	20	1	90	1
747	106	2	0.5	10	20	2	90	1
748	106	2	0.8	10	20	2	90	1
TDA 1034	76	0.1	13	200	20	0.5	100	100
CA 3130	110	1.5×10^3	30	10	5×10^{-4}	8	90	11

Exercises 5

5.1 The specification of a 741 op-amp includes the following: internal frequency compensation, short-circuit protection, offset voltage null capability, no latch-up. Explain the meaning and importance of each term.

5.2 Fig. 5.13 show the open-loop gain of an op-amp. The amplifier is to be used to provide a closed loop gain of 2000. Determine its bandwidth.

The amplifier is now compensated with a break frequency of 100 Hz. Draw the curve of the compensated gain and determine the bandwidth of the amplifier.

Fig. 5.13

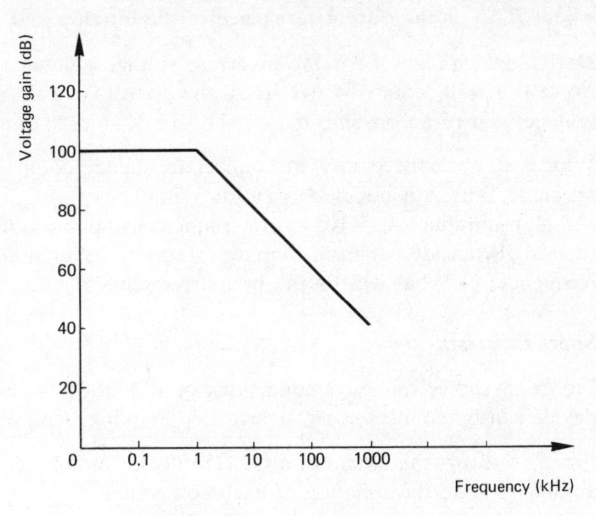

5.3 An op-amp has the following data: input resistance 50 kΩ, input capacitance 100 pF, output resistance 150 Ω. The voltage gain is 6000 into a 1 kΩ load with break frequencies at 2 MHz, 4 MHz and 30 MHz. Feedback is applied so that the overall gain is −100 into a 1 kΩ load. Calculate i) the required feedback components, ii) the input and output resistances.

Fig. 5.14

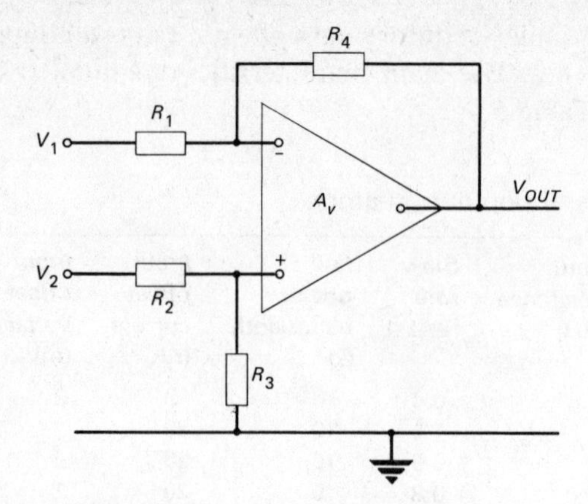

5.4 A 741 op-amp is to be used as an inverting amplifier with a voltage gain of 20 dB and break frequency at 800 Hz. The input impedance is to be 47 kΩ. Calculate suitable component values.

5.5 For the circuit given in Fig. 5.14 derive an expression for the value of R_2 which will make the gains V_0/V_1 and V_0/V_2 equal so that the circuit will act as a voltage comparator. Determine R_2 if $R_1 = 6.8$ kΩ, $R_3 = 7.5$ kΩ and $R_4 = 470$ kΩ. Also find the gain of the circuit to either input.

5.6 Show that the output resistance $R_{OUT(F)}$ of an op-amp connected to give a non-inverting gain is approximately

$$R_{OUT(F)} = R_{OUT} R_2 / A_v R_1$$

where R_{OUT} is the output resistance of the op-amp and A_v is the open loop gain.

5.7 Derive an expression for the inverting voltage gain of an op-amp. Show that this expression will reduce to $-R_2/R_1$ if the open loop gain is very large. Calculate the voltage gain of an op-amp if $A_v = 3 \times 10^4$, $R_1 = 12$ kΩ and $R_2 = 560$ kΩ.

5.8 Why is it often necessary to connect frequency compensation components to an op-amp? Why is it not necessary for a 741?

An op-amp has $A_v = 10^5$ at low frequencies falling to unity at 1 MHz with a slope of −20 dB/decade. *a*) Is this circuit internally compensated? *b*) Calculate its break frequency. *c*) What will be the break frequency if n.f.b. reduces the gain to 100?

Short Exercises

5.9 Fig. 5.15 shows the pin connections of a TL 081 BiFET op-amp. Show how this device can be connected as an inverting amplifier with a gain of 47.

5.10 Fig. 5.16 shows the circuit of a CA 3130 cmos op-amp connected as a non-inverting amplifier. State the function of each component.

5.11 Sketch the standard symbol for an op-amp and explain the meanings of the terms inverting and non-inverting gain, offset voltage, and open-loop gain.

5.12 An op-amp has an open-loop gain of 240 000. Calculate magnitude of the p.d. between the two inputs needed to produce the maximum output voltage of 30 V peak-to-peak.

Fig. 5.15

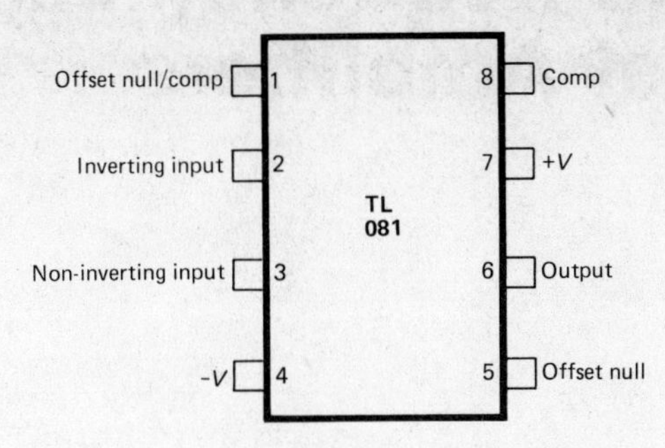

Fig. 5.16

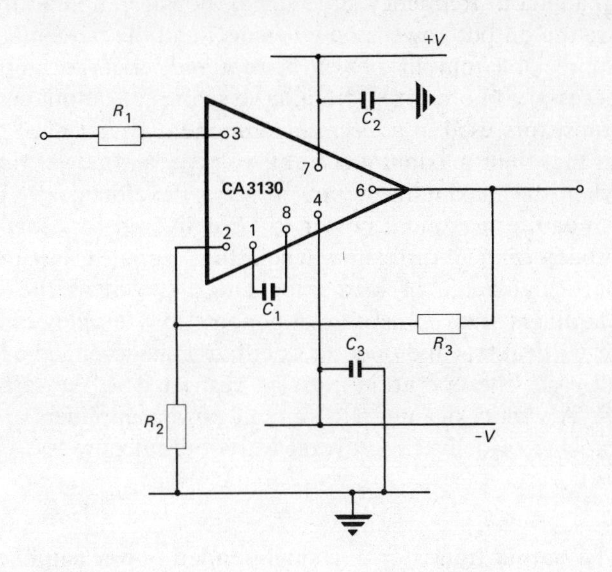

5.13 Explain clearly what is meant by the virtual earth concept when applied to an op-amp.

5.14 An op-amp has a compensating capacitor of 20 pF into which a maximum current of 25 μA can be delivered. Calculate the slew rate of the amplifier.

5.15 A non-inverting op-amp has $R_1 = 27\ \text{k}\Omega$, $R_2 = 150\ \text{k}\Omega$ and $A_v = 10^5$. Calculate the output voltage when 1 mV is applied to the circuit. Also calculate the voltage between the + and − terminals.

5.16 An op-amp has a slew rate of 5 V/μs. Calculate the maximum peak value of a sinusoidal output voltage at i) 1 MHz and ii) 3 MHz.

5.17 An uncompensated op-amp has $A_v = 90$ dB and break frequencies of 100 kHz and 1 MHz. Determine the smallest closed-loop gain possible without instability.

6 Audio-frequency Large-signal Amplifiers

Introduction

In an audio-frequency large signal (power) amplifier the main considerations are the output power, the efficiency, and the percentage distortion. When an appreciable output power is required, a large amplitude input signal is necessary in order to obtain large swings of output current and voltage. The transistors used in a discrete component circuit must be selected and biased so that their maximum current, voltage, and power ratings are not exceeded when the maximum output power is developed.

For the maximum power to be delivered to a load, without exceeding a predetermined distortion level, the output transistors must work into a particular value of load impedance, known as the optimum load. Power amplifiers using discrete components may employ either bipolar transistors or vertical fets and operate as either a single-ended or a push-pull circuit. In all cases the operation must be restricted to the safe operating area (page 9). A variety of integrated circuit power amplifiers are also available, some capable of delivering several watts output power.

Single-ended Power Amplifiers

The output transistor of a **single-ended power amplifier** should work into its optimum load impedance and usually this demands the use of an output transformer as shown in Fig. 6.1. The turns ratio of the output transformer should be equal to

$$\sqrt{[\text{Optimum load impedance/Actual load impedance}]}$$

D.C. stabilization of a transistor amplifier is best achieved by means of the potential divider bias circuit but to minimize d.c. power losses the emitter resistor should be of very low value, perhaps as small as 1 Ω. When such a low value of emitter resistance is used, the resistor is not decoupled because the required capacitance value would be very high. In high-power circuits where the emitter current may be several amperes, the emitter resistor may be omitted.

Fundamentally, a power amplifier is a converter of d.c. power taken from the power supplies into a.c. power delivered to the load. Usually, the amplifier parameter of the greatest importance is its efficiency; the **collector efficiency** η is given by

$$\eta = \frac{\text{a.c. power output to load}}{\text{d.c. power taken from power supply}} \times 100\% \qquad (6.1)$$

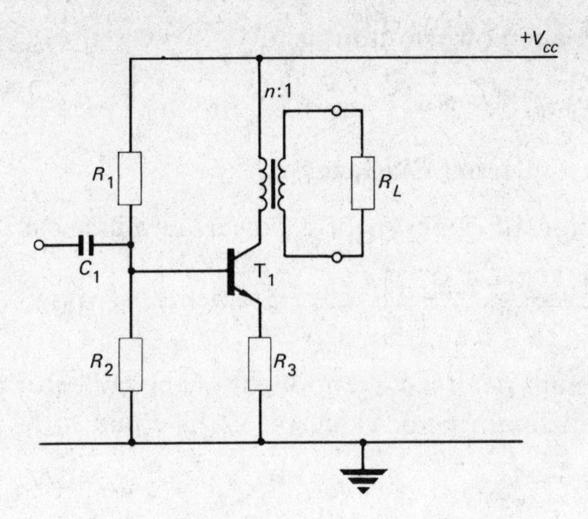

Fig. 6.1 Single-ended power amplifier.

The d.c. power P_{dc} taken from the power supply is equal to the product of the collector supply voltage and the d.c. component of the collector current, i.e.

$$P_{dc} = V_{cc} I_{c(dc)} \tag{6.2}$$

The d.c. power provides the a.c. power output plus various power losses within the amplifier itself. Therefore,

$$P_{dc} = P_{OUT(ac)} + \text{d.c. power losses} \tag{6.3}$$

D.C. power is lost within the amplifier because of the resistance of the primary winding of the output transformer, the emitter resistor, and dissipation at the collector of the transistor. Very often these losses, except for the last, are small enough to be neglected and then

$$P_{dc} = P_{OUT(ac)} + P_c \tag{6.4a}$$

where P_c is the collector dissipation.

Rearranging $\quad P_c = P_{dc} - P_{OUT(ac)} \quad$ (6.4b)

The d.c. power taken from the power supply is constant (but see page 137) and hence the collector dissipation attains its maximum value when the input signal is zero and there is no output power. Care must be taken to ensure that the maximum collector dissipation specified by the manufacturer is not exceeded.

The output power and efficiency of an a.f. power amplifier can be determined with the aid of an a.c. load line drawn on the output characteristics [EIII]. Since the collector dissipation is at its maximum value under quiescent conditions, the operating point must be chosen to lie on or under the maximum collector dissipation hyperbola.

When a signal is applied to the amplifier, the swings of collector current and voltage are

$$I_{c(max)} - I_{c(min)} \quad \text{and} \quad V_{ce(max)} - V_{ce(min)}$$

so that the a.c. power output is

$$P_{OUT(ac)} = \tfrac{1}{8}[I_{c(max)} - I_{c(min)}][V_{ce(max)} - V_{ce(min)}]\,\text{W} \tag{6.5}$$

Maximum Collector Efficiency

The collector efficiency η of a power amplifier can be written as

$$\eta = \frac{[I_{c(max)} - I_{c(min)}][V_{ce(max)} - V_{ce(min)}]}{8V_{cc}I_{c(dc)}} \times 100\% \tag{6.6}$$

The maximum peak a.c. component of the collector current is equal to $I_{c(dc)}$ and the maximum peak value of V_{ce} is equal to V_{cc}. Then

$$I_{c(max)} = 2I_{c(dc)} \qquad I_{c(min)} = 0 \qquad V_{ce(max)} = 2V_{cc} \qquad V_{ce(min)} = 0$$

Therefore,

$$\eta_{max} = \frac{2I_{c(dc)} \times 2V_{cc}}{8V_{cc}I_{c(dc)}} \times 100\% = 50\% \tag{6.7}$$

In practice, V_{ce} cannot be driven down to 0 V nor can I_c very nearly approach either $2I_{c(dc)}$ or 0 without considerable distortion being introduced. For this reason practical efficiencies are always considerably less than the theoretical maximum figure. Typical efficiencies are in the region of 40%.

Class A Push-pull Amplifiers

When the output power available from a single transistor is inadequate, or a small percentage distortion is required, the push-pull circuit shown in Fig. 6.2 can be employed. The operation of the circuit is fully described elsewhere [EIII].

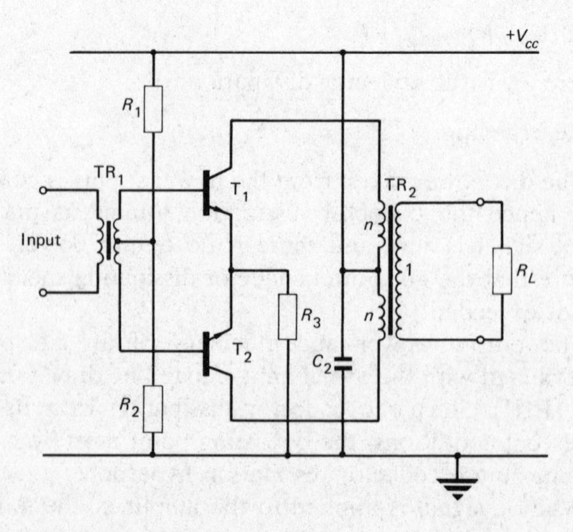

Fig. 6.2 Class A push-pull amplifier.

The mutual characteristic of a large-signal transistor can be represented by a power series:

$$I_c = I_{dc} + aV_{be} + bV_{be}^2 + cV_{be}^3 + \text{etc.} \tag{6.8}$$

where a, b, and c are constants and V_{be} is the voltage applied between the base and the emitter. Thus for T_1

$$V_{be} = V_S \sin \omega t$$

and for T_2

$$V_{be} = -V_S \sin \omega t$$

Analysis using this series shows that all even-order harmonics generated by the transistors cancel out if the transistors have identical characteristics but, in practice, they are not perfectly balanced. Generation of odd-order harmonics is small.

Example 6.1

A Class A push-pull amplifier uses two transistors whose mutual characteristics are

$$I_{c1} = 50 + 10V_{be} + 0.5V_{be}^2 \text{ mA}$$
$$I_{c2} = 55 + 11V_{be} + 0.6V_{be}^2 \text{ mA}$$

Calculate the percentage second harmonic distortion of the output signal if a 3 V peak sinusoidal signal is applied to each transistor.

Solution

$$I_{c1} = 50 + 30 \sin \omega t + 0.5 \times 9 \sin^2 \omega t$$
$$= 50 + 30 \sin \omega t + 2.25 - 2.25 \cos 2\omega t$$

[Note: % 2nd harmonic = 2.25/30 × 100 = 7.5%.]

$$I_{c2} = 55 - 33 \sin \omega t + 0.6 \times 9 \sin^2 \omega t$$
$$= 55 - 33 \sin \omega t + 2.7 - 2.7 \cos 2\omega t$$

[Note: % 2nd harmonic = 2.7/33 × 100 = 8.2%.]

The output signal is proportional to the *difference* between I_{c1} and I_{c2}.

$$I_{c2} - I_{c1} = 5.45 - 63 \sin \omega t - 0.45 \cos 2\omega t$$

% second harmonic distortion = 0.45/63 × 100 = 0.7% (*Ans*)

The circuit of a transformerless Class A push-pull amplifier is shown in Fig. 6.3.

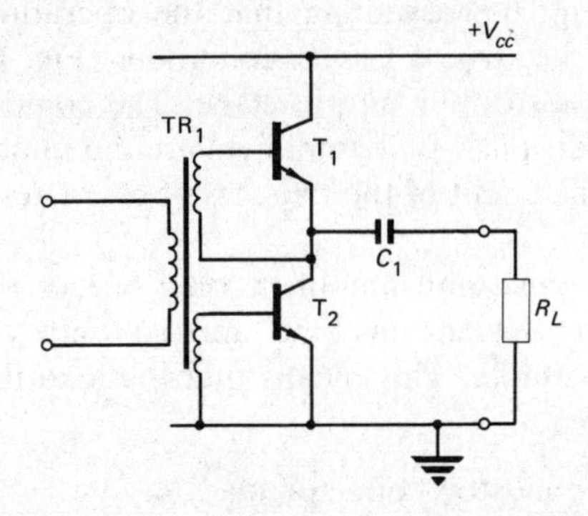

Fig. 6.3 Transformerless Class A push-pull amplifier.

Class B Push-pull Amplifiers

Nearly all push-pull amplifiers are operated under **Class B** conditions usi either the circuit shown in Fig. 6.2 or a complementary pair circuit. Whe sinusoidal input signal is applied to the circuit, one transistor condu during the positive half cycles and the other transistor conducts during negative half cycles. The collector current of each transistor therefore flc in a series of half-sinewave pulses and the two currents combine at output to produce a sinusoidal output waveform (Fig. 6.4).

The mutual characteristics of a transistor are non-linear for small val of collector current (Fig. 6.5*a*) and this gives rise to **cross-over distort** (see Fig. 6.5*b*). Cross-over distortion can be reduced by biasing b transistors to pass a small quiescent collector current.

The majority of Class B push-pull amplifiers are of the **complement pair type**, shown in Fig. 6.6. Two transistors, one p-n-p and the other n-p are slightly forward-biased and operated as a pair of emitter followers. T quiescent condition of the circuit is with the junction of the emitter resist at a potential of $V_{cc}/2$.

When an input signal is applied to the circuit, its positive half cycles dr T_1 into conduction and turn T_2 OFF. If the amplitude of the input signa large enough, T_1 is driven into saturation at the peak of the half cycle. T ON resistance of T_1 is very small and the OFF resistance of T_2 is very high that the circuit can be redrawn as shown in Fig. 6.7*a*. Current flows via and R_L to charge C_1 to V_{cc} volts and then the voltage across the load R V_{cc} volts. During the negative half cycles of the input signal voltage, T turned OFF and T_2 is saturated and the circuit can be represented by F 6.7*b*. Capacitor C_1 now discharges via R_4 and R_L and, when C_1 completely discharged, the load voltage is zero. The load voltage va about its mean value of $V_{cc}/2$, reaching a maximum positive value of volts and a minimum value of zero. The peak value of the a.c. componen the load voltage is $V_{cc}/2$ and the maximum output power is $V_{cc}^2/8R_L$ watts

The static output characteristics of the two transistors can be combinec obtain a dynamic characteristic for the push-pull amplifier. The out characteristics of one transistor are drawn in the usual manner but the ot set are drawn upside down and adjusted horizontally so that the collec supply voltages of the two characteristics coincide (see Fig. 6.8).

A composite load line can be drawn on the dynamic characteristic us the same method as for a single-ended circuit [EIII]. A typical load line been drawn in Fig. 6.8 assuming that the operating point is $I_{c1} = I_{c2} =$ $V_{ce1} = V_{ce2} = V_{cc}$, i.e. true Class B conditions. Fig. 6.8 assumes two n- transistors in a transformer output stage. The composite characteristics complementary pair push-pull amplifiers are the same as Fig. 6.8 except the horizontal adjustment of the two sets of characteristics is made to line the $V_{cc}/2$ points.

In the Class B push-pull amplifier, only one of the output transistor conducting at any instant in time and so only one-half of the out transformer is operative. This means that the load into which the transi works is equal to

$$n^2R_L \quad \text{or} \quad (\text{collector–collector load})/4$$

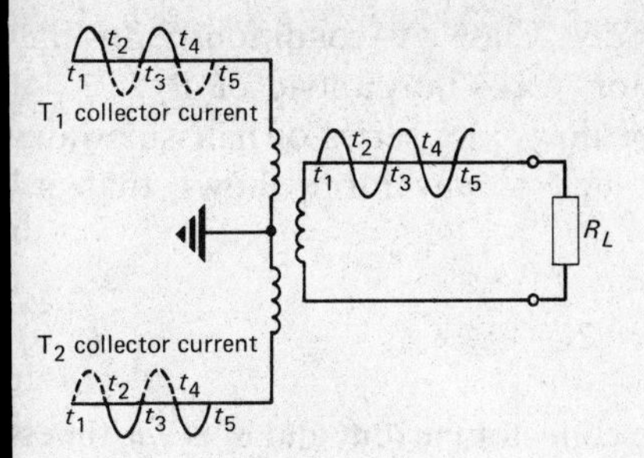

Fig. 6.4 Alternating currents in a Class B push-pull amplifier.

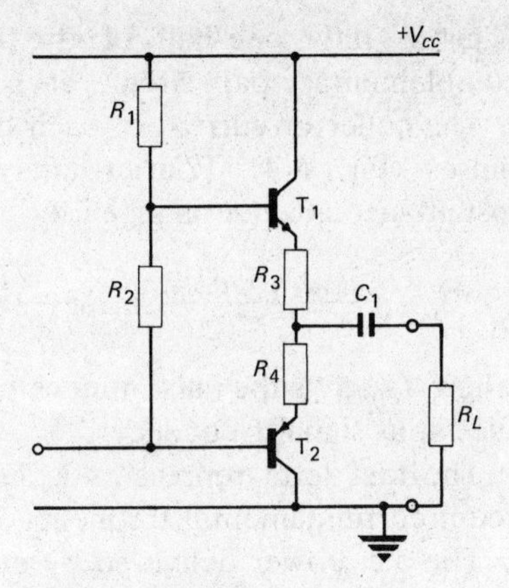

Fig. 6.6 Basic complementary pair Class B push-pull amplifier.

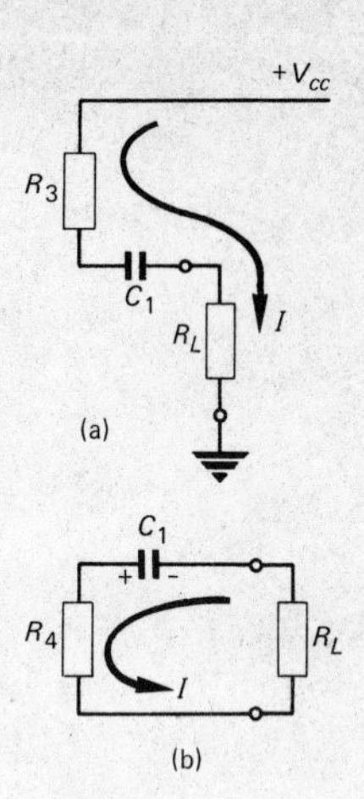

Fig. 6.7

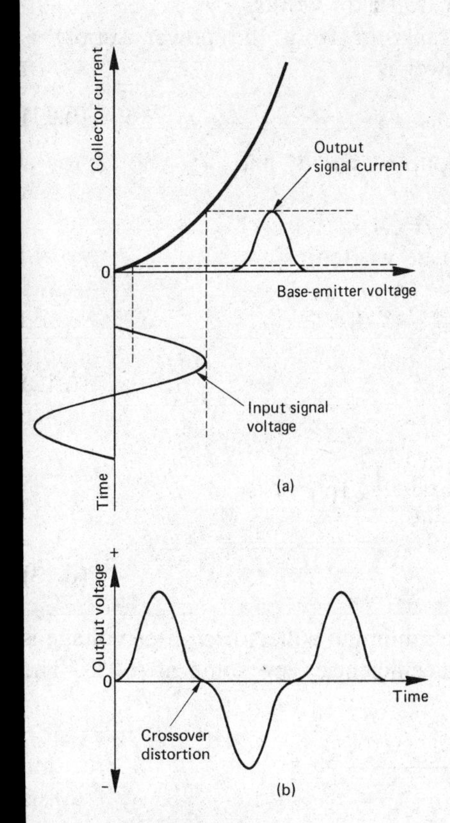

Fig. 6.5 Cross-over distortion.

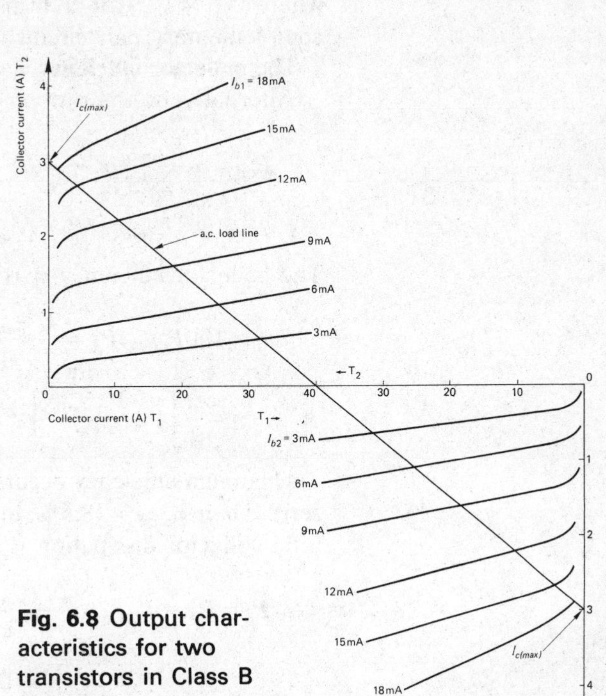

Fig. 6.8 Output characteristics for two transistors in Class B push-pull.

(This is only one-half of the load under Class A conditions.) In the complementary pair circuit, each transistor works into a load of R_L.

The collector current of each transistor flows in a series of half-sinewave pulses (Fig. 6.4). Fourier analysis of such a waveform shows that its instantaneous value is given by

$$i_c = \frac{I_{c(max)}}{\pi} + \frac{I_{c(max)}}{2} \sin \omega t - \frac{2I_{c(max)}}{3\pi} \cos 2\omega t + \cdots \tag{6.9}$$

where $I_{c(max)}$ is the maximum value of the collector current and ω is 2π times the input signal frequency.

The first term represents a d.c. component and the second term is the required fundamental frequency component.

The a.c. power delivered by each transistor to its load is

$$\left(\frac{I_{c(max)}}{2\sqrt{2}}\right)^2 R'_L$$

and so the *total* output power is

$$P_{OUT} = \tfrac{1}{4} I^2_{c(max)} R'_L \tag{6.10}$$

where R'_L is the load into which each transistor works.

The d.c. power taken by each transistor from the power supply is $I_{c(max)} V'_{cc}/\pi$ and so the *total* input power is

$$P_{dc} = 2I_{c(max)} V'_{cc}/\pi \tag{6.11}$$

where $V'_{cc} = V_{cc}$ for a transformer coupled circuit and $V'_{cc} = V_{cc}/2$ for a complementary pair circuit.

The **collector efficiency** η is $100P_{OUT}/P_{dc}$.

Alternatively, the output power can be written as

$$P_{OUT} = \frac{I_{c(max)}}{2\sqrt{2}} \times \frac{V_{ce(max)}}{\sqrt{2}} \times 2 = \frac{I_{c(max)} V_{ce(max)}}{2}$$

$$= \tfrac{1}{2} I_{c(max)} [V'_{cc} - V_{ce(min)}] \tag{6.12}$$

The collector efficiency η is

$$\eta = 100P_{OUT}/P_{dc} = \frac{I_{c(max)}[V'_{cc} - V_{ce(min)}]}{4I_{c(max)} V'_{cc}/\pi} \times 100$$

or $$\eta = \frac{\pi}{4}\left[1 - \frac{V_{ce(min)}}{V'_{cc}}\right] \times 100\% \tag{6.13}$$

Maximum efficiency occurs when the minimum collector/emitter voltage is zero. Then $\eta_{max} = 78.5\%$ but practical efficiencies are some 50–60%. The total collector dissipation is

$$P_c = P_{dc} - P_{OUT} = \frac{2I_{c(max)} V'_{cc}}{\pi} - \frac{I_{c(max)} V_{ce(max)}}{2}$$

$$= \frac{2V'_{cc}}{\pi} \cdot \frac{V_{ce(max)}}{R'_L} - \frac{V^2_{ce(max)}}{2R'_L} \tag{6.14}$$

To determine the maximum possible collector dissipation, differentiate P_c with respect to $V_{ce(max)}$ and equate the result to zero, i.e.

$$\frac{dP_c}{dV_{ce(max)}} = \frac{2V'_{cc}}{\pi R'_L} - \frac{V_{ce(max)}}{R'_L} = 0$$

Therefore, $2V'_{cc}/\pi = V_{ce(max)}$
Substituting this value of $V_{ce(max)}$ into equation (6.14) gives

$$P_{c(max)} = \frac{4(V'_{cc})^2}{\pi^2 R'_L} - \frac{2(V'_{cc})^2}{\pi^2 R'_L} = \frac{2(V'_{cc})^2}{\pi^2 R'_L} \qquad (6.15)$$

Maximum output power occurs when $V_{ce(max)} = V'_{cc}$ and is equal to $(V'_{cc})^2/2R'_L$ and therefore

$$P_{c(max)} = \frac{4P_{OUT(max)}}{\pi^2} \simeq 0.4P_{OUT(max)} \qquad (6.16)$$

If, for example, a Class B push-pull amplifier is required to deliver a power of 12 W to a load, then the maximum total collector dissipation is 4.8 W. This means that the transistors employed must have a power rating of at least 2.4 W.

Example 6.2

The transistors used in a Class B transformer push-pull amplifier have a maximum collector dissipation of 3.0 W. If the collector supply voltage is 18 V calculate the maximum power output and the peak collector current. If the transistors are driven so that the peak collector current is 0.8 times the maximum peak value and a quiescent current of 10 mA is used to reduce cross-over distortion, calculate the output power and the collector efficiency.

Solution

Maximum output power $= 5 \times 3 = 15$ W (*Ans*)

Therefore $I_{c(max)} V_{cc}/2 = 15$

$$I_{c(max)} = \frac{15 \times 2}{18} = 1.67 \text{ A} \quad (Ans)$$

The actual peak collector current is $0.8 \times 1.67 = 1.34$ A and

$$P_{OUT} = \frac{1.34}{2\sqrt{2}} \times \frac{0.8 \times 18}{\sqrt{2}} \times 2 = 9.65 \text{ W} \quad (Ans)$$

The mean collector current per transistor is $1.34/\pi$ A and so the total d.c. power taken from the supply is

$$P_{dc} = \frac{2 \times 1.34 \times 18}{\pi} + 10 \times 10^{-3} \times 18 = 15.54 \text{ W}$$

Therefore, the collector efficiency η in

$$\eta = \frac{9.65}{15.54} \times 100 = 62\% \quad (Ans)$$

Harmonic Distortion

The mutual characteristic of each transistor can be represented by a power series such as that given by equation (6.8). A similar analysis to that carried out for the Class A circuit is valid but there is now no bias voltage (true Class B) and only one transistor conducts at a time. Hence, considerable third harmonic content exists, and

$$I_{c1} = aV \sin \omega t + bV^2 \sin^2 \omega t + cV^3 \sin^3 \omega t$$
$$= aV \sin \omega t + \tfrac{1}{2}bV^2 - \tfrac{1}{2}bV^2 \cos 2\omega t + \tfrac{1}{4}3cV^3 \sin \omega t - \tfrac{1}{4}cV^3 \sin 3\omega t$$
$$= \tfrac{1}{2}bV^2 + (aV + \tfrac{1}{4}3cV^3) \sin \omega t - \tfrac{1}{2}bV^2 \cos 2\omega t - \tfrac{1}{4}cV^3 \sin 3\omega t$$

During the alternate half cycles of the input signal,

$$I_{c2} = -aV \sin \omega t + bV^2 \sin^2 \omega t - cV^3 \sin^3 \omega t$$
$$= \tfrac{1}{2}bV^2 - (aV + \tfrac{1}{4}3cV^3) \sin \omega t - \tfrac{1}{2}bV^2 \cos 2\omega t + \tfrac{1}{4}cV^3 \sin^3 \omega t$$

It is clear from these equations that even-order harmonics cancel out but odd-order harmonics do not and this means that the Class B circuit introduces considerably more distortion than does the Class A circuit.

The correct operation of the complementary pair circuit depends upon the quiescent potential at the junction of R_3 and R_4 being held constant at $V_{cc}/2$ volts. This requirement is usually satisfied by the use of d.c. negative feedback from the junction of the emitter resistors to the base of the driver transistor (see Fig. 6.9).

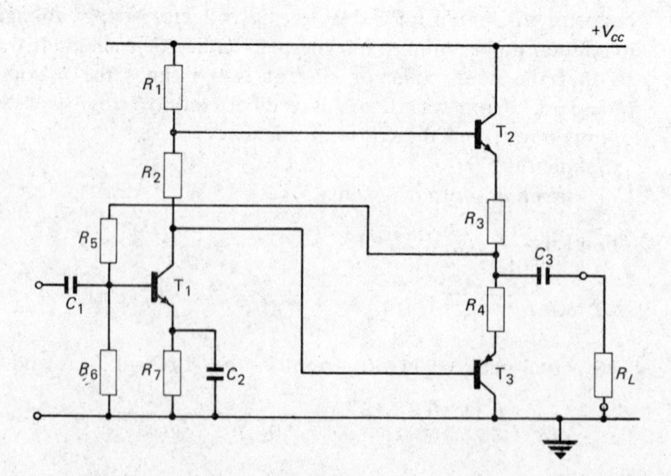

Fig. 6.9 Complementary pair Class B push-pull amplifier.

The base bias voltage for T_1 is obtained from the potential divider R_5 and R_6 connected between the output stage mid-point and earth. If the d.c. voltage at the mid-point should rise, the base bias voltage of T_1 will also rise and T_1 will conduct a larger current. This will make the voltages dropped across R_1 and R_2 increase, making the base potentials of T_2 and T_3 less positive. This, in turn, increases the resistance of T_2 and decreases the resistance of T_3 so that the voltage across T_2 rises while the voltage across T_3 falls. This action will tend to make the mid-point voltage move towards

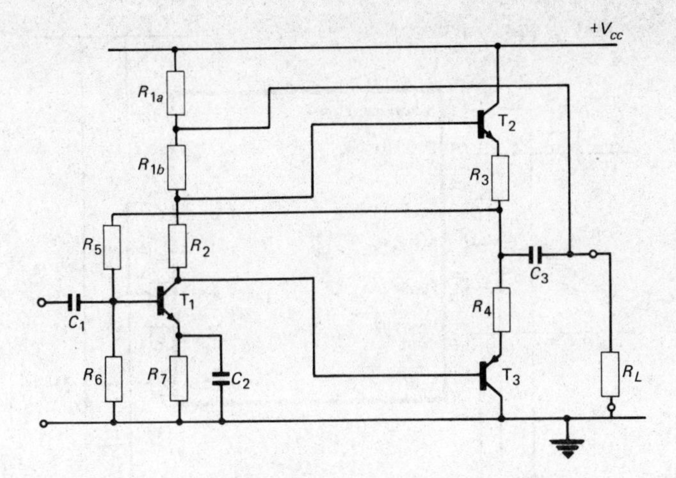

Fig. 6.10 The bootstrapped Class B push-pull amplifier.

$V_{cc}/2$ volts. The action of the d.c. feedback loop is equally effective in counteracting a fall in the mid-point voltage and so its effect is to stabilize the voltage at the desired value of $V_{cc}/2$ volts.

The d.c. component of the collector current of T_1 must be greater than the peak base current taken by T_2 and T_3. Because of this the maximum possible values of R_1 and R_2 are limited and this sets a limit to the gain of the driver stage. A considerable increase in the gain can be obtained if the stage is *bootstrapped*, as shown by Fig. 6.10. The resistor R_1 has been replaced by two resistors R_{1a} and R_{1b} and their junction connected to the top of the load resistor R_L. When a signal is applied to the circuit, the emitter potentials of T_2 and T_3 vary and so does the junction of R_{1a} and R_{1b}. This means that the signal voltages at either end of R_{1b} are very nearly equal. Hence the signal-frequency current that flows in R_{1b} is very small and so its effective a.c. resistance is very high. Since the a.c. voltage gain depends upon its a.c. collector load impedance, a large gain is made possible.

Integrated Circuit Power Amplifiers

A large number of **integrated circuit power amplifiers** are presently available from various sources giving output powers ranging from a few milliwatts to several watts. The most important parameters of such a device are its gain, input impedance, output impedance, quiescent supply current, supply voltage, bandwidth, distortion, sensitivity, and its maximum internal power dissipation. Information on these parameters and suggested circuits are provided by the manufacturer's data sheets. Some i.c.s must be mounted on a suitable heat sink and some are provided with short-circuit and/or thermal protection. The specific i.c. selected for a particular application depends upon the relative weights placed on the above parameters, the cost, and, of course, the ready availability from a convenient supplier. Some power amplifier i.c.s incorporate a pre-amplifier stage and/or an integral heat sink.

Fig. 6.11 shows a circuit using the LM 380 3 W i.c. power amplifier. It can be seen that very few external components are needed. The bypass capacitor

Fig. 6.11 LM 380 power amplifier.

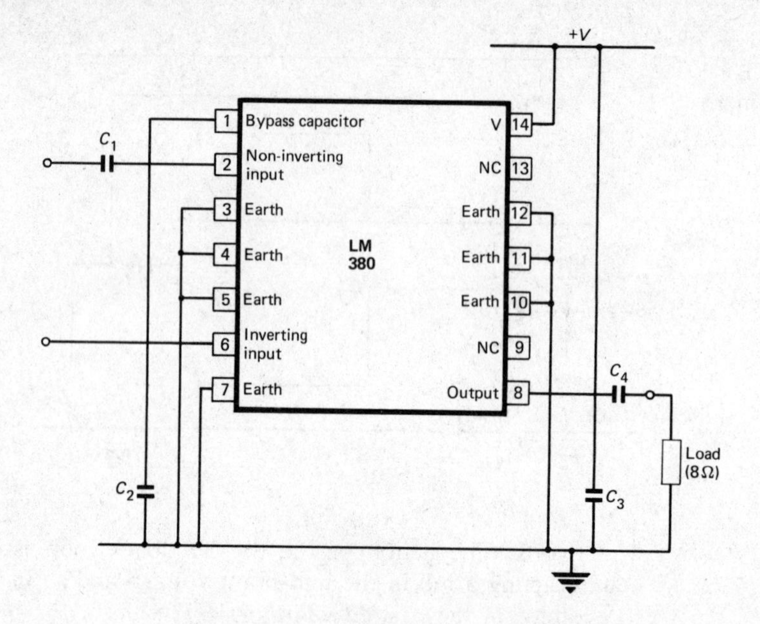

Fig. 6.12 TBA 810 power amplifier.

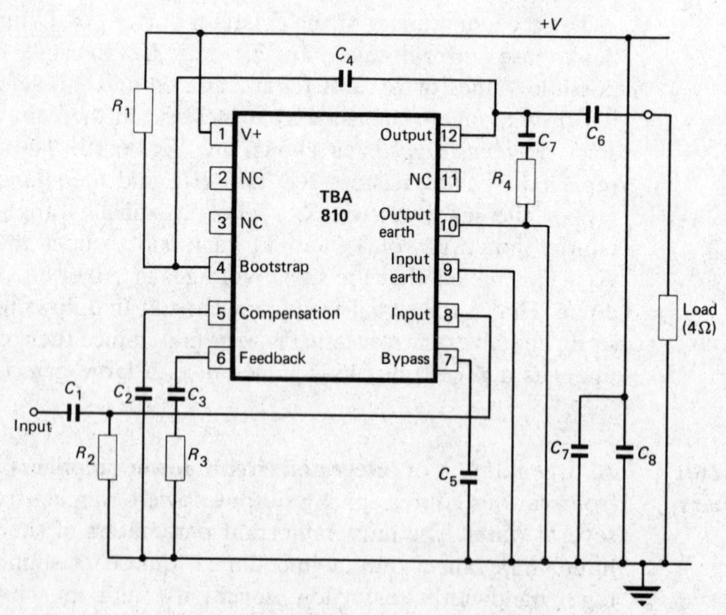

Table 6.1 I.C. power amplifiers

I.C.	*Power output* (W)	*Load resistance* (Ω)	*Voltage gain* (dB)	*Input resistance*
TDA 2610	4.5	15	37	45 kΩ
LM 389	0.325	8	26	50 kΩ
TBA 820	1.6	8	34	5 MΩ
CA 3131	5	8	48	200 kΩ

C_2 is not always used; when it is fitted, a value of some 4.7–10 μF should be used. Also $C_1 = C_3 = 0.1$ μF, and C_4 should be some hundreds of μF. The i.c. has a voltage gain of 50, an input resistance of 150 kΩ, and its maximum output voltage is equal to $V-4$ volts with 0.2% distortion.

Another example of an i.c. power amplifier is given by Fig. 6.12 which uses the TBA 810. This i.c., made by several manufacturers, can provide up to 7 W into a 4 Ω load with 10% distortion. The input resistance is about 5 MΩ. Because of the high power output great care is needed in the layout to avoid instability problems.

Table 6.1 lists some of the more important parameters of a number of other i.c. power amplifiers.

When a large output power is wanted it will be necessary to use a hybrid integrated circuit amplifier. The specification of one version includes: power output 25 W with 0.5% distortion into a 8 Ω load with a flat frequency response up to 100 kHz. An amplifier of this type will probably have an input impedance of some 70 kΩ and will need a heat sink. It will have the advantage of needing no external components, except perhaps an input and/or output coupling capacitor.

Measurements on Power Amplifiers

The arrangement used to measure the **output power** of an a.f. large-signal amplifier is shown in Fig. 6.13. The oscillator is set to the required test frequency and its output voltage is steadily increased until distortion of the output sinewave, displayed on the c.r.o. screen, is just observed. The voltage of the oscillator is then reduced slightly until no distortion of the signal can be observed. The r.m.s. voltage across the load resistance is then measured and the output power calculated.

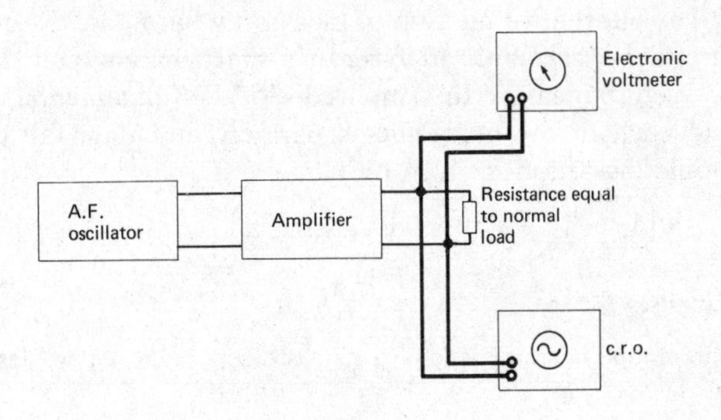

Fig. 6.13 Measurement of output power.

If a distortion factor meter is available, the output signal voltage can be varied until a designated value of percentage distortion is obtained.

If the **efficiency** of the amplifier is also to be measured it will be necessary to measure the direct current taken from the power supply by the amplifier when it is delivering its maximum output.

Measurement of Harmonic Distortion

The easiest way to measure the **harmonic distortion** produced by an amplifier is to use a **distortion factor meter**. This is essentially an instrument that measures the resultant voltage of any waveform that is applied across its input terminals. The instrument includes a tunable filter that can be switched

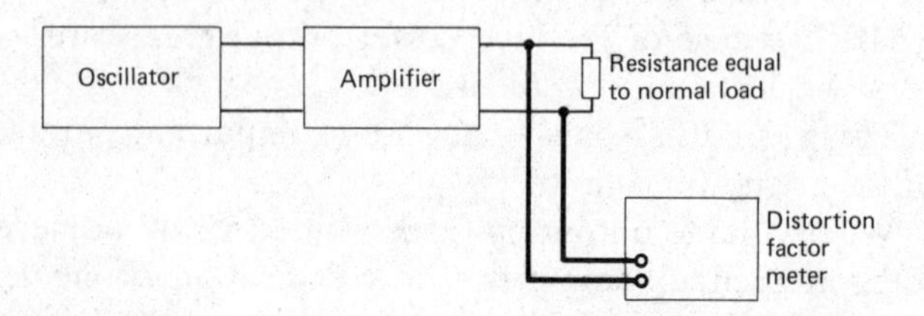

Fig. 6.14 Measurement of distortion.

into or out of a circuit. Fig. 6.14 shows the arrangement used for a distortion factor measurement. The frequency of the oscillator is set to the desired test value and its voltage is set to the value giving the rated output power of the amplifier (or the power value determined in the previous test). The reading of the distortion factor meter is noted and then the internal filter is switched into circuit and tuned to the frequency of the input signal provided by the oscillator. The output of the amplifier now consists of noise and distortion, the fundamental frequency component having been removed. The ratio of the two readings is then equal to the ratio

$$\frac{\text{r.m.s. (distortion plus noise) voltage}}{\text{r.m.s. (distortion plus noise plus signal) voltage}}$$

The percentage harmonic distortion is then taken as 100 times this ratio although this is not strictly true.

The alternative method of measuring harmonic distortion involves the use of either a **harmonic analyzer** or a **spectrum analyzer**. Either instrument can be used to measure the amplitudes of the fundamental frequency component and each of the harmonics separately and then the percentage total harmonic distortion is given by

$$\sqrt{[V_1^2 + V_2^2 + V_3^2 + \cdots]}/V_f \times 100\%$$

Exercises 6

6.1 The mutual characteristic of a transistor operated under Class A conditions is given by

$$I_c = I_{dc} + bV_{be} + cV_{be}^2$$

where I_c is the collector current and V_{be} is the base signal voltage. Prove that if two such transistors are connected in a push-pull circuit, even-order harmonic components are not present at the output. In a Class A push-pull amplifier the collector-collector load resistance is 100Ω. If the collector currents of the two transistors vary sinusoidally between 0.2 A and 1 A and the supply voltage is 24 V, calculate i) the output power and ii) the collector dissipation of each transistor.

Fig. 6.15

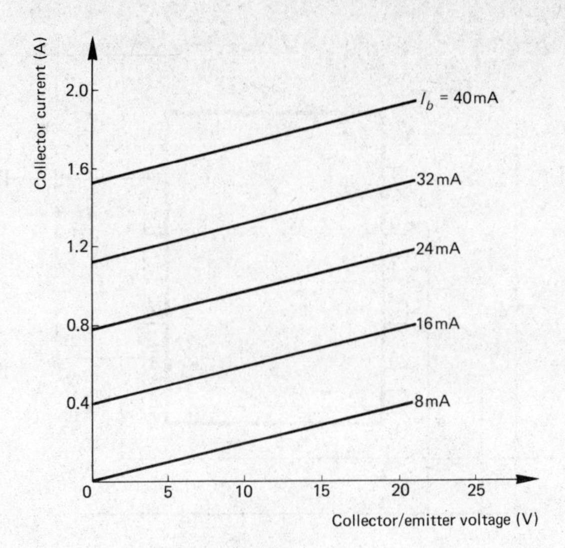

6.2 Draw the circuit of a Class B push-pull amplifier using complementary output transistors. Explain its operation. If the peak value of the collector current of each transistor is 1.6 A, calculate i) the output power in the 12Ω load, ii) the collector efficiency. Take $V_{cc} = 30$ V.

6.3 The output characteristics of a transistor are given by Fig. 6.15. Two such transistors are used in a Class B push-pull amplifier that is transformer-coupled to a 12.5 Ω load. The collector supply voltage is 20 V. The sinusoidal input signal provides a peak base current of 32 mA to each transistor. Using an a.c. load line drawn on the output characteristics, determine i) the output power, ii) the collector efficiency, and iii) the collector dissipation of each transistor.

6.4 Draw waveforms of the currents and voltages at various points in a Class B push-pull amplifier. Derive an expression for the efficiency of such a stage.

Calculate the power output and the efficiency of a Class B amplifier in which the peak collector current is 0.7 A and each transistor works into a load of 15 Ω and the supply voltage is 15 V.

6.5 Prove that the collector dissipation in a Class B push-pull amplifier is at its maximum value when the input sinusoidal signal voltage is $2/\pi$ times the maximum possible value. A Class B amplifier uses transistors whose collector dissipation power rating is 1.8 W. If the supply voltage is 16 V determine the maximum power output.

6.6 Determine the percentage increase in the collector dissipation of a Class B push-pull amplifier if the magnitude of the input signal is 50% of the value required to produce the maximum possible collector efficiency.

6.7 Data for an integrated circuit power amplifier includes the following:

Power output for 10% t.h.d. typically 4 W
Output current (repetitive peak value) 0.9 A
Input voltage for $P_{OUT} = P_{OUT(max)}$ 120 mV
Input impedance 45 kΩ
Frequency response up to 16 kHz

Explain the importance of each term.

Fig. 6.16

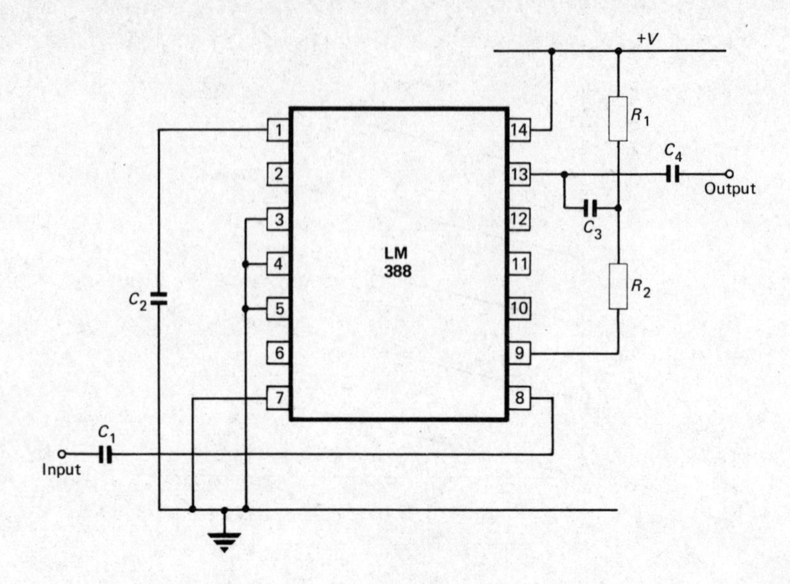

Fig. 6.17

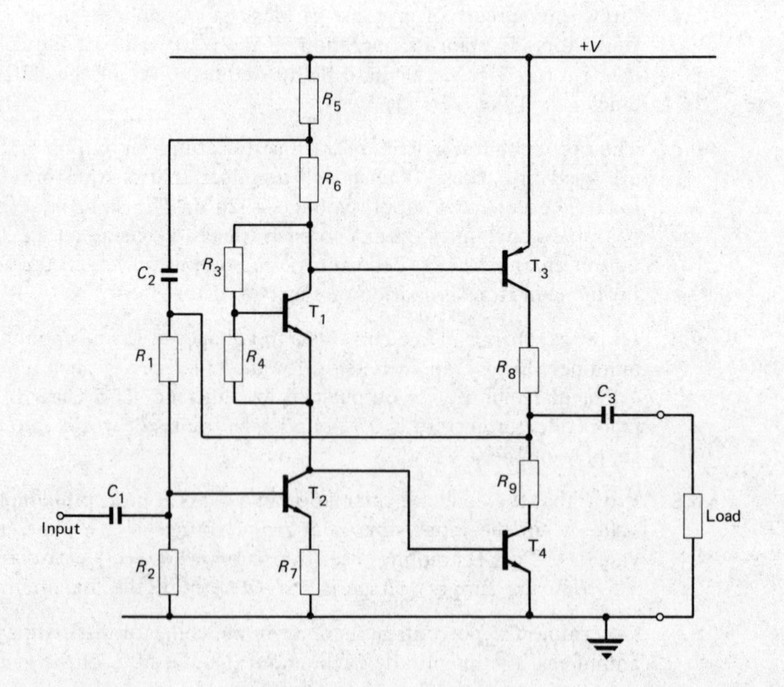

Short Exercises

6.8 Fig. 6.16 shows one way in which the LM 388 can be used as a power amplifier. Explain the function of each component shown.

6.9 Explain briefly why negative feedback has very little effect upon cross-over distortion.

6.10 By differentiating equation (6.14) with respect to $I_{c(max)}$ show that the maximum collector dissipation in a Class B push-pull amplifier occurs when the peak collector current of each transistor is 0.64 times the maximum permitted peak current.

6.11 Class B push-pull amplifiers often use a complementary pair output stage. Briefly explain i) how cross-over distortion is minimized, ii) why bootstrapping is necessary, iii) why n.f.b. is applied.

6.12 The load into which a Class B push-pull amplifier works is 12 Ω. What is the load on each transistor if i) an output transformer of total turns ratio 2:1 is used, ii) a complementary pair circuit is used?

6.13 A Class B complementary pair circuit is to deliver 10 W to a 8 Ω load. Determine the peak voltage across the load and the peak load current.

6.14 The ratings of the transistors used in a Class B push-pull circuit are $V_{ce(max)} = 20$ V, $I_{c(max)} = 1$ A. Calculate the maximum power that can be delivered to a 4 Ω load.

6.15 List the relative merits of i) single-ended, ii) Class A push-pull, iii) Class B push-pull operation of an audio-frequency power amplifier.

6.16 Explain the meaning of the term cross-over distortion and discuss how it can be minimized in a complementary pair circuit.

6.17 Two transistors, each having a maximum power dissipation of 1 W, are used in a Class B push-pull amplifier. Calculate the maximum possible output power. What is then the collector efficiency?

6.18 Explain the operation of the circuit shown in Fig. 6.17.

7 Sinusoidal Oscillators

Introduction

An **oscillator** is an electronic circuit whose function is to produce an alternating e.m.f. of a particular frequency and waveform. This chapter will be concerned solely with oscillators which generate an output of sinusoidal waveform, and other waveform generators will be discussed in Chapter 8.

All types of sinusoidal oscillator rely upon the application of positive feedback to a circuit that is capable of providing amplification, and they differ from one another mainly in the ways in which the feedback is applied. The basic block diagram of an oscillator is given in Fig. 7.1. When the oscillator is first switched on, a surge of current flows in the frequency-determining network and produces a voltage at the required frequency of oscillation across it. A fraction of this voltage is fed back, via the feedback network, to the input terminals of the amplifier and is then amplified to reappear across the frequency-determining network. A fraction of this larger voltage is fed back to the input, in phase with the input voltage, and is further amplified and so on. In this way the amplitude of the signal voltage builds up until the onset of non-linearity in the operation of the amplifier which reduces the loop gain to unity. The ways in which the loop gain is reduced have been discussed elsewhere [EIII].

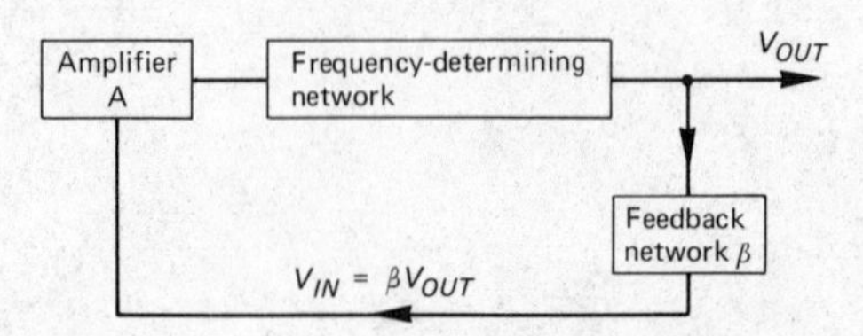

Fig. 7.1 The principle of an oscillator.

The frequency-determining network may consist of an *LC* circuit, or an *RC* circuit, or a piezo-electric crystal. The amplifying device may be a bipolar transistor, or a fet, or an operational amplifier.

The important characteristics of an oscillator are its

Frequency, or range of frequencies (if variable), of operation

Frequency stability

Amplitude stability

Percentage distortion of its output waveform.

The Generalized Oscillator

Referring to Fig. 7.1 the input voltage to the amplifying section is

$$V_{IN} = \beta V_{OUT} = \beta A_v V_{IN}$$

and so $V_{IN}(1-\beta A_v)=0$

The input voltage cannot be zero as an output voltage does exist and therefore $(1-\beta A_v)$ must be zero. Hence

$$\beta A_v = 1 \tag{7.1}$$

In general, both the gain A_v *and* the feedback factor β are complex and hence

$$|\beta|\underline{/\phi} \cdot |A_v|\underline{/\theta} = 1\underline{/0^\circ}$$

$$|\beta A_v|\underline{/\phi+\theta} = 1\underline{/0^\circ} \tag{7.2}$$

Equation (7.2) states that the two necessary requirements for oscillation to take place are that
the loop gain $|\beta A_v|$ must be unity
and the loop phase shift $\phi+\theta$ must be zero (or 360°)
This means that the frequency of oscillation of an oscillator circuit can be determined by equating the j parts of equation (7.2) to zero.

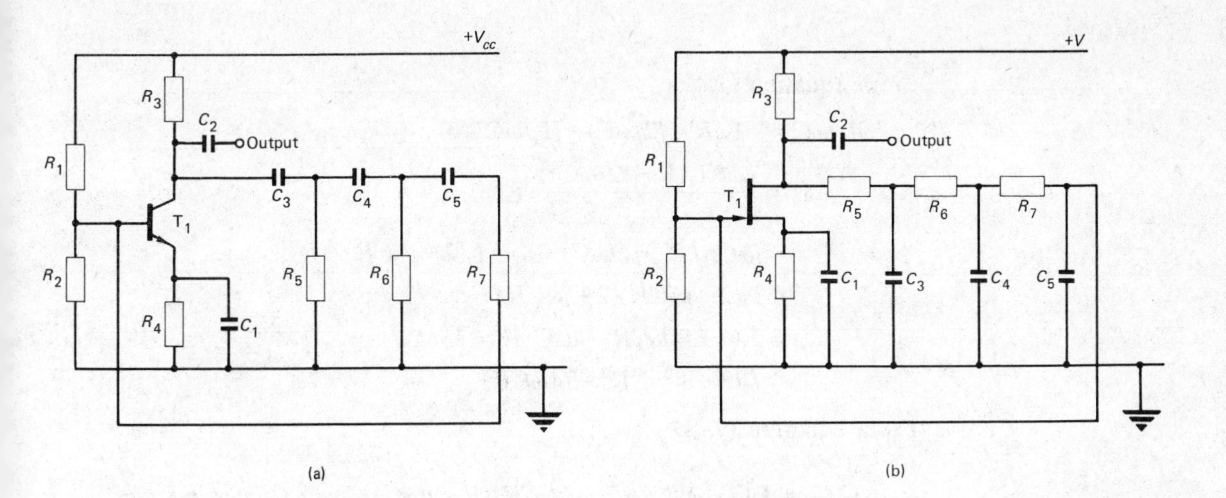

Fig. 7.2 Two *RC* oscillators.

Resistance-Capacitance Oscillators

Audio-frequency oscillators most often use a **resistance-capacitance** network to obtain the loop phase shift of 360° necessary for oscillations to take place. One type of *RC* oscillator uses a single stage amplifier and an *RC* ladder network, and two versions of it are given in Fig. 7.2.

1 Consider the circuit given in Fig. 7.2*b*. The a.c. equivalent circuit is shown by Fig. 7.3*a*. Applying Thevenin's theorem to the circuit gives Fig. 7.3*b* in which

$$V_{OUT} = g_m V_{gs} R_3$$
$$C_3 = C_4 = C_5 = C$$
$$R_6 = R_7 = R_5 + R_3 = R$$
$$V_{OUT} = I_1(R - j/\omega C) + jI_2/\omega C \tag{7.3}$$
$$0 = jI_1/\omega C + I_2(R - j2/\omega C) + jI_3/\omega C \tag{7.4}$$
$$0 = jI_2/\omega C + I_3(R - j2/\omega C) \tag{7.5}$$

From equation (7.5),

$$I_2 = -\frac{I_3(R - j2/\omega C)\omega C}{j} = jI_3(\omega CR - j2) = I_3(2 + j\omega CR)$$

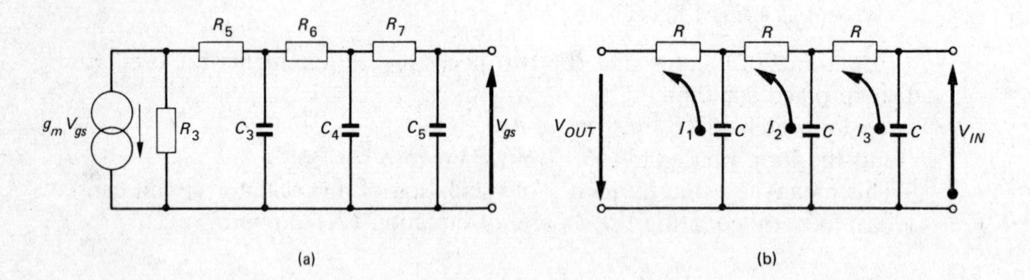

Fig. 7.3 (*a*) Equivalent circuit of Fig. 7.2*b*, (*b*) Thevenin version of (*a*).

From equation (7.4),

$$jI_1/\omega C = -I_2(R - j2/\omega C) - jI_3/\omega C$$
$$I_1 = \frac{-\omega C I_2(R - j2/\omega C)}{j} - I_3$$
$$= j\omega C I_2(R - j2/\omega C) - I_3 = I_2(2 + j\omega CR) - I_3$$
$$= I_3(2 + j\omega CR)(2 + j\omega CR) - I_3$$
$$= I_3(4 + j4\omega CR - \omega^2 C^2 R^2 - 1)$$
$$= I_3(3 - \omega^2 C^2 R^2 + j4\omega CR)$$

From equation (7.3),

$$V_{OUT} = I_3(3 - \omega^2 C^2 R^2 + j4\omega CR)(R - j/\omega C) + j\frac{I_3}{\omega C}(2 + j\omega CR)$$
$$= I_3\left(3R - \frac{j3}{\omega C} - \omega^2 C^2 R^3 + j\omega CR^2 + j4\omega CR^2 + 4R + \frac{j2}{\omega C} - R\right)$$
$$= I_3(6R - \omega^2 C^2 R^3 + j5\omega CR^2 - j/\omega C)$$
$$V_{IN} = -jI_3/\omega C = \frac{-jV_{OUT}}{(6R - \omega^2 C^2 R^3 + j5\omega CR^2 - j/\omega C)\omega C}$$
$$= \frac{-jV_{OUT}}{6\omega CR - \omega^3 C^3 R^3 + j5\omega^2 C^2 R^2 - j}$$

Therefore,

$$\beta = V_{IN}/V_{OUT} = \frac{1}{1-5\omega^2C^2R^2+j(6\omega CR-\omega^3C^3R^3)}$$

The circuit will oscillate at the frequency at which this ratio is wholly real, i.e. at which the sum of the j terms is equal to zero. Therefore,

$$6\omega_0 CR-\omega_0^3C^3R^3=0 \qquad \text{i.e. } \omega_0^2C^2R^2=6$$

$$\omega_0=\sqrt{6}/CR \quad \text{and} \quad f_0=\sqrt{6}/2\pi CR \text{ Hz} \tag{7.6}$$

At this frequency the ratio $\beta = V_{IN}/V_{OUT}$ is

$$V_{IN}/V_{OUT}=\frac{1}{1-5\times 6}=-1/29$$

Since the loop gain of the circuit must be, initially, greater than unity, the amplifier must have an inverting gain in excess of 29.

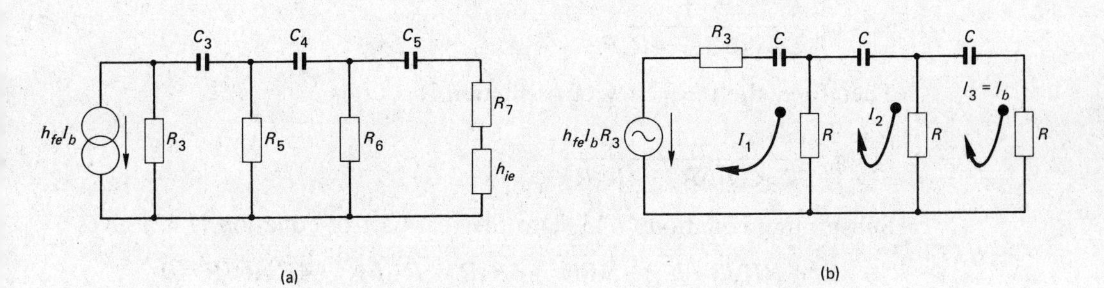

Fig. 7.4 (*a*) Equivalent circuit of Fig. 7.2*a*, (*b*) Thevenin version of (*a*).

2 When the alternative circuit is used (Fig. 7.2*a*) the a.c. equivalent circuit is given by Fig. 7.4*a* modified by the use of Thevenin's theorem to Fig. 7.4*b*. In this last case $C_3=C_4=C_5=C$ and $R_5=R_6=R_7+h_{ie}=R$.

From Fig. 7.4*b*

$$I_b(2R-j/\omega C)-I_2R=0 \tag{7.7}$$

$$I_2(2R-j/\omega C)-I_bR-I_1R=0 \tag{7.8}$$

$$I_1(R_1+R_3-j/\omega C)-I_2R=-h_{fe}I_bR_3 \tag{7.9}$$

From equation (7.7),

$$I_2=I_b(2R-j/\omega C)/R$$

Substituting into equations (7.8) and (7.9) gives

$$I_b\left[\frac{1}{R}(2R-j/\omega C)(2R-j/\omega C)-R\right]=I_1R \tag{7.10}$$

$$-I_b[2R-j/\omega C-h_{fe}R_3]=-I_1[R+R_3-j/\omega C] \tag{7.11}$$

Cross-multiplying equations (7.10) and (7.11),

$$[(2-j/\omega CR)(2R-j/\omega C)-R][R+R_3-j/\omega C]$$
$$=R[2R-h_{fe}R_3-j/\omega C]$$

$$(3R-1/\omega^2C^2R-j4/\omega C)(R+R_3-j/\omega C)$$
$$=2R^2-h_{fe}RR_3-jR/\omega C$$

$$3R^2+3RR_3-j3R/\omega C-1/\omega^2C^2-R_3/\omega^2C^2R$$
$$+j/\omega^3C^3R-j4R/\omega C-j4R_3/\omega C-4/\omega^2C^2$$
$$=2R^2-h_{fe}RR_3-jR/\omega C$$

$$R^2+RR_3(3+h_{fe})-j6R/\omega C-5/\omega^2C^2$$
$$-R_3/\omega^2C^2R+j/\omega^3C^3R-j4R_3/\omega C=0 \qquad (7.12)$$

Equating the j terms to zero:

$$1/\omega_0^3C^3R=6R/\omega_0C+4R_3/\omega_0C$$

$$\omega_0=\frac{1}{C\sqrt{[6R^2+4R_3R]}} \qquad (7.13)$$

Therefore, the frequency of oscillation is

$$f_0=\frac{1}{2\pi C\sqrt{[6R^2+4R_3R]}} \qquad (7.14)$$

Substituting equation (7.13) into the real part of equation (7.12) gives

$$R^2+RR_3(3+h_{fe})-5(6R^2+4R_3R)-R_3(6R^2+4R_3R)/R=0$$
$$R^2+3RR_3+RR_3h_{fe}-30R^2-20R_3R-6R_3R-4R_3^2=0$$
$$-29R^2-23RR_3+RR_3h_{fe}-4R_3^2=0$$

Therefore

$$h_{fe}\geqslant\frac{29R}{R_3}+23+\frac{4R_3}{R} \qquad (7.15)$$

Differentiating equation (7.15) with respect to R shows that the value of R_3/R that gives the smallest possible value of h_{fe} is 2.69. For this ratio $h_{fe(min)}=44.54$. This means that the transistor used must have an h_{fe} of at least 45 if the circuit is to oscillate.

Example 7.1

Calculate the frequency of oscillation and the minimum h_{fe} required for a phase-shift oscillator of the kind shown in Fig. 7.2*a* if $R=1200\ \Omega$, $C=0.01\ \mu F$ and $R_3=2200\ \Omega$.

Solution From equation (7.14)

$$f_0=1/2\pi\times10^{-8}\sqrt{[6\times1200^2+4\times1200\times2200]}=3632\text{ Hz} \quad (Ans)$$

From equation (7.15)

$$h_{fe(min)} = \frac{29 \times 1200}{2200} + 23 + \frac{4 \times 2200}{1200} = 46.2 \quad (Ans)$$

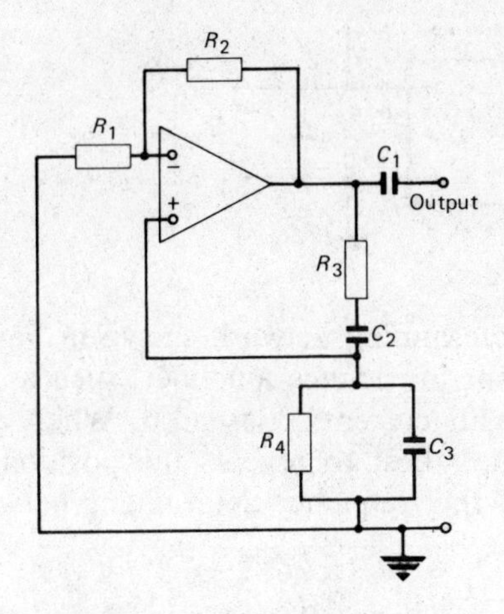

Fig. 7.5 Wien bridge oscillator.

Wien Bridge Oscillator

Another kind of RC oscillator is known as the **Wien bridge** circuit and is shown in Fig. 7.5. The voltage applied to the non-inverting terminal is

$$V_+ = \frac{\dfrac{V_{OUT}R_4}{1+j\omega C_3R_4}}{\dfrac{R_4}{1+j\omega C_3R_4} + R_3 + \dfrac{1}{j\omega C_2}} = \frac{V_{OUT}R_4}{\dfrac{C_3R_4}{C_2} + R_3 + R_4 + j\left(\omega R_3R_4C_3 - \dfrac{1}{\omega C_2}\right)}$$

V_+ will be in phase with V_{OUT}, and the circuit will oscillate, when the j terms sum to zero, i.e. when

$$\omega_0 R_3R_4C_3 = 1/\omega_0 C_2$$

$$\omega_0 = \frac{1}{\sqrt{[R_3R_4C_2C_3]}} \quad \text{and} \quad f_0 = \frac{1}{2\pi\sqrt{[R_3R_4C_2C_3]}}\,\text{Hz} \tag{7.16}$$

At this frequency

$$V_{OUT}/V_+ = 1 + \frac{R_3}{R_4} + \frac{C_3}{C_2} \tag{7.17}$$

and this is the amplifier gain necessary for a loop gain of unity. If $R_3 = R_4$ and $C_2 = C_3$ the amplifier gain needed is 3 and this is easily obtained by making $R_2 = 2R_1$.

The oscillations build up in amplitude until amplifier non-linearities cause the loop gain to fall to unity.

If a bipolar or field effect transistor is used to provide the required amplification, a two-stage circuit is necessary to obtain the necessary 360° phase shift. The required voltage gain is obtained by applying overall n.f.b. to the circuit.

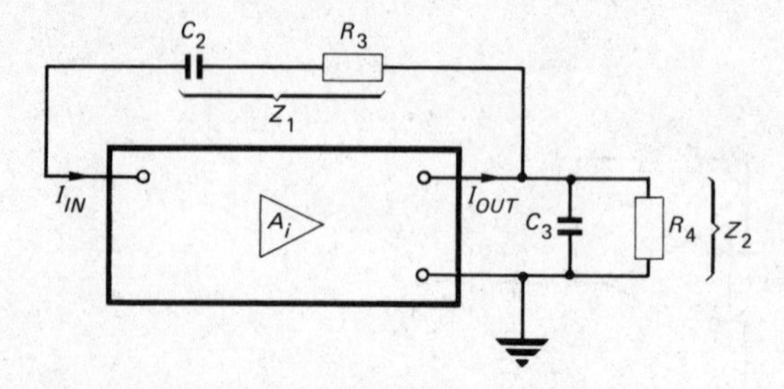

Fig. 7.6 Current-feed Wien bridge oscillator.

The frequency-determining network shown in Fig. 7.5 is best suited for use with a high input impedance amplifier, such as an op-amp, a fet or a bipolar transistor with current-series n.f.b. When a low-input impedance amplifier is used, it is best to reverse the position of the series and the parallel elements of the frequency-determining network (Fig. 7.6) so that a current is fed back.

$$I_{IN}=\frac{I_{OUT}Z_2}{Z_1+Z_2}=\frac{I_{IN}\dfrac{R_4}{1+j\omega C_3R_4}}{\dfrac{R_4}{1+j\omega C_3R_4}+R_3+\dfrac{1}{j\omega C_2}}$$

This is the same equation as before and will therefore lead to the same equations for the oscillation frequency and the required gain.

Example 7.2

If in Fig. 7.6, $R_3=47\ \text{k}\Omega$, $R_4=1200\ \Omega$, $C_2=1$ nF and $C_3=0.01\ \mu$F calculate i) the frequency of oscillation, ii) the necessary current gain of the amplifier.

Solution From equation (7.16)

$$f_0=\frac{1}{2\pi\sqrt{[47\times10^3\times1200\times10^{-9}\times10^{-8}]}}=6702\ \text{Hz}\quad(\textit{Ans})$$

From equation (7.17)

$$A_{i(\min)}=1+\frac{47\times10^3}{1200}+\frac{0.01}{0.001}=50.2\quad(\textit{Ans})$$

For both types of circuit, frequency variation can be achieved by the simultaneous variation of the ganged capacitors C_2 and C_3. If required, range switching can be obtained by switching in new values for R_3 and R_4 (usually equal).

Usually, the amplitude of the output waveform is limited by the use of a negative temperature coefficient (n.t.c.) resistor, such as a thermistor, in the negative feedback path, establishing the gain of the amplifier section of the circuit. For example, in Fig. 7.5, R_2 can be a thermistor; as the oscillation amplitude builds up, the voltage across R_2 increases also and the higher power dissipation raises the temperature of and therefore reduces the resistance of R_2. The reduction in the ratio R_2/R_1 reduces the gain of the amplifier and so limits the oscillation amplitude.

The Wien oscillator is preferred to the ladder *RC* oscillator when a variable frequency output is wanted since there are fewer components to be simultaneously controlled.

Inductance-Capacitance Oscillators

Resistance-capacitance oscillators are not suited to use at frequencies in excess of some tens of kilohertz because the required values of *R* and *C* become impracticably small. For the generation of higher frequencies, the frequency-determining network is provided by an **inductance-capacitance** network.

A number of different configurations are possible and several have been given elsewhere [EIII] but perhaps the most popular are the tuned-collector or tuned-drain oscillators, the Hartley oscillator, and Colpitts oscillator.

Fig. 7.7 Tuned-collector oscillator.

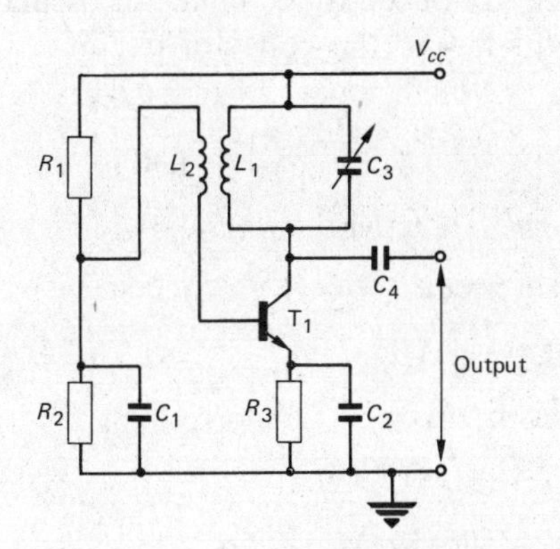

Fig. 7.7 shows the circuit of a **tuned-collector** oscillator in which R_1, R_2 and R_3 are bias components while C_1 and C_2 are decoupling components. Variable capacitor C_3 tunes the circuit to the desired frequency of oscillation and C_4 is a d.c. blocker. The action of the circuit is as follows. When the collector supply voltage is first switched on, the resulting surge in the d.c. current causes a minute oscillatory current to flow in the tuned circuit. This current flows in the inductor L_1 and induces a voltage at the same frequency into the inductor L_2. This voltage is then applied to the base of the transistor. The transistor introduces a 180° phase shift between its base and its collector terminals, and the mutual inductance between L_1 and L_2 must be such that the loop phase shift is zero. The amplified voltage causes a larger oscillatory current to flow in L_1 and a larger e.m.f. is induced into L_2 and so on. Provided the loop gain is greater than unity, the amplitude of the oscillations builds up until the point is reached where the transistor is driven into cut-off and saturation. The loop gain is then reduced to unity and the oscillation amplitude remains constant.

The derivation of the maintenance conditions required for steady-state sinusoidal oscillations to occur in an LC oscillator will be based upon the simplified h-parameter a.c. equivalent circuit. If the oscillation frequency is well above the audio range, the h-parameter circuit is not really valid (Chapter 1) and the hybrid-π circuit should be used. However, the analysis would then be too complicated for this book. The results obtained in the following pages are approximately correct and are commonly used as a basis for design.

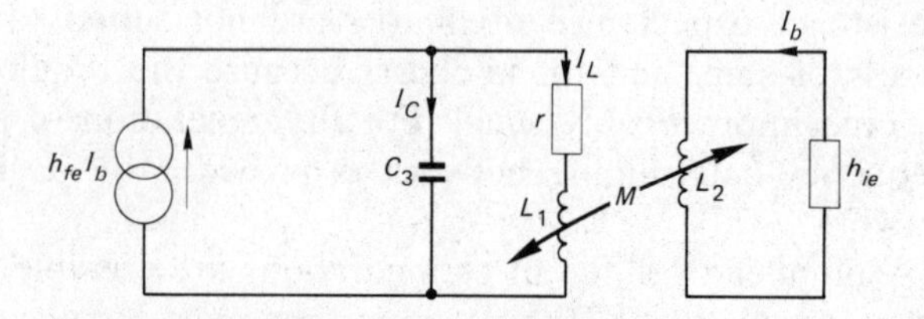

Fig. 7.8 h-parameter equivalent circuit of Fig. 7.7.

The equivalent circuit of the tuned collector oscillator is shown in Fig. 7.8. Applying Kirchhoff's law to the collector circuit

$$0 = I_L(r + j\omega L_1) + j\omega M I_b + (h_{fe}I_b - I_L)\frac{1}{j\omega C_3}$$

$$I_L(j/\omega C_3 - r - j\omega L_1) = I_b(j\omega M - jh_{fe}/\omega C_3) \tag{7.18}$$

And, from the base circuit,

$$0 = I_b(h_{ie} + j\omega L_2) + j\omega M I_L$$

$$-I_L j\omega M = I_b(h_{ie} + j\omega L_2) \tag{7.19}$$

Divide equation (7.18) by (7.19),

$$\frac{-r + j(1/\omega C_3 - \omega L_1)}{-j\omega M} = \frac{j(\omega M - h_{fe}/\omega C_3)}{h_{ie} + j\omega L_2}$$

and cross-multiplying

$$-rh_{ie} - j\omega r L_2 + jh_{ie}(1/\omega C_3 - \omega L_1) - L_2/C_3 + \omega^2 L_1 L_2$$
$$= \omega^2 M^2 - h_{fe}M/C_3 \tag{7.20}$$

For oscillations to occur the j part of equation (7.20) must be equal to zero, hence

$$-\omega_0 r L_2 + h_{ie}(1/\omega_0 C_3 - \omega_0 L_1) = 0$$
$$-\omega_0^2 r L_2 C_3 + h_{ie} - h_{ie}\omega_0^2 L_1 C_3 = 0$$
$$\omega_0^2 = h_{ie}/(L_1 C_3 h_{ie} + L_2 C_3 r)$$

Therefore,

$$f_0 = \frac{1}{2\pi\sqrt{[L_1 C_3 + (L_2 C_3 r/h_{ie})]}} \tag{7.21}$$

$$f_0 \simeq \frac{1}{2\pi\sqrt{(L_1C_3)}} \quad \text{if} \quad h_{ie} \gg r \tag{7.22}$$

Equating the real parts of equation (7.20) gives

$$-rh_{ie} - \frac{L_2}{C_3} + \omega_0^2 L_1 L_2 = \omega_0^2 M^2 - h_{fe}\frac{M}{C_3}$$

$$h_{fe} = \frac{1}{M}[rh_{ie}C_3 + L_2 + \omega_0^2 C_3(M^2 - L_1L_2)]$$

$$\simeq \frac{rh_{ie}C_3}{M} + \frac{L_2}{M} + \frac{C_3}{L_1C_3M}(M^2 - L_1L_2)$$

$$h_{fe(min)} = \frac{rh_{ie}C_3}{M} + \frac{M}{L_1} \tag{7.23}$$

Equations (7.21) and (7.23) ignore the effects of h_{oe} and the impedance into which the oscillator output is delivered. These have the effect of decreasing the calculated value for ω_0 and increasing the calculated required minimum value of h_{fe}.

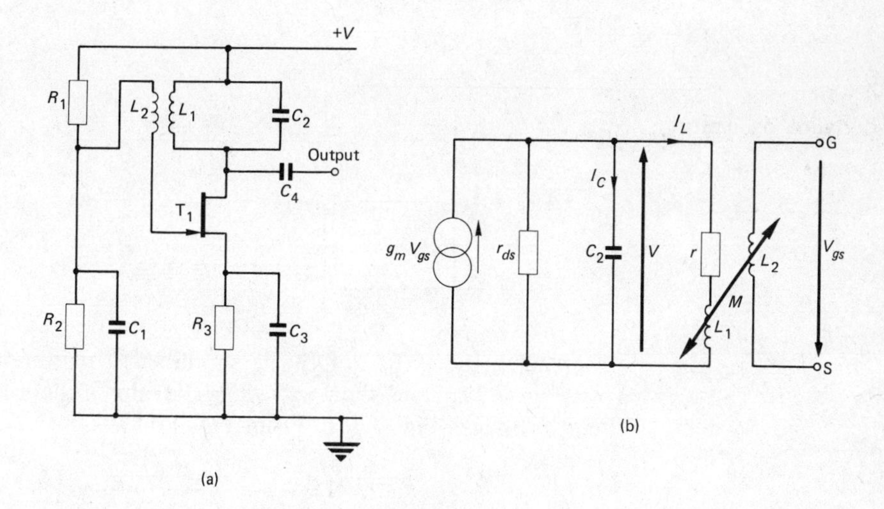

Fig. 7.9 (*a*) Tuned-drain oscillator, (*b*) equivalent circuit of (*a*).

The circuit of a **tuned-drain** oscillator is shown in Fig. 7.9*a*. Bias for the jfet is provided by a potential divider circuit. The a.c. equivalent circuit of the amplifier is given in Fig. 7.9*b*. From this

$$I_L = \frac{V}{r + j\omega L_1} \quad \text{and} \quad V_{gs} = j\omega M I_L$$

$$\frac{jg_m\omega MV}{r + j\omega L_1} = \frac{V}{r_{ds}} + jV\omega C_2 + \frac{V}{r + j\omega L_1}$$

$$jg_m r_{ds}\omega M = r + j\omega L_1 + jrr_{ds}\omega C_2 - r_{ds}\omega^2 L_1 C_2 + r_{ds}$$

Equating the real terms to zero,

$$0 = r + r_{ds} - r_{ds}\omega_0^2 L_1 C_2$$

$$\omega_0^2 = \frac{r + r_{ds}}{L_1 C_2 r_{ds}} = \frac{1}{L_1 C_2}\left(1 + \frac{r}{r_{ds}}\right)$$

$$f_0 = \frac{1}{2\pi\sqrt{(L_1 C_2)}}\sqrt{\left(1 + \frac{r}{r_{ds}}\right)} \qquad (7.24)$$

Equating the j terms to zero,

$$g_m r_{ds}\omega_0 M = \omega_0 L_1 + r_{ds} r \omega_0 C_2$$

$$M = \frac{L_1 + r_{ds} r C_2}{g_m r_{ds}} \qquad (7.25)$$

Colpitts and Hartley Oscillators

Two other types of LC oscillator that are often used in electronic equipment are the Colpitts and Hartley oscillators. These oscillators can be represented by the generalized diagram shown in Fig. 7.10 in which X_1 is the reactance connected between base and collector, X_2 is the reactance between the base and the emitter, and X_3 is the reactance between the collector and the emitter.

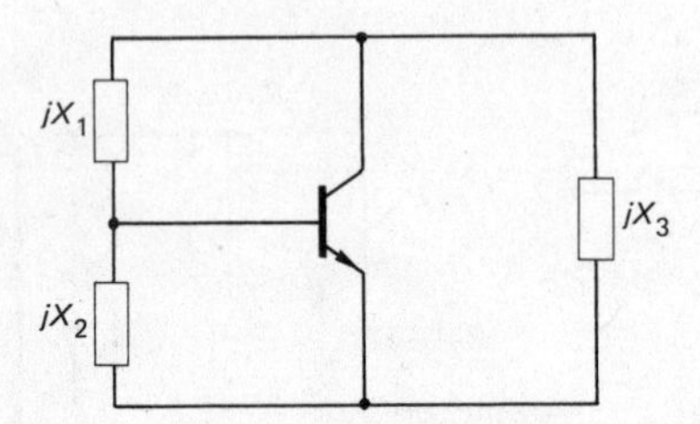

Fig. 7.10 Three-impedance oscillator.

Replacing the transistor by its simplified h-parameter equivalent circuit gives Fig. 7.11a and then applying Thevenin's theorem to the left of the circuit produces Fig. 7.11b. From Fig. 7.11b,

$$I_b = \frac{-IjX_2}{jX_2 + h_{ie}} \quad \text{and} \quad I = \frac{jh_{fe} I_b X_3}{jX_1 + jX_3 + \dfrac{jX_2 h_{ie}}{jX_2 + h_{ie}}}$$

$$I = \frac{\dfrac{-h_{fe} I jX_2 jX_3}{jX_2 + h_{ie}}}{jX_1 + jX_3 + \dfrac{jX_2 h_{ie}}{jX_2 + h_{ie}}}$$

$$= \frac{-h_{fe} I jX_2 jX_3}{jX_1 jX_2 + jX_1 h_{ie} + jX_2 jX_3 + jX_3 h_{ie} + jX_2 h_{ie}}$$

Hence,

$$jX_2 jX_3 + jX_1 jX_2 + jh_{ie}(X_1 + X_2 + X_3) = -h_{fe} jX_2 jX_3$$

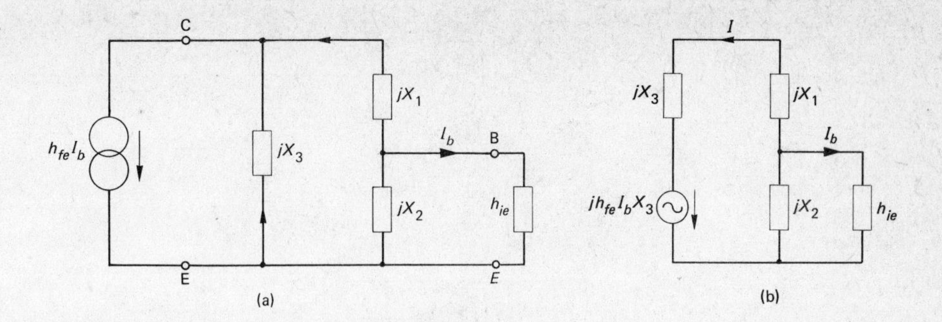

Fig. 7.11 (*a*) Equivalent circuit of Fig. 7.10, (*b*) Thevenin version of (*a*).

Equating real parts:

$$0 = X_2X_3(1 + h_{fe}) + X_1X_2$$

$$0 = X_3(1 + h_{fe}) + X_1$$

Hence the reactances X_1 and X_3 must be of opposite sign. For oscillations to be maintained

$$1 + h_{fe} = -X_1/X_3 \tag{7.26}$$

Equating the j parts

$$h_{ie}(X_1 + X_2 + X_3) = 0 \quad \text{or} \quad X_1 + X_2 + X_3 = 0 \tag{7.27}$$

Now, from equation (7.26), $|X_1| > |X_3|$ so X_2 must have the same sign as X_3.

For a *Colpitts oscillator*, X_2 and X_3 are capacitive reactances and X_1 is an inductive reactance.

For a *Hartley oscillator*, X_2 and X_3 are inductive and X_1 is capacitive.

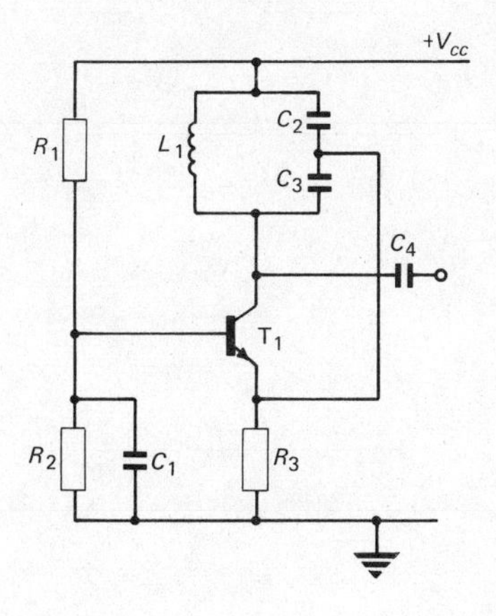

Fig. 7.12 Colpitts oscillator.

Colpitts Oscillator

The circuit diagram of a **Colpitts oscillator** is shown in Fig. 7.12. The frequency of oscillation is easily obtained from equation (7.27), by putting $jX_1 = j\omega L_1$, $jX_2 = 1/j\omega C_2$, $jX_3 = 1/j\omega C_3$ to give

$$\frac{1}{j\omega_0 C_2}+\frac{1}{j\omega_0 C_3}+j\omega_0 L_1=0$$

$$\frac{1}{C_2}+\frac{1}{C_3}=\omega_0^2 L_1$$

$$\omega_0^2=\frac{1}{L_1}\left(\frac{1}{C_2}+\frac{1}{C_3}\right)=\frac{1}{L_1}\left(\frac{C_2+C_3}{C_2C_3}\right)$$

$$f_0=\frac{1}{2\pi\sqrt{\left(L_1\dfrac{C_2C_3}{C_2+C_3}\right)}}\text{ Hz} \tag{7.28}$$

The minimum value for the current gain h_{fe} if oscillations are to be maintained is, from equation (7.26),

$$h_{fe(min)}=-1-\frac{j\omega_0 L_1}{1/j\omega_0 C_3}=-1+\omega_0^2 L_1 C_3$$

$$h_{fe(min)}=-1+\frac{C_2+C_3}{L_1C_2C_3}\cdot L_1C_3=-1+1+\frac{C_3}{C_2}=\frac{C_3}{C_2} \tag{7.29}$$

Usually C_3 is made equal to C_2 and then there is ample gain for oscillations to start. Equation (7.28) only gives an approximation to the actual frequency of oscillation, partly because of the use of the simplified h-parameter equivalent circuit and partly because the self-resistance of the inductance has been neglected.

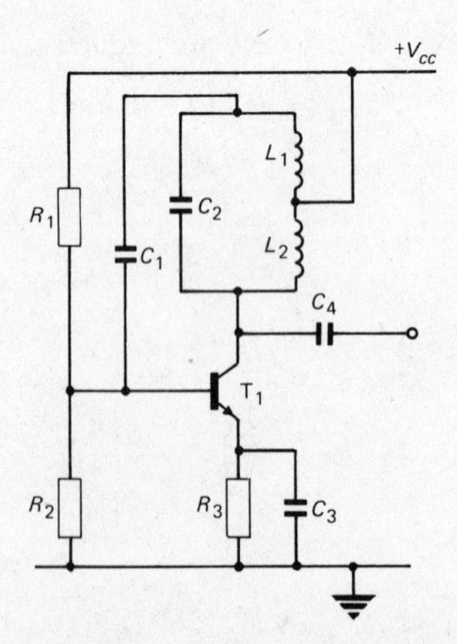

Fig. 7.13 Hartley oscillator.

Hartley Oscillator

Fig. 7.13 shows the circuit of a **Hartley oscillator**. Now, $X_1=1/j\omega C_2$ $X_2=j\omega(L_1+M)$, and $X_3=j\omega(L_2+M)$. Substituting into equation (7.27 gives

$$\frac{1}{j\omega_0 C_2}+j\omega_0(L_1+M)+j\omega_0(L_2+M)=0$$

$$\frac{1}{C_2}=\omega_0^2(L_1+L_2+2M) \qquad \omega_0^2=\frac{1}{C_2(L_1+L_2+2M)}$$

$$f_0=\frac{1}{2\pi\sqrt{[C_2(L_1+L_2+2M)]}} \tag{7.30}$$

From equation (7.26)

$$h_{fe(min)}=-1-\frac{1/j\omega_0 C_2}{j\omega_0(L_2+M)}=-1-\frac{1}{\omega_0^2 C_2(L_2+M)}=-1+\frac{C_2(L_1+L_2+2M)}{C_2(L_2+M)}$$

$$h_{fe(min)}=-1+1+\frac{L_1+M}{L_2+M}=\frac{L_1+M}{L_2+M} \tag{7.31}$$

Frequency Stability

The frequency stability of an oscillator is the amount by which its frequency drifts away from the desired value. It is desirable that the frequency drift should be small, and the maximum allowable change in frequency is usually specified as so many parts per million, e.g. ± 1 part in 10^6 would mean a maximum frequency drift of ± 1 Hz if the frequency of oscillation were 1 MHz but of ± 100 Hz if the frequency were 100 MHz.

Factors that affect the frequency stability of an oscillator are
the load into which the oscillator works,
the parameters of the active device,
the supply voltage,
stability of the components used in the frequency-determining network.

Load on the Oscillator

The frequency of oscillation is not independent of the **load** into which the oscillatory power is delivered. If the load resistance should vary, the frequency of oscillation will not be stable. Variations in the external load can be effectively removed by connecting a buffer amplifier between the oscillator and its load.

Variations in the Supply Voltage

The parameters of an active device, such as the current or voltage gain and the input and output capacitances, are functions of the quiescent collector current and hence of the supply voltage. Any changes in the **supply voltage** will cause one or more parameters to vary and the oscillation frequency to drift. This cause of frequency instability can be minimized by the use of adequate power supply stabilization.

Circuit Components

Changes in the temperature of the **circuit components** will produce changes in inductance and capacitance and hence in the oscillation frequency. Steps that can be taken to minimize this effect include: keeping the frequency-determining components well clear of any heat producing sources; and mounting the components inside a thermostatically controlled oven.

Crystal Oscillators

The best frequency stability that can be achieved with an *LC* oscillator is limited to about ±10 parts in 10^6 per °C and if a better frequency stability is needed a **crystal oscillator** must be employed. A crystal oscillator is one in which the frequency-determining network is provided by a piezo-electric crystal.

Piezo-electric Crystals

A **piezo-electric crystal** is a material, such as quartz, having the property that, if subjected to a mechanical stress, a potential difference is developed across it, and if the stress is reversed a p.d. of the opposite polarity is produced. Conversely, the application of a potential difference to a piezo-electric crystal causes the crystal to be stressed in a direction depending on the polarity of the applied voltage.

In its natural state, quartz crystal is of hexagonal cross-section with pointed ends. If a small thin plate is cut from a crystal, the plate will possess a particular natural frequency and, if an alternating voltage at this frequency is applied across it, the plate will vibrate vigorously. The natural frequency of a crystal depends upon its dimensions, the mode of vibration, and its original position or *cut* in the crystal. The important characteristics of a particular cut are its natural frequency and its temperature coefficient. One cut, the **GT cut,** has a negligible temperature coefficient over a temperature range of 0°C to 100°C. The **AT cut** has a temperature coefficient that varies from about +10 p.p.m./°C at 0°C to 0 p.p.m. at 40°C and about +20 p.p.m./C° at 90° C. Crystal plates are available with fundamental natural frequencies from about 4 kHz up to about 10 MHz or so. For higher frequencies the required plate thickness is very small and the plate becomes very fragile; however, a crystal can be operated at a harmonic of its fundamental frequency and such *overtone* operation raises the possible upper frequency to about 100 MHz.

The electrical equivalent circuit of a piezo-electric crystal is shown in Fig. 7.14. The inductance L represents the inertia of the mass of the crystal when it is vibrating; the capacitance C_1 represents the reciprocal of the stiffness of the plate; and the resistance R represents the frictional losses of the vibrating plate. The capacitance C_2 is the actual capacitance of the crystal. The $L-C_1-R$ branch of the equivalent circuit is series resonant at a frequency f_s when $\omega_s L_1 = 1/\omega_s C_1$. Hence

Fig. 7.14 Electrical equivalent circuit of a piezo-electric crystal.

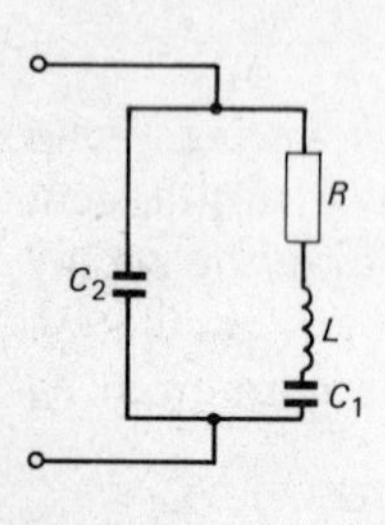

$$f_s = \frac{1}{2\pi\sqrt{(LC_1)}}\ \text{Hz} \tag{7.32}$$

At higher frequencies $\omega L > 1/\omega C_1$ and the effective reactance of the series branch is inductive. At some frequency f_p the effective inductive reactance of the series branch has the same magnitude as the reactance of the shunt capacitor C_2. At this frequency f_p, **parallel resonance** occurs.

$$\omega_p C_2 = \frac{1}{\omega_p L - 1/\omega_p C_1} \qquad \omega_p C_2(\omega_p L - 1/\omega_p C_1) = 1$$

$$\omega_p^2 L C_2 - C_2/C_1 = 1 \qquad \omega_p^2 = \frac{1}{LC_2} + \frac{1}{LC_1}$$

$$\omega_p^2 = \frac{1}{LC_1}(1 + C_1/C_2) = \omega_s^2(1 + C_1/C_2)$$

$$f_p = f_s\sqrt{(1 + C_1/C_2)} \tag{7.33}$$

Since $C_1 \ll C_2$ there is very little difference between the series and parallel resonant frequencies and usually f_p is about $1.01f_s$.

The resonant frequencies of a piezo-electric crystal are *very* stable and this means that such an element can be used as the frequency-determining section of an oscillator.

Crystals manufactured to be resonant at frequencies below 1 MHz have, typically, $L = 100$ H, $C_1 = 0.4$ pF, and $R = 2$ kΩ. Higher-frequency crystals have smaller dimensions and typical values are $L = 20$ mH, $C = 0.2$ pF, and $R = 50$ Ω. The value of the shunt capacitance is generally some 5–10 pF at all frequencies but this value is augmented by the various stray capacitances that appear in parallel with the crystal when it is fitted into an oscillator circuit. However, even a large change in the circuit capacitance has very little effect upon the self-resonant frequencies of a crystal as is illustrated by the following example.

Example 7.3

A piezo-electric crystal has a series capacitance C_1 of 0.040 pF and a parallel capacitance of 10 pF. The crystal is connected into an oscillator circuit and it then has a total shunt capacitance of 15 pF connected across its terminals. Determine the percentage change in i) the series resonant frequency, ii) the parallel resonant frequency if, due to a change of transistor, the shunt capacitance is reduced from 15 pF to 12 pF.

Solution i) The series resonant frequency is unaffected by the value of the shunt capacitance C_2. Hence,

% change in frequency = 0% (*Ans*)

ii) From equation (7.33)

$$f_{p1} = f_s\sqrt{\left(1 + \frac{0.04}{25}\right)} \tag{7.34}$$

and after the change in transistor

$$f_{p2} = f_s\sqrt{\left(1 + \frac{0.04}{22}\right)} \tag{7.35}$$

Dividing equation (7.35) by equation (7.34)

$$\frac{f_{p2}}{f_{p1}} = \sqrt{\left(\frac{1+0.04/22}{1+0.04/25}\right)} \geqslant 1.00011$$

The percentage change in the parallel resonant frequency is

$$\frac{1.00011 f_{p1} - f_{p1}}{f_{p1}} \times 100\% = 0.011\% \quad (Ans)$$

Crystal Oscillator Circuits

When the power supply is first connected to a **crystal oscillator circuit** a voltage pulse is applied to the crystal and causes it to vibrate at its resonant frequency. An alternating voltage at the resonant frequency is then developed between the terminals of the crystal. If this voltage is applied to the base of a transistor, it will be amplified to appear at the collector load. If some fraction of this amplified voltage is fed back to the crystal in the correct phase, it will cause the crystal to vibrate more vigorously. A larger alternating voltage will then appear across the crystal and will be amplified and fed back to the crystal and so on. The circuit will therefore oscillate at either the series or the parallel resonant frequency of the crystal.

A number of crystal oscillator circuits have been developed and several examples of them are shown in [EIII]. Fig. 7.15 shows two examples, one using a bipolar transistor and the other using an op-amp as the active device.

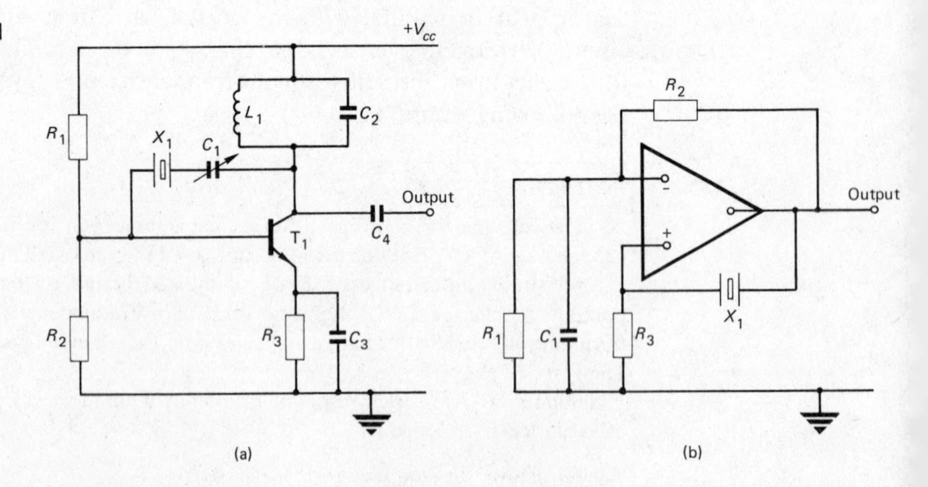

Fig. 7.15 Two crystal oscillator circuits.

Exercises 7

7.1 Draw the electrical equivalent circuit for a piezo-electric crystal and explain why it will resonate at two different frequencies.

A crystal is mounted in a holder which has a capacitance of 6 pF. Calculate the two resonant frequencies of the crystal if the parameters of the crystal are $L = 10$ H, $C_1 = 0.06$ pF and $R = 3000\ \Omega$.

7.2 Derive expressions for i) the minimum value of mutual inductance to maintain oscillations in a Hartley oscillator, ii) the frequency at which the circuit then oscillates.

Determine the minimum value of h_{fe} required for oscillations to take place at 1 MHz if $h_{ie} = 1000\ \Omega$, $L_1 = 220\ \mu$H, $L_2 = 20\ \mu$H, $C = 100$ pF, and $r = 25\ \Omega$.

7.3 Draw the circuit of a Wien bridge *RC* oscillator and explain its operation. Derive an expression for the minimum current (or voltage) gain of the amplifier and the frequency of oscillation. Calculate if the feedback network components are $R_3 = R_4 = 10$ kΩ and $C_3 = C_2 = 0.01\ \mu$F.

7.4 Use the generalized three-impedance concept to derive expressions for the frequency of oscillation and the maintenance condition of an oscillator,

A Colpitts oscillator is to operate at 4 MHz using an inductor of 80 μH and a transistor whose minimum h_{fe} is 50. Calculate suitable values for the tuned-circuit capacitors.

7.5 Draw the circuit of a Wien bridge oscillator. State the parts of the circuit that determine i) the frequency of oscillation, ii) whether the circuit will oscillate.

Quote a typical frequency range for an oscillator of this type and explain why it is not used at radio frequencies.

Fig. 7.16

82 kΩ 1 nF 12 kΩ 10 nF A_v V_{OUT}

7.6 Fig. 7.16 shows the circuit of an *RC* oscillator. Determine the minimum gain of the amplifier for oscillations to take place. Also calculate the oscillation frequency.

7.7 Draw and explain the operation of a crystal oscillator.

In a particular circuit the crystal has a shunt capacitance of 12 pF and a series capacitance of 0.06 pF. The crystal is connected to a fet whose input capacitance is 20 pF. Calculate the percentage change in the oscillation frequency when a capacitor of 10 pF is connected across the crystal.

Short Exercises

7.8 Draw the circuit of an op-amp phase-shift oscillator. State the minimum voltage gain which the op-amp must provide.

7.9 In the circuit shown in Fig. 7.2*a*, $C_3 = C_4 = C_5 = 0.01\ \mu F$ and $R_5 = R_6 = 1500\ \Omega$.
i) If $h_{ie} = 800\ \Omega$ choose a suitable value for R_7; ii) If the oscillation frequency is 3 kHz calculate R_3 and $h_{fe(min)}$.

7.10 A jfet has $g_m = 4$ mS and $r_{ds} = 60$ kΩ and is to be used in a Colpitts oscillator. The oscillator is to work with an inductance of 20 μH at a frequency of 6 MHz. Determine the values of the two frequency-determining capacitors.

7.11 A crystal has $C_1 = 0.05$ pF and $C_2 = 4$ pF. Express the parallel resonant frequency of the crystal as a percentage of its series resonant frequency.

7.12 A piezo-electric crystal has $L = 50$ mH, $R = 10$ kΩ and $C_1 = 0.02$ pF. Calculate its series resonant frequency.

7.13 Draw and explain the operation of a tuned-drain oscillator using an enhancement-mode mosfet.

7.14 Draw and explain the operation of a Hartley oscillator using a depletion-mode mosfet.

7.15 Draw the circuit of an *RC* oscillator using a junction fet.

8 Non-sinusoidal Waveform Generators

Introduction

Waveform generators find frequent application in both analogue and digital circuitry whenever there is a need for a rectangular, a sawtooth, or (less often) a triangular waveform. Rectangular waveforms are produced by a class of oscillator known as a **multivibrator**. In a multivibrator, positive feedback is applied to a circuit to drive two transistors alternately ON and OFF so that the devices act as switches rather than as amplifiers. There are three main categories of multivibrator: the *monostable*, the *astable*, and the *bistable* or *flip-flop*.

An **astable** multivibrator operates continuously to provide a rectangular waveform with particular values of pulse repetition frequency and mark/space ratio. A **monostable** circuit has one stable and one unstable state; normally the circuit rests in its stable state but it can be switched into its alternative state by the application of a trigger pulse, where it will remain for a time determined by its component values. Lastly, the **bistable** or **flip-flop** circuit has two stable states and it will remain in either one until switching is initiated by a trigger pulse.

Multivibrators can be designed using bipolar transistors, fets, operational amplifiers, logic elements such as NAND and NOR gates [DT&S] and are also available as integrated circuits.

A sawtooth voltage consists of a voltage that rises linearly with time, known as a **ramp**, until its maximum value is reached when it then falls rapidly to zero. Immediately the voltage has fallen to zero, it begins another ramp, and so on. Sawtooth waveforms are employed whenever a voltage or current that increases linearly with time is required. More rarely a triangle-shaped waveform may be needed. Sawtooth and triangle generators can be fabricated using discrete components/devices, op-amps, and purpose-built integrated circuits.

Bistable Multivibrators

A bistable multivibrator, or flip-flop, is a circuit having two stable states. The circuit will remain in one state or the other until a trigger pulse is applied to switch the circuit to its other state. It will remain in the second state until another trigger pulse is applied to switch the circuit back to its original condition. A number of variants of the flip-flop circuit are in common usage: the SR, JK, T and D flip-flops, and these have all been described in another volume [DT&S].

The circuit of a **discrete component bistable multivibrator** is shown in Fig. 8.1. Suppose the circuit is in one of its two stable states with T_2 ON and T_1 OFF. The stable state bias conditions of the circuit can be calculated by the use of the circuit of Fig. 8.2*a*. Apply Thevenin's theorem to the left of the point marked as X. The open-circuit voltage V_{oc} is then

$$V_{oc} = \frac{(V_{bb}+V_{cc})R_6}{R_2+R_3+R_6} - V_{bb}$$

and the open-circuit resistance is

$$R_{oc} = R_6(R_2+R_3)/(R_2+R_3+R_6)$$

and so Fig. 8.2*a* can be redrawn as shown by Fig. 8.2*b*. From Fig. 8.2*b*,

$$I_{b2} = (V_{oc} - V_{be})/R_{oc}$$

and, unless this base current drives T_2 into saturation,

$$I_{c2} = h_{FE} I_{b2}$$

Saturation of T_2 can easily be checked since, if T_2 *is* saturated, the product $I_{c2}R_7$ will be calculated as being greater than the supply voltage V_{cc}. Clearly, this cannot be the case and means that I_{c2} must be less than its determined value. Hence T_2 is saturated and its collector/emitter voltage is the saturation voltage $V_{CE(SAT)}$ of the transistor. The collector current of T_2 is then (see Fig. 8.2*c*),

$$I_{c2} = \frac{V_{cc} - V_{CE(SAT)}}{R_7} - \frac{V_{bb} + V_{CE(SAT)}}{R_4 + R_5} \qquad (8.1)$$

The base/emitter voltage V_{be1} of the non-conducting transistor T_1 can be found from Fig. 8.2*d*.

$$V_{be1} = \frac{(V_{CE(SAT)} + V_{bb})R_4}{R_4 + R_5} - V_{bb} \qquad (8.2)$$

and this voltage must be *negative* to take the n-p-n transistor T_1 into non-conduction.

Example 8.1

The transistors in the circuit of Fig. 8.3 have $h_{FE} = 50$. Calculate i) the base and collector currents of a conducting transistor, ii) the base/emitter and collector/emitter voltages of a non-conducting transistor.

Assume for each transistor $V_{CE(SAT)} = 0.1$ V and $V_{be} = 0.6$ V.

Solution

i) $$R_{oc} = \frac{6.8(4.7+1.2)}{6.8+4.7+1.2} = 3.159\ \text{k}\Omega$$

$$V_{oc} = \frac{(12+6)\times 6.8}{6.8+4.7+1.2} - 6 = 3.638\ \text{V}$$

Therefore $$I_b = \frac{3.638-0.6}{3.159\times 10^3} = 0.962\ \text{mA} \quad (Ans)$$

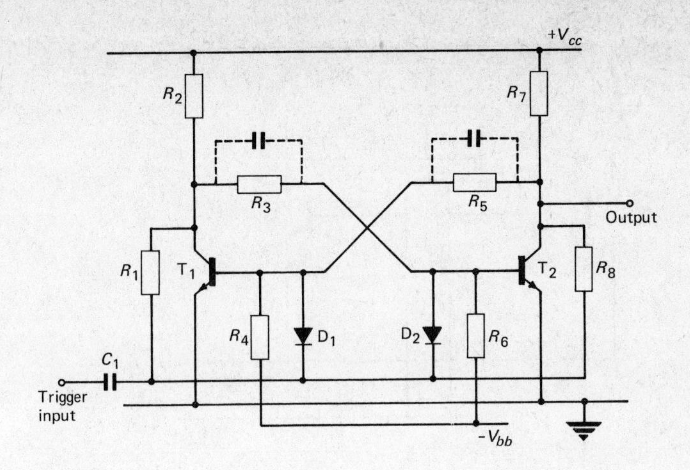

Fig. 8.1 Bistable multivibrator.

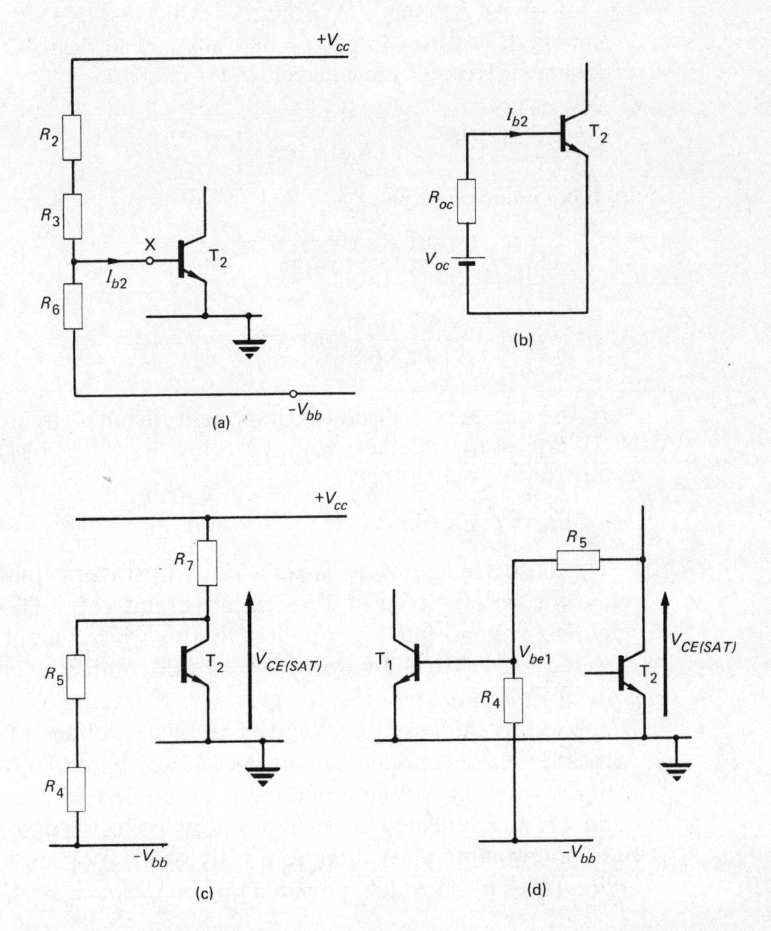

Fig. 8.2

Fig. 8.3

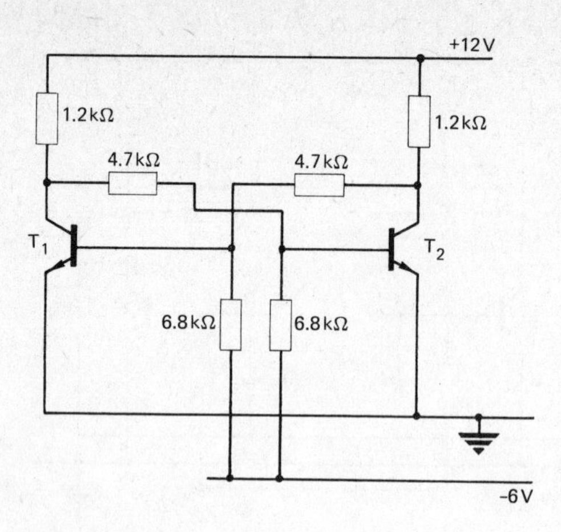

Alternatively $I_b = \dfrac{12-0.6}{(1.2+4.7)\times 10^3} - \dfrac{6+0.6}{6.8\times 10^3} = 0.962\text{ mA}$ (*Ans*)

Now $h_{FE}I_b = 50\times 0.962\times 10^{-3} = 48.1\text{ mA}$ and so clearly the conducting transistor *is* saturated. Hence, from equation (8.1)

$$I_c = \frac{12-0.1}{1.2\times 10^3} - \frac{6+0.1}{(4.7+6.8)\times 10^3} = 9.39\text{ mA} \quad (Ans)$$

ii) From equation (8.2)

$$V_{BE} = \frac{(6+0.1)\times 6.8\times 10^3}{(4.7+6.8)\times 10^3} - 6 = -2.39\text{ V} \quad (Ans)$$

Also $V_{CE} = \dfrac{12\times(4.7+6.8)}{1.2+4.7+6.8} - 6 = 4.87\text{ V}$ (*Ans*)

In the design of a discrete component bistable circuit, equations (8.1) and (8.2) are used together with ensuring that the base current of the ON transistor is greater than

$$(V_{cc} - V_{CE(SAT)})/h_{FE}R_2 \quad (\text{or } R_7)$$

The base voltage of T_2 is sufficiently positive to bias D_2 into conduction while the base voltage of T_1 is negative and keeps D_1 non-conducting. If a negative voltage pulse is applied to the trigger input terminal, it will be directed via D_2 to the base of T_2. T_2 will conduct a smaller current and its collector potential will rise towards V_{cc}. A fraction of this change in voltage is applied to the base of T_1 and, if the base voltage of T_1 is taken positive, will cause T_1 to conduct. Then the collector potential of T_1 starts to fall and a negative-going voltage increment is applied to the base of T_2. T_2 now conducts an even smaller current and its collector potential rises, taking the base potential of T_1 even more positive and so on. In this way a cumulative action takes place which, *provided the loop gain is greater than unity*, turns T_1

Fig. 8.4

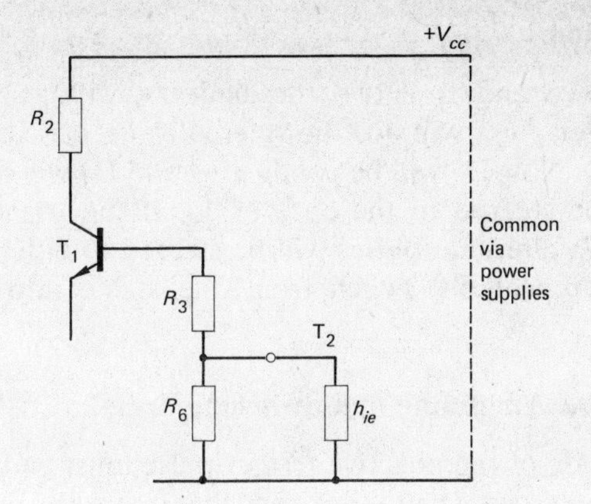

ON and T_2 OFF. Thus the circuit has been switched from one stable state to the other by the application of a trigger pulse.

The **loop gain** of the circuit is the product of the voltage gains of the two transistors *and* the fraction of the collector voltage changes that appear at the base/emitter terminals of the other transistor. For each transistor the voltage gain A_v is approximately equal to

$$h_{fe}R_{L(eff)}/h_{ie}$$

where $R_{L(eff)}$ is the effective collector load. From Fig. 8.4 (taken from Fig. 8.1), if $R = R_6h_{ie}/(R_6 + h_{ie})$ then

$$R_{L(eff)} = R_2(R_3 + R)/(R_2 + R_3 + R) \tag{8.3}$$

Also, from Fig. 8.4 the fraction β of V_{CE} that is applied to the base of the other transistor is

$$\beta = R/(R_3 + R_4) \tag{8.4}$$

The loop gain of the circuit of Fig. 8.1 is hence

$$\beta A = \left(\frac{R}{R_3 + R}\right)\left(\frac{h_{FE}R_{L(eff)}}{h_{ie}}\right) \times \left(\frac{R'}{R' + R_5}\right)\left(\frac{h_{FE}R'_{Lfeff}}{h_{ie}}\right) \tag{8.5}$$

where $R' = R_4h_{ie}/(R_4 + h_{ie})$ and $R'_{L(eff)} = \dfrac{R_7(R_5 + R')}{R_7 + R_5 + R'}$

Example 8.2

Calculate the loop gain of the circuit shown in Fig. 8.3 if $h_{ie} = 2000\ \Omega$.

Solution Since the circuit is symmetrical $R = R'$ and $R_{L(eff)} = R'_{L(eff)}$.

$$R = 6.8 \times 2/(6.8 + 2) = 1.545\ \text{k}\Omega$$

$$R_{L(eff)} = \frac{1.2(4.7 + 1.545)}{1.2 + 4.7 + 1.545} = 1\ \text{k}\Omega$$

Therefore, the loop gain is

$$\beta A = \left(\frac{1.545}{4.7 + 1.545}\right)^2 \times \left(\frac{50 \times 1}{2}\right)^2 = 38.25 \quad (Ans)$$

Toggle Action

When T_1 is ON and T_2 is OFF, the diode D_1 will be conducting and D_2 will not. A trigger pulse will now be steered to the base of T_1 to turn it OFF and hence T_2 ON. Now D_2 will be conducting and D_1 will not and the next trigger pulse will be steered to the base of T_2. If the trigger pulses are obtained from a *clock*, alternate pulses will be steered to different transistors and the circuit will repeatedly switch from one stable into the other, i.e. it will **toggle**.

Trigger Pulse Amplitude and Switching Speed

The amplitude of the negative trigger pulse must be large enough to initiate the switching process and this means that it must be greater than the positive bias voltage at the base of the ON transistor. The trigger pulses are of rectangular waveform but circuit capacitances will tend to differentiate the pulses with the result shown in Fig. 8.5. Clearly the negative-going trigger pulses are reduced in width and care must be taken to ensure that the pulse duration is long enough to initiate the switching process.

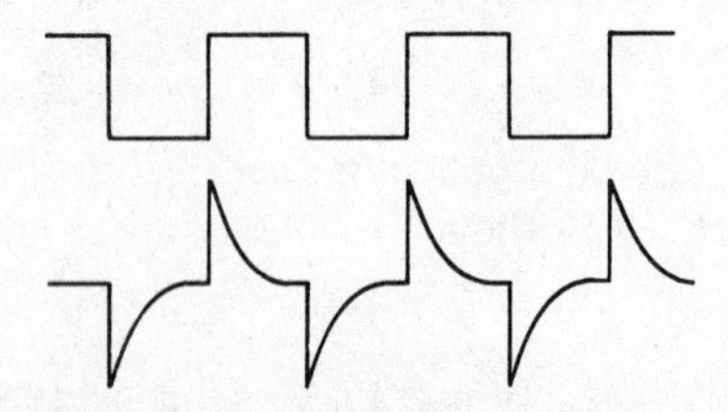

Fig. 8.5 Effect of circuit capacitances on trigger pulses.

The **switching speed** of the circuit is determined by a number of factors, perhaps the most important of which is the switching speed of the transistors themselves. When a transistor is used as a switch, it is cut off, or non-conducting, in one state, and saturated or ON in the other state. The active region, in which the transistor will act as an amplifier, is rapidly passed through and is not of much interest. When the transistor is OFF, *both* of its p-n junctions are reverse-biased and only leakage current flows. When the transistor is ON, *both* junctions are forward biased and the collector/emitter voltage is very small, typically 0.1 V.

When a transistor is OFF and a voltage is applied to its base to turn it ON, there will be a time delay t_d, during which charge is supplied to the base and the transistor capacitances are charged, before the collector current starts to increase. Once the transistor starts to conduct current, more charge is needed to establish the normal charge density at the base/emitter junction and for this reason the collector current rises relatively slowly, taking a time t_r to reach its maximum value. When the saturation value of the collector current is reached

$$V_{CE} = V_{CE(SAT)} \quad \text{and} \quad I_c = \frac{V_{cc} - V_{CE(SAT)}}{R_L}$$

This collector current is equal to $h_{FE}I_b$. If I_b is larger than $I_{C(SAT)}/h_{FE}$ the collector current will rise more rapidly but its final value will not be greater.

When a saturated transistor is turned OFF, the excess charge stored in the base region must be removed and the collector current remains constant at its saturated value for a time t_s. Once the excess charge has been removed, the collector current will fall and will take a time t_f to reach zero (see Fig. 8.6).

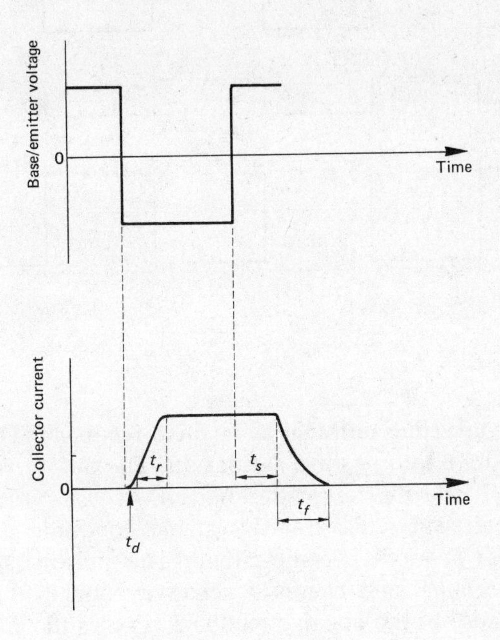

Fig. 8.6 Transistor switching.

If a capacitor C, generally known as a **speed-up capacitor**, is connected in parallel with R_3 and with R_5, the initial *change* in collector voltage is transferred to the appropriate base more rapidly than otherwise. In the ON state, the voltage across C is I_bR_3 and the stored charge is I_bR_3C coulombs. If C is of such a value that the charge stored is greater than the total excess charge stored in the base, then when the base voltage falls to zero, C takes an impulse current that rapidly removes the excess charge.

The waveforms at various points in the circuit of Fig. 8.3 are given by Fig. 8.7.

Operational Amplifier Bistable Multivibrator

The bistable multivibrator can also be fabricated using an **operational amplifier**, the circuit being shown in Fig. 8.8. When the output of the op-amp is at its negative saturation value $V^-_{o(SAT)}$, diode D_2 is ON. A positive trigger pulse applied to the circuit is then steered to the non-inverting terminal via R_3 and not to the inverting terminal. A positive voltage appears at the output terminals and a fraction of this voltage is fed back, via R_6 and C_2,

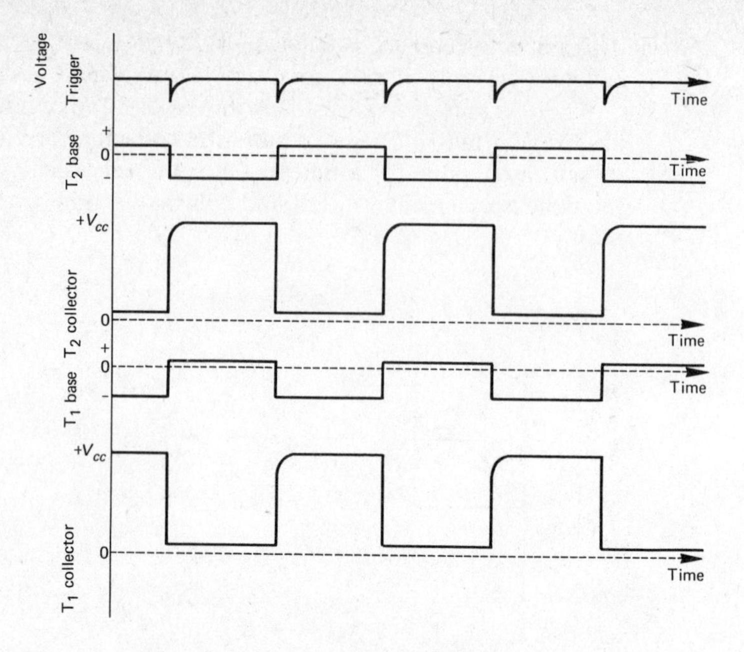

Fig. 8.7 Bistable multivibrator waveforms.

to the non-inverting terminal as positive feedback. This causes the op-amp to rapidly drive into positive saturation. The output voltage of the circuit is then $V^{+}_{o(SAT)}$ and this keeps the diode D_2 OFF.

The circuit will remain in this stable condition until the next positive trigger pulse is applied to the circuit. This pulse is steered, via R_2, to the inverting terminal and causes a negative voltage to appear at the output which, because of the positive feedback, causes the circuit to switch back to its negative saturation stable state. The circuit is said to toggle when a clock waveform is applied to the trigger input terminal.

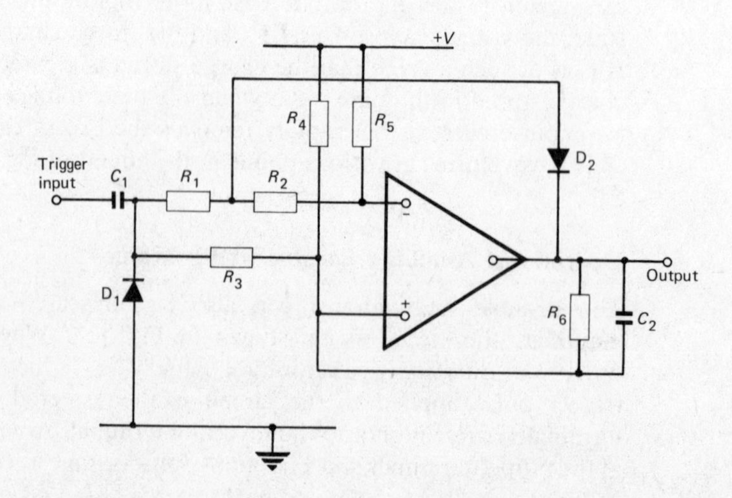

Fig. 8.8 Op-amp bistable multivibrator.

Integrated NAND/NOR Bistable Multivibrators

The bistable multivibrator can also be made by connecting two NOR gates or four NAND gates together as shown in Fig. 8.9*a* and *b*.

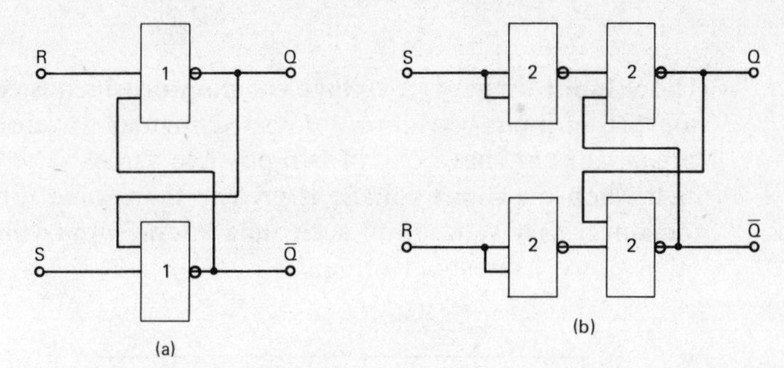

Fig. 8.9 (*a*) NOR gate bistable multivibrator, (*b*) NAND gate bistable multivibrator.

The output of a **NOR gate** is 1 only when both of its inputs are at 0. If one or both inputs are at 1, the output of the NOR gate is 0.

1 Suppose $Q=1$ and $\bar{Q}=0$. If $S=1$ and $R=0$, both inputs to the lower gate are at 1 and so $\bar{Q}$ remains at 0. The two inputs to the upper gate are at 0 and so Q stays at 1.

2 When $S=0$ and $R=1$, the upper gate has one input at 0 and the other at 1 and hence $Q=0$. Both inputs to the lower gate are now at 0 and so $\bar{Q}$ becomes 1. Thus the circuit has changed state from $Q=1$, $\bar{Q}=0$ to $Q=0$, $\bar{Q}=1$.

3 If now $S=1$ and $R=0$ again, then the lower gate has one input at 1 and the other at 0. Then $\bar{Q}$ becomes 0 and so both inputs to the upper gate will be at 0, and Q will switch back to the 1 state.

4 If the condition $S=R=1$ should exist, the operation of the circuit cannot be predicted.

The output of a **NAND gate** is at 0 only when both of its inputs are at 1, otherwise its output is at logical 1.

1 Suppose that initially $Q=1$, $\bar{Q}=0$, $S=R=0$. Then the inputs to the upper gate are 1 and 0 and this gives $Q=1$. The two inputs to the lower gate are both 1 and so $\bar{Q}=0$.

2 The circuit should change state to $Q=0$, $\bar{Q}=1$ if $S=0$, $R=1$. When R becomes 1, the lower output gate will have one input at 1 and the other at 0 and this means that $\bar{Q}$ will switch to 1. Now the upper output gate has both of its inputs at 1 and Q becomes 0.

3 If now $S=1$, $R=0$, the upper output gate will have one input at 0 and the other at 1 and so Q becomes 1. Now the lower output gate has both inputs at 1 and $\bar{Q}$ switches to 0.

The NAND and NOR gates will almost certainly be members of either the t.t.l. (including Schottky) or the cmos logic families [DT&S] and all the gates needed will be within the one package. Examples are the 4 001 (cmos) and 7 402 (t.t.l.) quad 2-input NOR gates, and the 4 011 (cmos) and 7 400 (t.t.l.) quad 2-input NAND gates.

Integrated Bistable Multivibrators

A variety of bistable multivibrators are readily available in both the cmos and t.t.l. families and are specifically one of the SR, JK, or D versions of the basic circuit [DT&S].

Schmitt Trigger

The **Schmitt trigger** is a **voltage comparator** which is commonly employed to convert an input waveform into a rectangular waveform. The output of the circuit can only have one of two possible values. The output voltage will be high when the input voltage is greater than some threshold value and will remain at that value until such time as the input voltage falls below some other, lower, threshold value.

Fig. 8.10 Schmitt trigger.

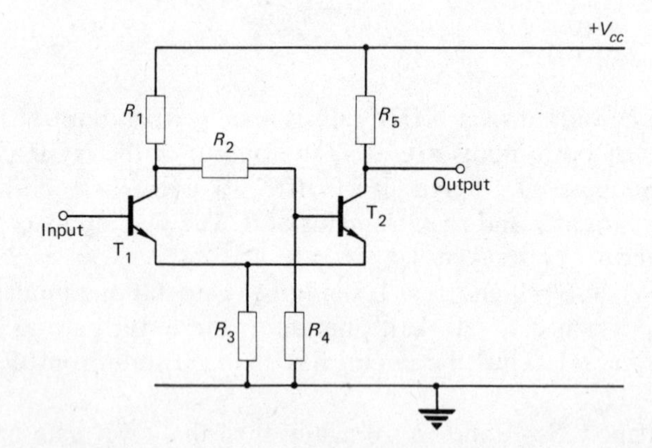

The circuit of a discrete component Schmitt trigger is shown in Fig. 8.10. When the input voltage is zero, T_2 will be biased ON by the potential $V_{cc}R_4/(R_1+R_2+R_4)$ at its base. The emitter current of T_2 flowing through the common emitter resistor R_3 will develop a voltage that will hold the emitter of T_1 positive with respect to its base so that T_1 is cut off. The output voltage is now equal to

$$(V_{CE(SAT)}+I_{E2}R_3) \text{ volts}$$

If now a voltage is applied to the base of T_1 that takes it more positive than the voltage across R_3, then T_1 will start to conduct and its collector potential will become less positive. This, in turn, will reduce the positive potential at the base of T_2 and so T_2 will conduct a smaller current. The voltage across R_3 will fall and this will increase the forward bias voltage V_{be} of T_1 so that T_1 will conduct more current. The collector potential of T_1 will fall still further and so on. A cumulative effect takes place (provided the loop gain is greater than unity) and T_1 will be turned ON and T_2 OFF. The output voltage is now V_{cc} volts.

If now the input voltage is reduced, the point will be reached where T_1 comes out of its saturated condition and conducts less current. The collector potential of T_1 then increases in the positive direction and this positive voltage increment causes T_2 to conduct. The positive voltage dropped across the emitter resistor R_3 increases in value and so reduces the forward bias applied to T_1. Therefore T_1 conducts a smaller current and its collector potential becomes more positive and so on; a cumulative effect takes place which results in the condition T_1 OFF and T_2 ON.

The difference between the two input voltages V_1 and V_2 at which switching takes place is known as the **hysteresis** of the circuit. Generally, this should be as small as possible but it is typically in the region of 1 V.

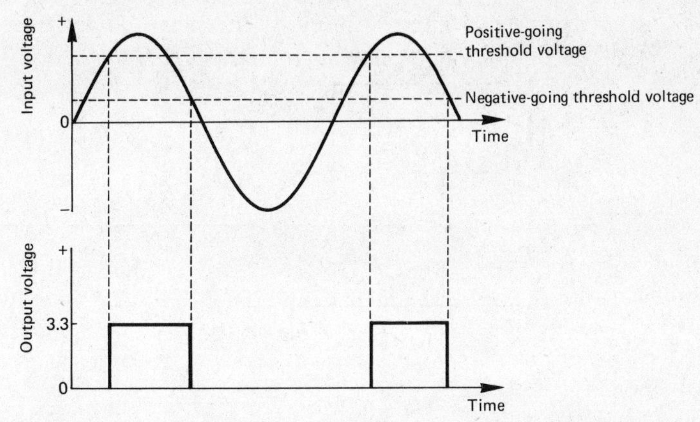

Fig. 8.11 Operation of a Schmitt trigger.

Fig. 8.11 shows typical waveforms to be expected from a Schmitt trigger circuit. When the sinusoidal input voltage becomes more positive than the upper threshold voltage, the circuit switches to give an output voltage of +3.3 V. The output voltage will remain at this value until the input voltage falls below the lower threshold voltage and at this point the output voltage suddenly switches to approximately 0 V. The output voltage will now remain at 0 V until the input voltage again exceeds the upper threshold voltage.

The upper threshold voltage V_1 at which the circuit switches from its normal state of T_1 OFF and T_2 ON is equal to the base/emitter voltage V_{be} of T_1 plus the voltage developed across R_3. That is

$$V_1 = V_{be} + I_{e2}R_3 \simeq V_{be} + I_{c2}R_3$$

With T_1 OFF and T_2 ON, Thevenin's theorem can be applied to the base of T_2 to give Fig. 8.12 in which

$$V_{oc} = V_{cc}R_4/(R_1+R_2+R_4) \quad \text{and} \quad R_{oc} = R_4(R_1+R_2)/(R_1+R_2+R_4)$$

Fig. 8.12

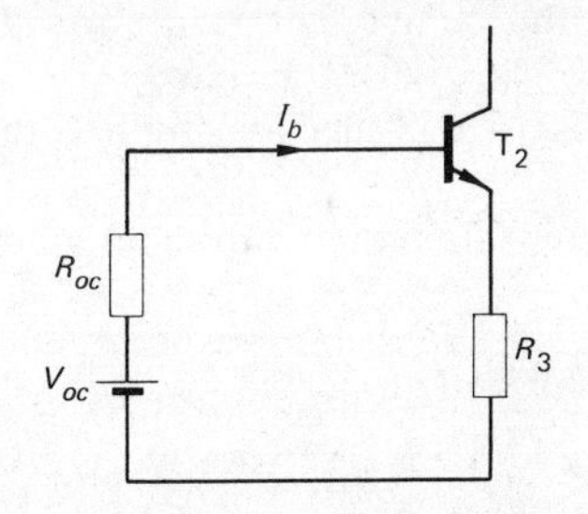

From Fig. 8.12,

$$V_{oc} - V_{be} = I_b R_{oc} + (1 + h_{fe}) I_b R_3 \simeq I_b (R_{oc} + h_{fe} R_3)$$

Therefore $I_b = (V_{oc} - V_{be})/(R_{oc} + h_{fe} R_3)$ and

$$V_1 = V_{be} + h_{fe} R_3 (V_{oc} - V_{be})/(R_{oc} + h_{fe} R_3) \tag{8.6}$$

Fig. 8.13

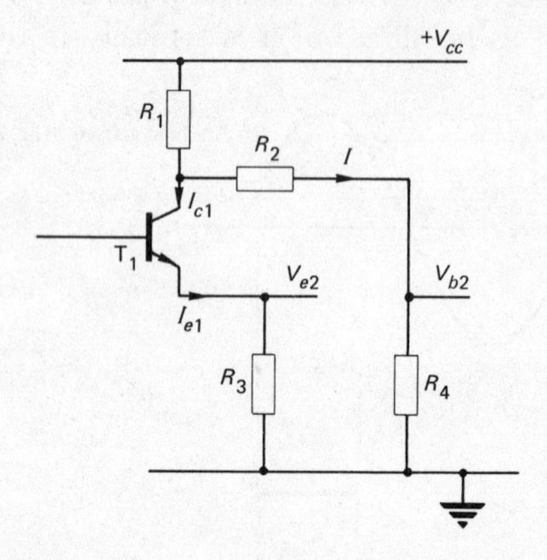

The lower threshold voltage V_2 is the value at which the circuit will be switched back to its original state of T_1 OFF and T_2 ON. For T_2 to start conducting current, its base potential must become more positive than its emitter potential. Consider Fig. 8.13,

$$V_{cc} = R_1(I_{c1} + I) + (R_2 + R_4)I$$

T_2 will switch when $V_{e2} = V_{b2}$ or

$$I_{c1} R_3 = V_{be} + I_4 R_4 = \frac{R_4(V_{cc} - R_1 I_{c1})}{R_1 + R_2 + R_4}$$

$$I_{c1}[R_3(R_1 + R_2 + R_4) + R_1 R_4] = V_{cc} R_4$$

$$I_{c1} = \frac{V_{cc} R_4}{R_3(R_1 + R_2 + R_4) + R_1 R_4} \tag{8.7}$$

and $V_2 = V_{be} + I_{c1} R_3$ (8.8)

Example 8.3

In a Schmitt trigger circuit the component values are $R_1 = 1200\ \Omega$, $R_2 = 27\ \text{k}\Omega$, $R_3 = 1000\ \Omega$ and $R_4 = 33\ \text{k}\Omega$. The transistors have $h_{fe} = 100$ and $V_{be} = 0.6$ V while the collector supply voltage is 12 V. Calculate the two threshold voltages and the hysteresis of the circuit.

Solution From equation (8.6) the upper threshold voltage V_1 is

$$V_1 = 0.6 + \frac{100 \times 1000(6.47 - 0.6)}{15.21 \times 10^3 + 100 \times 1000}$$

$$= 5.7\ \text{V} \quad (Ans)$$

and from equation (8.8) the lower threshold voltage V_2 is

$$V_2 = \frac{12 \times 1 \times 10^3 \times 33 \times 10^3}{10^3 \times 1.2 \times 10^3 + 27 \times 10^3 \times 10^3 + 10^3 \times 33 \times 10^3 + 1.2 \times 10^3 \times 33 \times 10^3} + 0.6$$

$$= 4.53 \text{ V} \quad (Ans)$$

The hysteresis of the circuit is

$$5.7 - 4.53 = 1.17 \text{ V} \quad (Ans)$$

Hysteresis provides some measure of noise protection to a circuit. Once the input signal voltage has passed through a threshold voltage and the circuit has switched, the output voltage will remain constant even though the input voltage may have noise voltage superimposed upon it. Always provided, of course, that the noise voltages are not so large that they cause the input voltage to reach the other threshold value. Clearly, the larger the hysteresis, the greater the noise protection afforded.

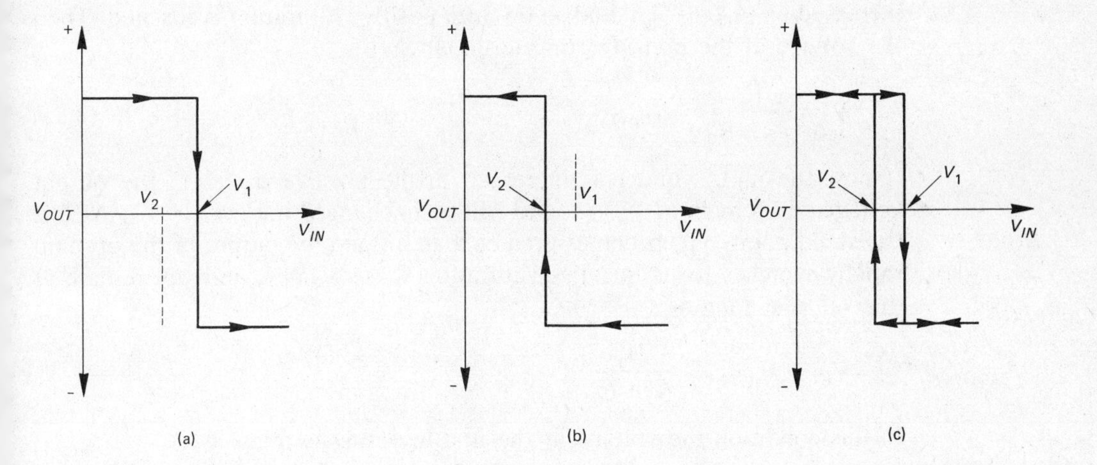

Fig. 8.14 Transfer characteristics of a Schmitt trigger.

The **transfer characteristic** of a Schmitt trigger circuit is shown by Fig. 8.14 and is a plot of output voltage to a base of input voltage. Fig. 8.14*a* shows how the output voltage changes from one value to another as the input voltage is increased to the upper threshold voltage V_1 volts. Similarly, Fig. 8.14*b* shows how the output voltage suddenly reverts to its original value when the input voltage is reduced below the lower threshold voltage of V_2 volts. Finally, Fig. 8.14*c* shows the complete transfer characteristic. Note that the transfer characteristic drawn shows the output voltage of the trigger varying between ± values, but for many circuits this would be between either $\pm V_{OUT}$ and approximately zero volts.

An **operational amplifier** can be connected to act as a Schmitt trigger, the circuit being shown by Fig. 8.15*a*. The input voltage is applied to the inverting terminal of the op-amp and positive feedback to its non-inverting terminal.

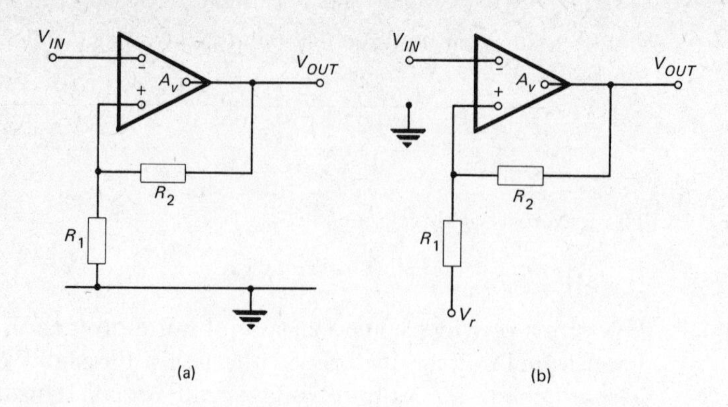

Fig. 8.15 Op-amp Schmitt triggers: (*a*) inverting, (*b*) inverting with reference voltage.

The feedback factor is $\beta = R_1/(R_1+R_2)$ and the loop gain is βA_v.

If the output voltage V_{OUT} increases by say δV_{OUT} then the voltage fed back to the non-inverting terminal is $\beta\, \delta V_{OUT}$ and so V_{OUT} will be further increased by $A_v\beta\, \delta V_{OUT}$ and so on until positive saturation is reached. Then the voltage at the non-inverting terminal is

$$V_1 = \frac{R_1}{R_1+R_2} \cdot V^+_{o(SAT)} \tag{8.9a}$$

If now the input voltage is increased in the positive direction the output voltage will remain at $V^+_{o(SAT)}$ and will be unchanged until $V_{IN} = V_1$. At this threshold point, V_{IN} becomes greater than V_1 and the output of the op-amp rapidly switches to its negative saturation value $V^-_{o(SAT)}$ and will remain at this value as long as

$$V_{IN} > V^-_{o(SAT)} \cdot \frac{R_1}{R_1+R_2}$$

In this condition the voltage at the non-inverting terminal is

$$V_2 = R_1 V^-_{o(SAT)}/(R_1+R_2) \tag{8.9b}$$

V_1 and V_2 are the threshold voltages and $(V_1 - V_2)$ is the hysteresis of the circuit and is small. The op-amp employed should have a fairly high slew rate and for most purposes the 741 is likely to prove inadequate.

If R_1 is taken to a reference voltage V_r instead of to earth (Fig. 8. 15*b*), the threshold voltages equations are modified thus:

$$V_1 = V_r + \frac{R_1}{R_1+R_2}(V^+_{o(SAT)} - V_r) \tag{8.10}$$

$$V_2 = V_r - \frac{R_1}{R_1+R_2}(V^-_{o(SAT)} + V_r) \tag{8.11}$$

Example 8.4

An op-amp Schmitt trigger has $R_1 = 560\ \Omega$, $R_2 = 12\ \text{k}\Omega$, and $V_r = 1$ V. For $V^+_{o(SAT)} =$

$V^-_{o(SAT)} = 8$ V, calculate the hysteresis of the circuit.

Solution From equations (8.10) and (8.11)

$$V_1 = 1 + \frac{560}{12560}(8-1) = 1.312 \text{ V}$$

$$V_2 = 1 - \frac{560}{12560}(8+1) = 0.599 \text{ V}$$

Therefore the hysteresis is $V_1 - V_2 = 0.713$ V (*Ans*)

With the op-amp connected as shown in Fig. 8.15, the output rectangular waveform is inverted relative to the input signal. A non-inverting circuit can easily be obtained by merely applying the reference voltage to the inverting terminal and the input signal to the non-inverting terminal.

Integrated Schmitt Trigger Circuits

The Schmitt trigger circuit is also available as an **integrated circuit** device and examples in the t.t.l. and cmos logic families are

t.t.l. 7413 dual 4-input NAND Schmitt trigger
7414 hex inverter Schmitt trigger
74132 quad 2-input NAND Schmitt trigger.

Each of these devices has a positive-going threshold voltage of 1.7 V and a negative-going threshold voltage of 0.9 V, i.e. a hysteresis of 0.8 V.

c.m.o.s. 4093 quad 2-input NAND Schmitt trigger

This has threshold voltages of 4 V and 6 V.

The pin connections of these devices are given in Fig. 8.16.

A NAND Schmitt trigger can be used as either a NAND gate *or* as an inverting Schmitt trigger circuit, the necessary connections, referring to one "gate" in the 4093 i.c., being shown in Fig. 8.17. Fig. 8.18 shows how the 4093 can be used as a sine-to-square wave convertor or as a level detector.

Fig. 8.16 Pin connections of four integrated Schmitt triggers.

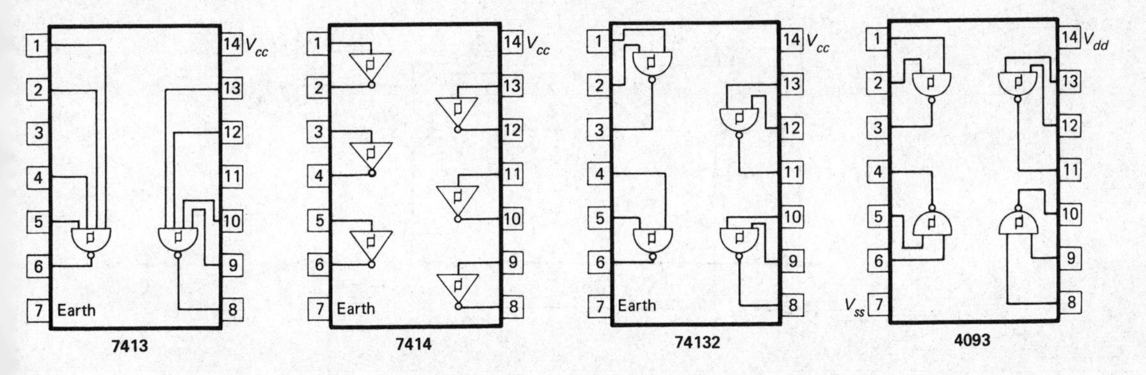

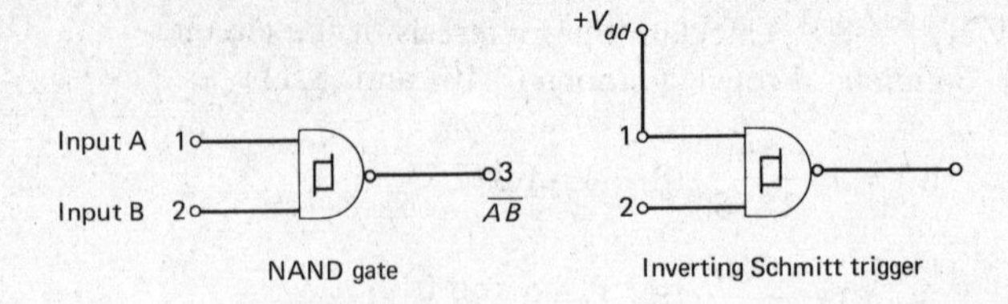

Fig. 8.17 Applications of the 4093 Schmitt trigger.

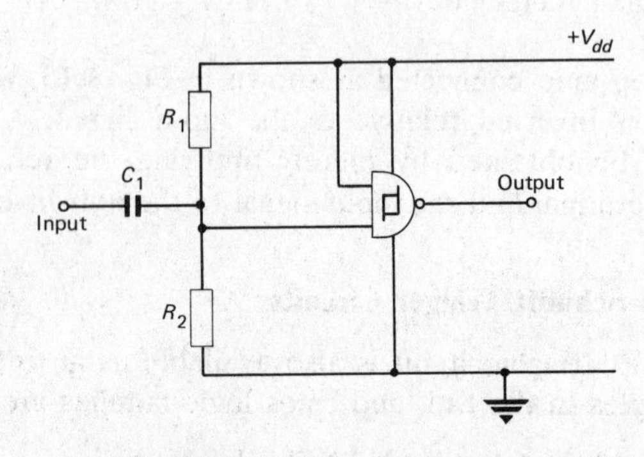

Fig. 8.18 Use of the 4093 as a level detector.

Monostable Multivibrators

The **monostable multivibrator** is a circuit that has one stable state and one unstable state. When the power supply is switched on, the circuit will settle in its stable state and remain there until a trigger pulse is applied to the circuit to initiate switching to the unstable state. The circuit will then remain in this unstable state for a time that is determined by its timing components and will then revert to its stable condition.

The circuit of a discrete component monostable multivibrator is shown in Fig. 8.19. The stable state is T_2 ON and T_1 OFF because of the voltages applied to the respective base terminals. Usually T_2 is saturated so that its collector potential is $V_{ce(SAT)} \simeq 0.1$ V.

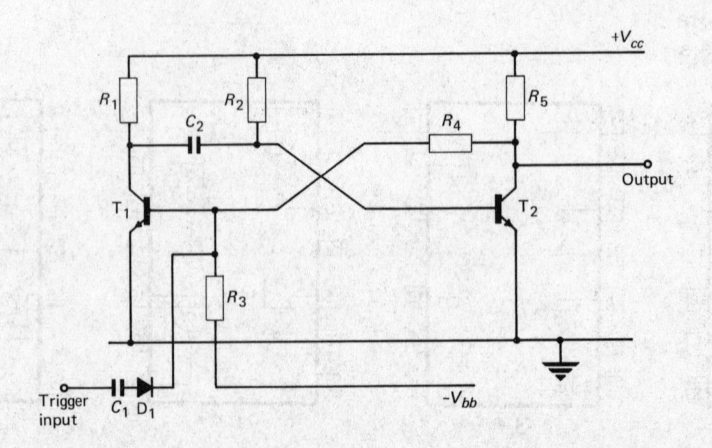

Fig. 8.19 Monostable multivibrator.

Fig. 8.20

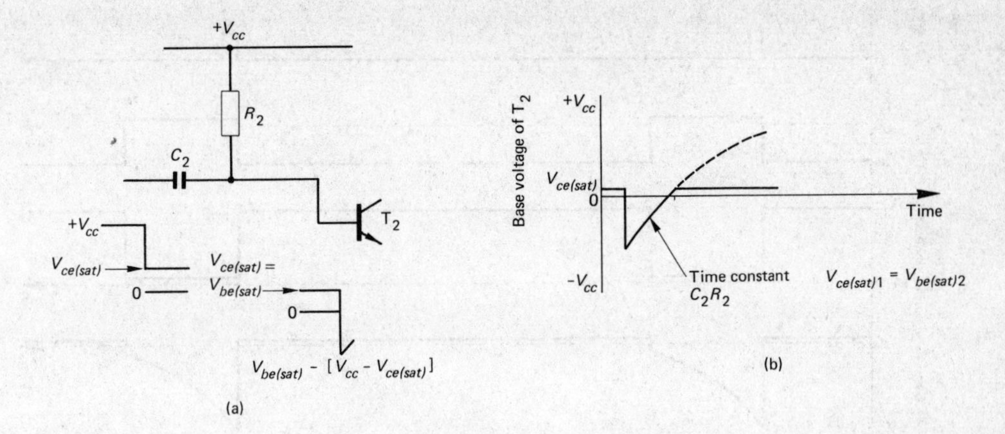

If a positive trigger pulse is applied to the circuit, it will drive T_1 into its saturated condition and the collector potential of T_1 will fall suddenly from $+V_{cc}$ volts to very nearly 0 V (actually $V_{ce(SAT)}$). A negative-going voltage pulse is passed, via C_2, to the base of T_2 and this takes the base potential of T_2 from 0 V to $-V_{cc}$ volts, to turn T_2 OFF. The collector voltage of T_2 rises suddenly from 0 V to $+V_{cc}$ volts. Capacitor C_2 now has a p.d. of $-V_{cc}$ volts across its plates (see Fig. 8.20*a*) and starts to charge, with time constant C_2R_2, towards $+V_{cc}$ volts (Fig. 8.20*b*). It should be noted that the total voltage applied across the C_2R_2 circuit is $2V_{cc}$ volts. Immediately the voltage across C_2 and hence the base voltage of T_2 passes through 0 V, T_2 starts to conduct and its collector potential falls. A negative-going voltage pulse is now applied to the base of T_1 causing it to conduct less current. The collector potential of T_1 rises and so on; a regenerative action takes place which rapidly switches the circuit back to its stable state, i.e. T_2 ON and T_1 OFF.

Taking $V_{ce(SAT)} \simeq V_{be(SAT)} \simeq 0$, then

$$V_{c2} = 2V_{cc}(1 - e^{-t/C_2R_2})$$

Immediately the base voltage of T_2 passes through 0 V (V_{cc} volts on the curve of Fig. 8.20*b*), T_2 will start to conduct. At this time T seconds

$$V_{cc} = 2V_{cc}(1 - e^{-T/C_2R_2})$$

$$T = 0.69C_2R_2 \tag{8.12}$$

Fig. 8.21 shows the waveforms at different points in the circuit. The output waveform can be improved by the use of a speed-up capacitor connected in parallel with R_4.

Use of NAND/NOR Gates

Integrated circuit NAND and NOR gates are readily available and provide a convenient and economic means of fabricating a monostable multivibrator circuit. A number of both types of gate are in both the t.t.l. and the cmos

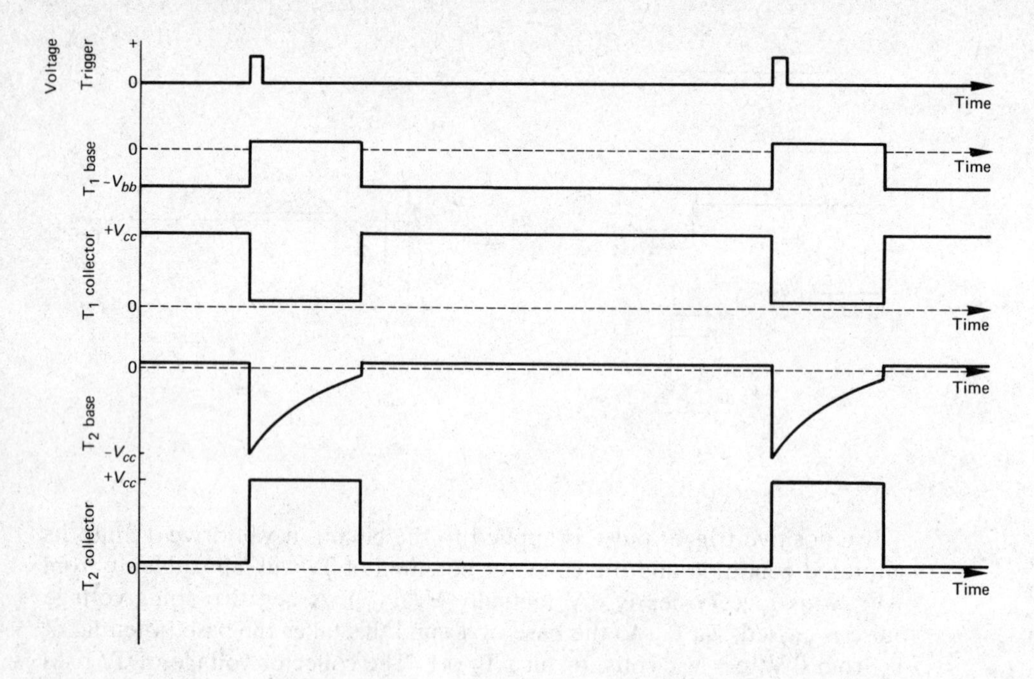

Fig. 8.21 Monostable multivibrator waveforms.

families, providing four 2-input gates in one package. A monostable multivibrator requires the use of two NOR or NAND gates as shown in Fig. 8.22*a* and *b*. In both circuits the relaxation time is determined by the time constant C_1R_2.

The **NOR circuit** has a stable state with a low ($\simeq 0$ V) output voltage; hence in the absence of a trigger pulse the output of the left-hand gate is high ($\simeq +5$ V). When a positive trigger pulse is applied to the circuit, the output of the input gate switches to 0 V and the change in potential is passed through C_1 to take the input of the right-hand gate to 0 V also. The output of this gate then switches high ($\simeq 5$ V). The left-hand plate of C_1 is now at 0 V and the capacitor charges up towards $+V_{cc}$ with a time constant of C_1R_2. When the gate input voltage reaches the threshold value, the output gate switches back to 0 V. Hence equation (8.12) applies with a different time constant. For t.t.l. gates the threshold voltage is 2 V and for cmos gates it is about $V_{dd}/2$.

The **NAND** monostable multivibrator works in a similar manner except that the stable state is a high ($\simeq 5$ V) output and triggering is accomplished by a negative-going voltage pulse.

Schmitt Trigger Monostable Multivibrator

A **Schmitt trigger** i.c. can be used as a monostable multivibrator, the necessary connections being shown in Fig. 8.23. In the stable condition, the n-channel mosfet is non-conducting and C_1 is fully charged. Both inputs to the trigger are then high and so the stable state is a low output.

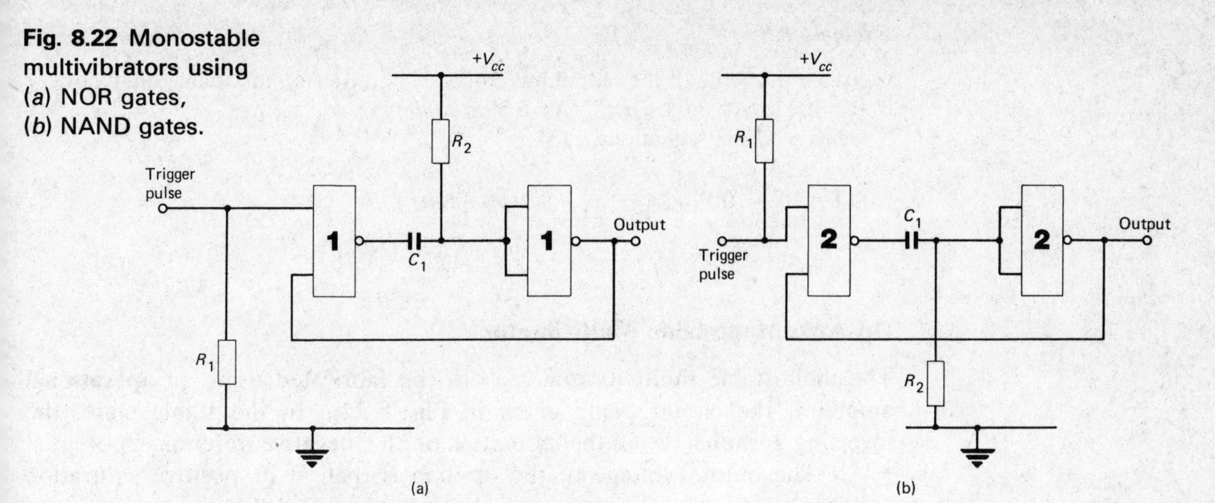

Fig. 8.22 Monostable multivibrators using (*a*) NOR gates, (*b*) NAND gates.

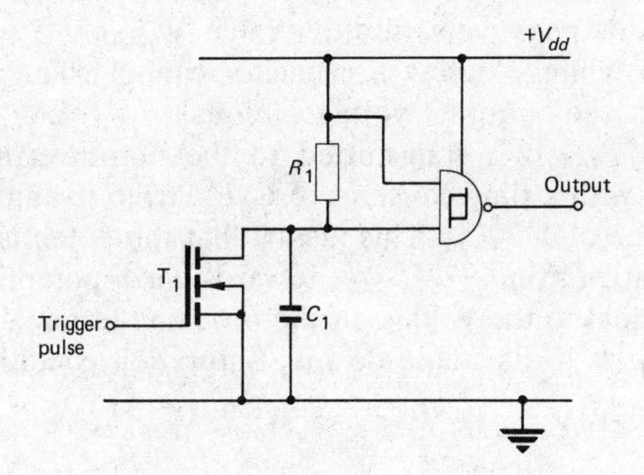

Fig. 8.23 Schmitt trigger monostable multivibrator.

When a positive voltage pulse is applied to the trigger input, T_1 is turned ON and C_1 is rapidly discharged. One input to the trigger is now low and so its output is switched to its high value. Capacitor C_1 now charges up via resistor R_1 and when the voltage across its terminals reaches the threshold value of the Schmitt trigger the output voltage switches back to the low value.

The expression for the voltage across C_1 is

$$v_c = V_{dd}(1 - e^{-t/R_1C_1})$$

The circuit will change states when this voltage reaches the threshold value V_1. This occurs after a time T, therefore,

$$V_1 = V_{dd}(1 - e^{-T/C_1R_1}) \qquad V_1/V_{dd} = 1 - e^{-T/C_1R_1}$$

$$-\frac{T}{C_1R_1} = \log_e\left(\frac{V_{dd} - V_1}{V_{dd}}\right) \qquad T = C_1R_1 \log_e\left(\frac{V_{dd}}{V_{dd} - V_1}\right) \tag{8.13}$$

Example 8.5

Calculate the width of the output pulse of a Schmitt trigger monostable multivibrator if $R = 100\ \text{k}\Omega$, $C = 0.01\ \mu\text{F}$, $V_{dd} = 5\ \text{V}$ and $V_1 = 3\ \text{V}$.

Solution From equation (8.13),

$$T = 10^5 \times 10^{-8} \log_e \left[\frac{5}{5-3}\right] = 916\ \mu\text{s} \quad (Ans)$$

Op-Amp Monostable Multivibrator

The monostable multivibrator can also be fabricated using an **operational amplifier**, the circuit being given in Fig. 8.24*a*. In the stable state, the inverting terminal is at the potential of the negative reference potential $-V_{ref}$. The output voltage of the op-amp is then at its positive saturation voltage $V^+_{o(SAT)}$.

When a positive trigger voltage, of magnitude greater than V_{ref}, is applied via C_1, the inverting terminal goes positive and the output of the circuit switches to its negative saturation value $V^-_{o(SAT)}$.

Since the voltage across a capacitor cannot change instantaneously, the change in the output voltage, equal to $-2V_{o(SAT)}$ (assuming that $V^+_{o(SAT)} = V^-_{o(SAT)}$), is transmitted to the non-inverting terminal. C_2 then discharges, with a time constant of C_2R_2, from its initial value of $V^+_{o(SAT)}$ to its new value of $V^-_{o(SAT)}$. This means that the potential at the non-inverting terminal varies from $-2V_{o(SAT)}$ towards earth potential. When this voltage becomes equal to the voltage at the inverting terminal, i.e. $-V_{ref}$, the circuit rapidly reverts to its stable positive saturation condition. Thus,

$$-V_{ref} = -2V_{o(SAT)}e^{-T/C_2R_2} \qquad V_{ref}/2V_{o(SAT)} = e^{-T/C_2R_2}$$

$$T = C_2R_2 \log_e \left[\frac{2V_{o(SAT)}}{V_{ref}}\right] \tag{8.14}$$

The circuit waveforms are shown in Fig. 8.24*b*.

Integrated Circuit Monostable Multivibrators

A number of monostable multivibrators are available in both the t.t.l. and cmos logic families. Examples of each are shown in Fig. 8.25. The **74121** is a monostable multivibrator in which the pins labelled $\bar{A}_1$ and $\bar{A}_2$ are negative-edge triggered logic inputs which will trigger the monostable when either or both A_1 and A_2 go to logical 0 and pin 5 (B) is at logical 1. Pin 5 is one input to a positive Schmitt trigger (the other input being connected to the output of the gate to which $\bar{A}_1$ and $\bar{A}_2$ connect). The function of this trigger circuit is to provide level detection. Without an external capacitor or resistor the width of the output pulse is 30 ns. Pin 9 should be connected to pin 14. For other pulse widths, an external capacitor C_1 should be connected between pins 10 and 11 and an external resistor should be connected

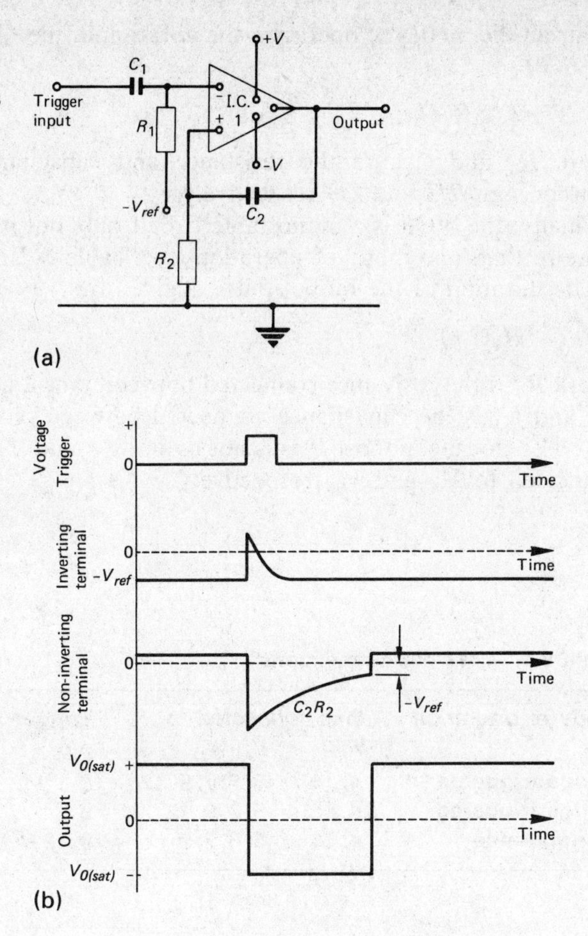

Fig. 8.24 (*a*) Op-amp monostable multivibrator, (*b*) waveforms in an op-amp monostable multivibrator.

between pins 11 and 14, and pin 9 left unconnected. The output pulse width is then given by

$$T = 0.695 R_1 C_1 \tag{8.15}$$

The **74122** is a retriggerable monostable multivibrator which provides an output pulse width given by

$$T = 0.32 R_1 C_1 (1 + 0.7/R_1) \tag{8.16}$$

where C_1 is the external capacitor connected between pins 11 and 13. R_1 is the external resistor connected between pins 13 and 14. If a 10 kΩ resistance is required this can be provided internally by shorting pin 9 to pin 14 instead.

The **4047** is a cmos device that can be used as either a monostable or an astable multivibrator. When used as a monostable it can be operated as either a positive-edge or a negative-edge triggered device and can also be made retriggerable. The necessary pin connections for each mode of operation are given in Table 8.1.

For all the modes of operation the duration of the Q output pulse is given by

$$T = 2.48 R_1 C_1 \tag{8.17}$$

where R_1 and C_1 are the resistance and capacitance values connected between pins 2/3 and 1/3 respectively.

Finally, the **4098** is a monostable circuit only but it can also be operated in more than one mode of operation (see Table 8.2).

The duration of the output pulse is given by

$$T = R_1 C_1 / 2 \tag{8.18}$$

where R_1 is the resistance connected between pins 2 and 16 (or pins 14 and 16) and C_1 is the capacitance connected between pins 1 and 2 (or 14 and 15). If one of the monostables is not used, its + and − trigger inputs must be connected to V_{ss} and V_{dd} respectively.

Table 8.1 4047 pin connections

Mode of operation	*Pins connected to* V_{dd}	V_{ss}	*Trigger input pin*	*Output pins*
+ edge triggered	4, 14	5, 6, 7, 9, 12	8	10, 11
− edge triggered	4, 8, 14	5, 7, 9, 12	6	10, 11
Retriggerable	4, 14	5, 6, 7, 9	8, 12	10, 11

Table 8.2 4098 pin connections

Mode of operation	*Pins connected to* V_{dd}	V_{ss}	*Trigger input pin*	*Short together pins*
Leading-edge trigger	3/13	—	4/12	5–7/9–11
Leading-edge trigger and retriggerable	3–5/11–13	—	4/12	—
Trailing-edge trigger	3/13	—	5/11	4–6/10–12
Trailing-edge trigger and retriggerable	3/13	4/12	5/11	—

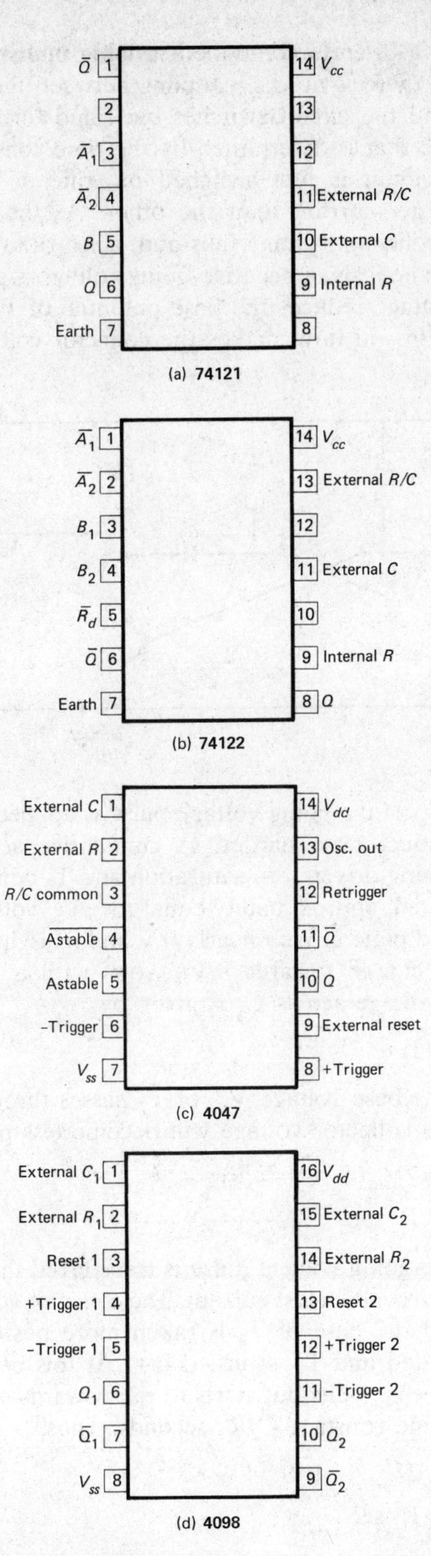

Fig. 8.25 Pin connections of four i.c. monostable multivibrators.

Astable Multivibrators

The circuit of a discrete component **astable multivibrator** is shown in Fig. 8.26. Because there is no d.c. coupling between the transistors, there is no stable state and the circuit switches back and forth between two unstable states at a rate that is determined by the time constants C_1R_2 and C_2R_3.

When the circuit is first switched on, one of the transistors, say T_1, conducts a larger current than the other. As the collector current of T_1 increases, its collector voltage falls and, since the voltage across C_1 cannot change instantaneously, a negative-going voltage is passed via C_1 to the base of T_2. This voltage reduces the base potential of T_2 and makes T_2 conduct less current. This, in turn, makes the collector voltage of T_2 become more

Fig. 8.26 Astable multivibrator.

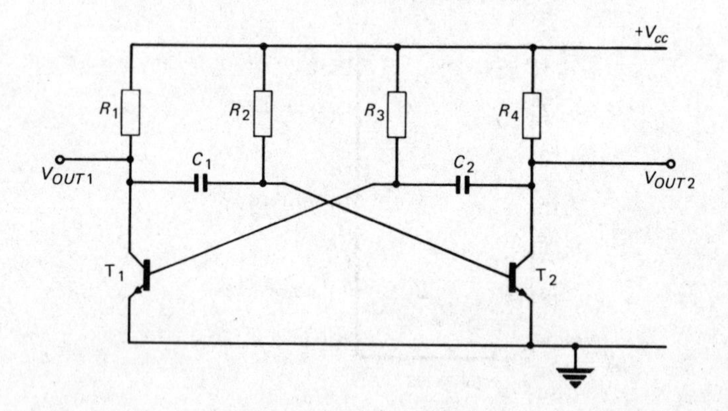

positive and a positive-going voltage pulse is applied to the base of T_1. This makes T_1 conduct even harder. A cumulative action takes place which results in T_1 being driven into saturation and T_2 being cut-off. T_1 now has a negative potential, approximately equal to $-V_{cc}$ volts, at its base, and so C_1 has its left-hand plate at very nearly 0 V and its right-hand plate at $-V_{cc}$. C_1 now starts to charge towards $+V_{cc}$ volts with a time constant of C_1R_2 seconds. The voltage across C_1 is given by

$$v_{c1} = 2V_{cc}(1-e^{-t/C_1R_2}) \tag{8.19}$$

Immediately the base voltage V_{b2} of T_2 passes through 0 V, T_2 will start to conduct and its collector voltage will become less positive. At this instant

$$v_{c1} = V_{cc} = 2V_{cc}(1-e^{-t_1/C_1R_2})$$

$$t_1 = 0.69C_1R_2 \text{ sec} \tag{8.20}$$

Now a negative-going voltage pulse is transferred through C_2 to the base of T_1 to cause T_1 to conduct less current. The collector voltage of T_1 rises towards $+V_{cc}$ volts and the base of T_2 is taken more positive and so on until T_2 becomes saturated and T_1 is turned OFF. At this point the base of T_2 is at approximately $-V_{cc}$ volts but starts to rise towards $+V_{cc}$ as the capacitor C_2 charges with time constant C_2R_3 seconds. Thus,

$$v_{c2} = V_{cc} = 2V_{cc}(1-e^{-t_2/C_2R_3})$$

$$t_2 = 0.69C_2R_3 \text{ sec} \tag{8.21}$$

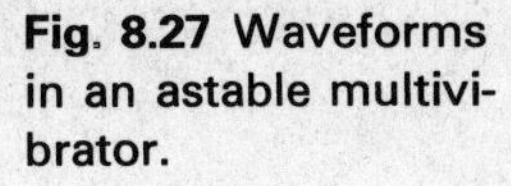

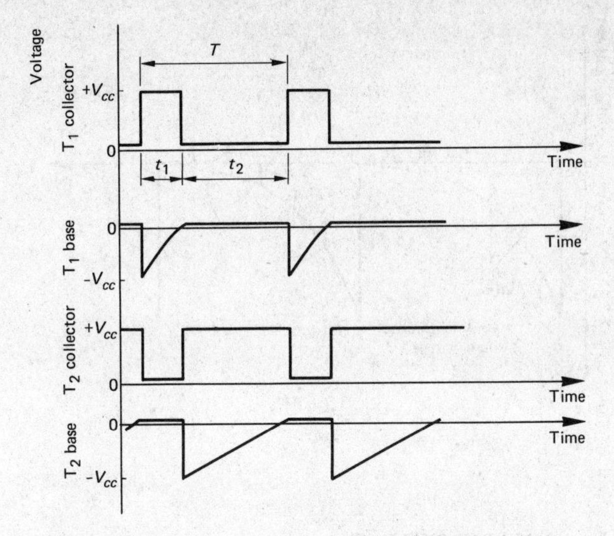

Fig. 8.27 Waveforms in an astable multivibrator.

Fig. 8.27 shows the waveforms at various points in an astable multivibrator.

The periodic time T of the output rectangular waveform is the sum of t_1 and t_2, i.e.

$$T = 0.69(C_1R_2 + C_2R_3) \text{ sec} \tag{8.22}$$

If $t_1 = t_2$ the output voltage is of square waveform with a periodic time of

$$T = 1.38CR \text{ sec} \tag{8.23}$$

Synchronization of an Astable Multivibrator

The frequency stability of an astable multivibrator depends not only on the timing components C_1, C_2, R_2, and R_3, but also upon the transistor parameters and the stability of the power supplies. The inherent frequency stability of an astable multivibrator can be improved by inserting a suitable **synchronizing signal** into the base of one of the transistors.

The synchronization pulses will be superimposed upon the normal base voltage of the transistor and will have no effect when the transistor is ON. When the transistor is OFF its base voltage is charging exponentially from $-V_{cc}$ volts towards $+V_{cc}$ volts and the transistor switches ON immediately the base voltage reaches zero. The superimposed synchronizing pulses will take the base voltage to zero at a moment in advance of the normal time and so ensure that the transistor is always switched ON at the same time. Fig. 8.28 illustrates the principle. The time taken for the transistor to switch ON is reduced from t_1 to t_2. The multivibrator adjusts its frequency so that the ratio of the frequencies of the synchronizing waveform and of the unsynchronized multivibrator is an exact ratio of a whole number. This means that the synchronized multivibrator will act as a **frequency divider**.

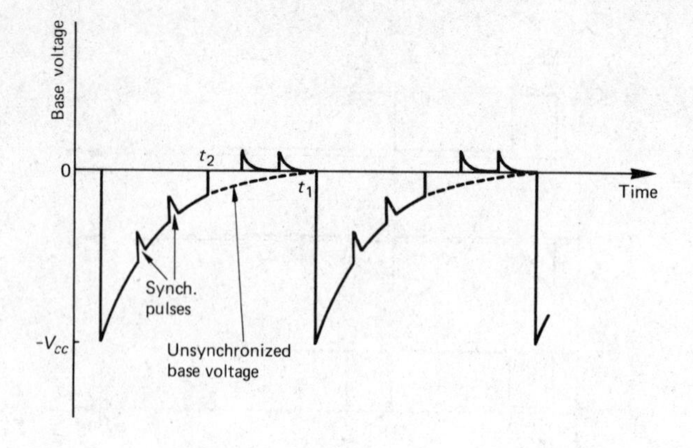

Fig. 8.28 Synchronization of an astable multivibrator.

Use of NAND/NOR Gates

Both NAND and NOR gates can be used to produce an astable multivibrator circuit; the circuit is the same for each type of gate since they are employed as invertors. Fig. 8.29*a* shows the circuit of a **NOR astable multivibrator**. If cmos devices are used, another resistor (R_2) is often connected in series with the input to gate A to protect its input protective diodes. The output of either type of gate is high when both its inputs are low, and low when both its inputs are high.

Suppose that the circuit has *just* switched into the state with the output of gate B high at V_H volts. ($V_H = 3.6$ V for t.t.l. and $V_{dd}/2$ for cmos.) The input to gate B and thus the output of gate A is low, approximately 0 V, and the input to gate A must be at the *threshold voltage* V_1 of the gate, i.e. the voltage at which the output of the gate changes state (2 V for t.t.l. and $V_{dd}/2$ for cmos). The sudden increase in gate B output voltage from 0 V to $+V_H$ volts is passed through C_1 to the input of gate A. This makes the input voltage of gate A equal to $V_1 + V_H$ volts. C_1 now has one plate at $+V_H$ volts and the other connected via R_1 to 0 V and so it commences to charge up at a rate determined by the time constant C_1R_1 seconds.

As the capacitor voltage increases, the gate A input voltage falls (see Fig. 8.29*b*). When the input voltage has fallen to the threshold value V_1, the gate switches back to its high output state and this causes gate B to change state also. The output of gate B therefore falls abruptly from $+V_H$ volts to 0 volts. This negative-going voltage pulse is transferred through C_1 to the input of gate A so that its input voltage suddenly falls from V_1 to $V_1 - V_H$ volts. Now C_1 has its right-hand plate connected to 0 V and its left-hand plate connected via R_1 to V_H volts and so it commences to charge with the opposite polarity to before. The input voltage to gate A increases towards V_1 and immediately it reaches this voltage the gate changes state and so on. The expression for the input voltage to gate A is

$$v = V_H - (V_H + V_1)e^{-t/C_1R_1}$$

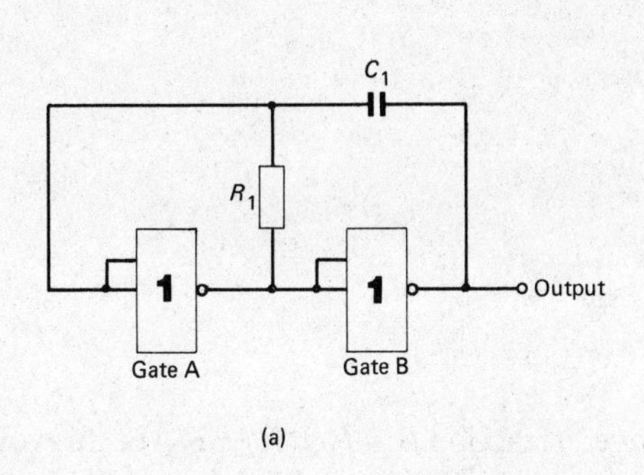

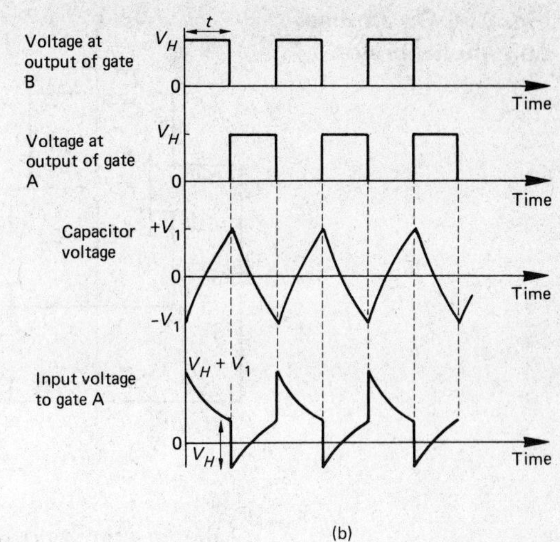

Fig. 8.29 (*a*) NOR gate astable multivibrator, (*b*) circuit waveforms.

The circuit changes state when $v = V_1$ hence

$$\frac{V_H - V_1}{V_H + V_1} = e^{-t_1/C_1R_1}$$

$$t_1 = C_1R_1 \log_e \left(\frac{V_H + V_1}{V_H - V_1}\right) \tag{8.24}$$

Therefore, the periodic time T of the output waveform is

$$T = 2t_1 = 2C_1R_1 \log_e \left(\frac{V_H + V_1}{V_H - V_1}\right) \tag{8.25}$$

For cmos gates, $V_H = V_{dd}$ and $V_1 = V_{dd}/2$ and then

$$T = 2C_1R_1 \log_e \left(\frac{3/2}{1/2}\right) = 2.2C_1R_1 \text{ secs} \tag{8.26}$$

Example 8.6

Determine the value of C_1 for a cmos NAND gate astable multivibrator which is to operate at a frequency of 15 kHz if $R_1 = 10\ \text{k}\Omega$.

Solution From equation (8.26),

$$C_1 = \frac{1}{15 \times 10^3 \times 2.2 \times 10^4} = 3\ \text{nF} \quad (Ans)$$

Operational Amplifier Astable Multivibrator

The circuit of an astable multivibrator using an **op-amp** is shown in Fig. 8.30. Positive feedback is applied to the op-amp by the potential divider $R_2 + R_3$ connected across the output terminals. Suppose the output voltage

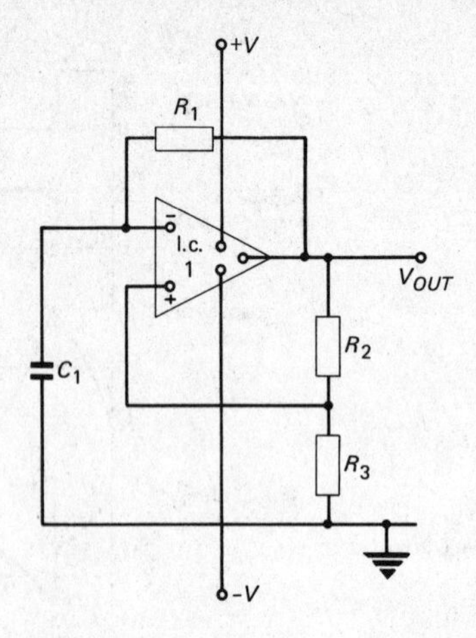

Fig. 8.30 Op-amp astable multivibrator.

of the op-amp is positive; a fraction $\beta = R_3/(R_2+R_3)$ of this voltage is fed back to the non-inverting terminal and the amplifier is rapidly driven into positive saturation. The saturated output voltage $V^+_{o(SAT)}$ is applied across the series R_1C_1 circuit and C_1 is charged at a rate determined by the time constant C_1R_1. The voltage across C_1 is applied to the inverting terminal of the op-amp and when, after a time t_1, this voltage becomes more positive than $V^+_{o(SAT)}R_3/(R_2+R_3)$, the op-amp will switch to have a negative output voltage. The positive feedback will then ensure that the op-amp is rapidly driven into its negative saturation condition. Its output voltage is then $V^-_{o(SAT)}$. Capacitor C_1 now starts to charge up with the opposite polarity to before and, when after a time t_2 seconds its voltage exceeds $V_{o(SAT)}-$ $R_3/(R_2+R_3)$, the circuit switches back to its positive saturated condition and so on.

At the moment of switching from $V^-_{o(SAT)}$ to $V^+_{o(SAT)}$, C_1 is charged to $\beta V^-_{o(SAT)}$ volts and so the voltage at the inverting terminal suddenly increases to

$$-\beta V^-_{o(SAT)} + V^+_{o(SAT)} \text{ volts}$$

Hence at time t_1

$$\beta V^+_{o(SAT)} = V^+_{o(SAT)} - (V^+_{o(SAT)} - \beta V^-_{o(SAT)})e^{-t_1/C_1R_1} \tag{8.27}$$

$$V^+_{o(SAT)}(1-\beta) = (V^+_{o(SAT)} - \beta V^-_{o(SAT)})e^{-t_1/C_1R_1}$$

$$t_1 = C_1R_1 \log_e \left[\frac{V^+_{o(SAT)} - \beta V^-_{o(SAT)}}{V^+_{o(SAT)}(1-\beta)}\right] \tag{8.28}$$

Similarly,

$$t_2 = C_1R_1 \log_e \left[\frac{V^-_{o(SAT)} - \beta V^+_{o(SAT)}}{V^-_{o(SAT)}(1-\beta)}\right] \tag{8.29}$$

The periodic time of the output waveform is $T = t_1 + t_2$. If the positive and negative saturation voltages $V^+_{o(SAT)}$ and $V^-_{o(SAT)}$ are equal then equation (8.27) can be written as

$$\beta V_{o(SAT)} = V_{o(SAT)} - V_{o(SAT)}(1+\beta)e^{-t_1/C_1R_1}$$

which corresponds to equation (8.26) with $\beta V_{o(SAT)} = V_1$, the threshold value.

Then, $1-\beta = (1+\beta)e^{-t_1/C_1R_1}$ and

$$t_1 = t_2 = C_1R_1 \log_e \left[\frac{1+\beta}{1-\beta}\right] = C_1R_1 \log_e \left[\frac{1+R_3/(R_2+R_3)}{1-R_3(R_2+R_3)}\right]$$

$$t_1 = t_2 = C_1R_1 \log_e [1+2R_3/R_2] \tag{8.30}$$

The periodic time T of the output waveform is

$$T = t_1 + t_2 = 2C_1R_1 \log_e [1+2R_3/R_2] \tag{8.31}$$

Example 8.7

The op-amp used in an astable multivibrator circuit has $V^+_{o(SAT)} = 13.5$ V and $V^-_{o(SAT)} = -12$ V. Calculate the frequency of the output waveform if $R_1 = 12$ kΩ, $R_2 = 20$ kΩ, $R_3 = 51$ kΩ and $C_1 = 0.1$ μF. Calculate also the output frequency if the output saturation voltages are assumed to be equal.

Solution

$$\beta = R_3/(R_2+R_3) = 0.72$$

From equation (8.28)

$$t_1 = 1.2\times10^{-3} \log_e \left[\frac{13.5+0.72\times12}{13.5(1-0.72)}\right] = 2.12 \text{ ms}$$

and from equation (8.29)

$$t_2 = 1.2\times10^{-3} \log_e \left[\frac{-12-0.72\times13.5}{-12(1-0.72)}\right] = 2.24 \text{ ms}$$

$$T = t_1 + t_2 = 4.36 \text{ ms}$$

Therefore the frequency $= 1/T \simeq 229$ Hz (*Ans*)

If $V^+_{o(SAT)}$ and $V^-_{o(SAT)}$ are assumed to be equal, equation (8.31) will apply. Therefore,

$$T = 2.4\times10^{-3} \log_e [1+102/20] = 4.34 \text{ ms}$$

and $f = 1/T \simeq 230$ Hz (*Ans*)

Integrated Circuit Astable Multivibrators

One of the devices quoted earlier, the cmos **4047** (page 189) can also be connected to operate as an astable multivibrator. Fig. 8.31 shows the circuit; if gated operation is desired the gate pulse should be connected to pin 5. the frequency of the output waveform is given by equation (8.32), i.e.

$$f = 1/4.4R_1C_1 \text{ Hz} \tag{8.32}$$

Fig. 8.31 4047 astable multivibrator.

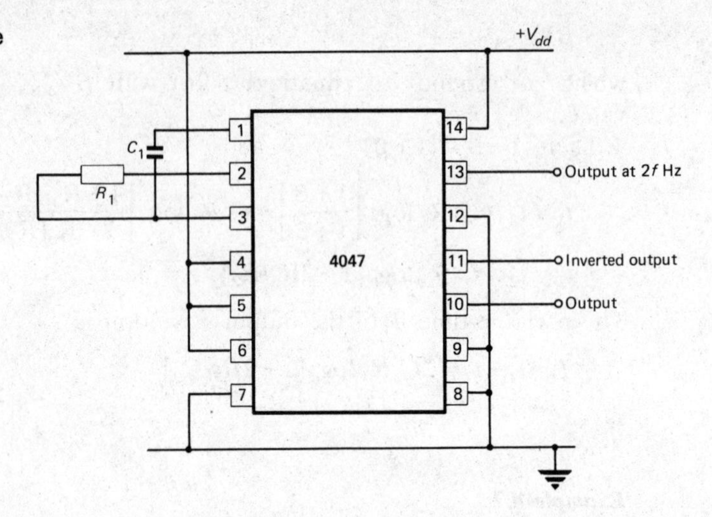

Fig. 8.32 4093 Schmitt trigger astable multivibrator.

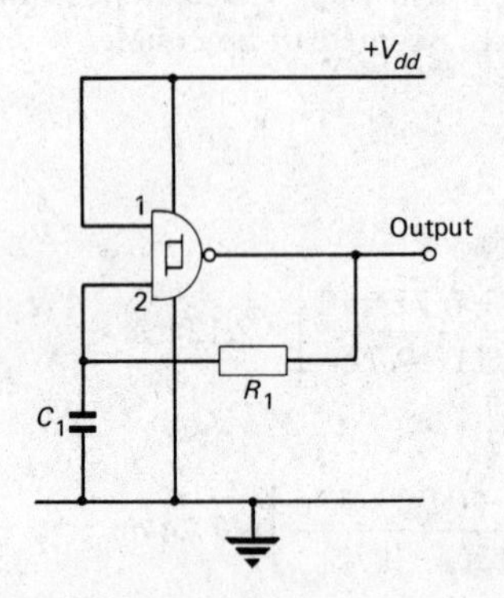

A **4093** Schmitt trigger i.c. can be connected as an astable multivibrator (Fig. 8.32). Suppose the i.c. has just switched to have its output high. This means that input 2 must be low and hence C_1 is discharged. C_1 now commences to charge at a rate determined by the time constant C_1R_1 seconds. When the voltage across the capacitor reaches the threshold voltage V_1 of the device, both of its inputs are then high and the trigger switches to its low output state. The capacitor then discharges at the same rate until its voltage reaches the value at which the circuit again switches and so on. The periodic time of the output waveform is

$$T = C_1R_1 \log_e \left[\frac{V_{dd}}{V_{dd} - V_1}\right] + C_1R_1 \log_e \left[\frac{V_{dd}}{V_{dd} - V_2}\right] \qquad (8.33)$$

where V_1 and V_2 are the threshold voltages of the Schmitt trigger.

Sawtooth and Triangle Waveform Generators

The sawtooth waveform, shown in Fig. 8.33*a*, is widely used as the timebase waveform in c.r.o.s and television receivers but it has many other applications as well. On the other hand the triangle waveform (Fig. 8.33*b*) is much less often used.

For both waveforms the most important parameters are the frequency and the linearity of the *ramp*. The linearity is generally expressed as the percentage of *slope error*, i.e.

$$\text{Slope error} = \frac{\text{Slope at start—Slope at end}}{\text{Slope at start}} \times 100\% \tag{8.34}$$

The basic principle of a **sawtooth generator** is illustrated by Fig. 8.34*a*. A capacitor C is charged through a resistor R from a constant voltage supply V and the capacitor voltage increases with increase in time according to

$$v_c = V(1 - e^{-t/CR}) \tag{8.35}$$

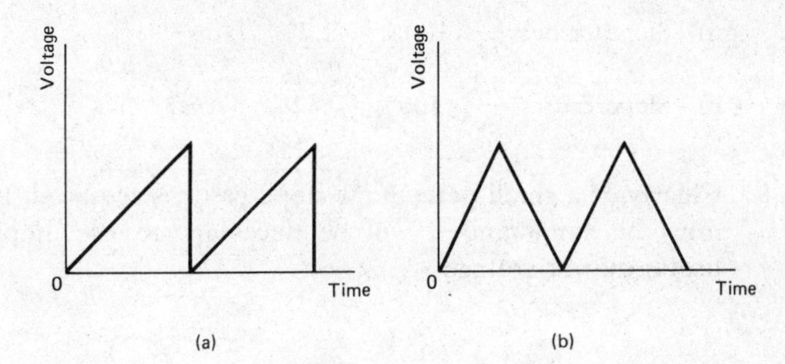

Fig. 8.33 (*a*) Sawtooth, (*b*) triangular waveforms.

Initially the slope dv_c/dt of the waveform is fairly linear but it becomes increasingly exponential as time increases. If the switch is closed to discharge the capacitor before the markedly non-linear part of the waveform is reached, and is then opened again when the capacitor voltage reaches 0 V, a sawtooth waveform can be generated (see Fig. 8.34*b*).

The departure from linearity is determined using equations (8.34) and (8.35):

$$\text{Slope} = dv_c/dt = Ve^{-t/CR}$$

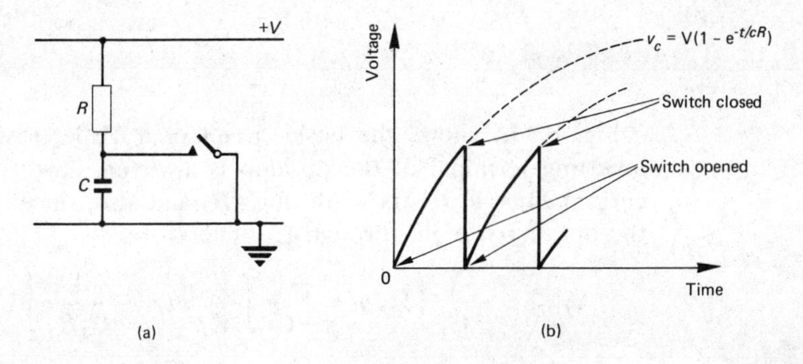

Fig. 8.34 Principle of a sawtooth waveform generator.

Thus the slope at the start of the ramp, $t=0$, is V and the slope at the end of the ramp, $t=t_1$, is $Ve^{-t_1/CR}$.

Therefore,

$$\text{Slope error} = \frac{V - Ve^{-t_1/CR}}{V} \times 100\%$$

$$= \frac{\text{Peak voltage of sawtooth}}{\text{Supply voltage}} \times 100\% \qquad (8.36)$$

This result means that for good linearity the peak voltage of the sawtooth waveform should be much smaller than the supply voltage.

Example 8.8

A simple sawtooth generator operates from a 24 V supply. Determine the percentage slope error if the peak ramp is i) 10 V and ii) 0.1 V.

Solution From equation (8.36),

i) $\text{Slope error} = \frac{10}{24} \times 100\% = 41.7\%$ (*Ans*)

ii) $\text{Slope error} = \frac{0.1}{24} \times 100\% = 0.42\%$ (*Ans*)

Clearly, if a small percentage slope error is required, the peak ramp voltage must be small and it will be necessary to use amplification to obtain a useful output voltage.

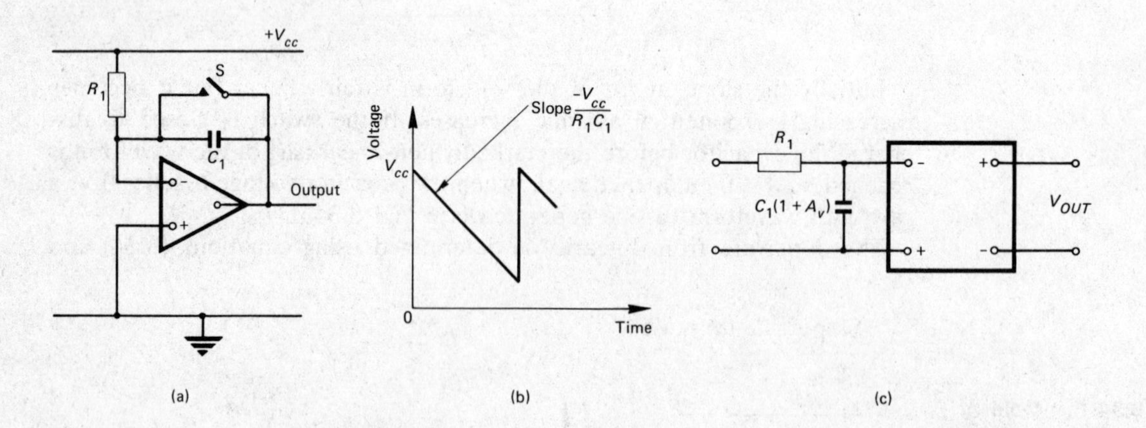

Fig. 8.35 Basic Miller sawtooth waveform generator.

Fig. 8.35*a* shows the basic circuit of a **Miller sawtooth generator.** The inverting terminal of the op-amp is a virtual earth point and so an input current $I_{IN} = V_{cc}/R_1$ flows through R_1 and also, since the input impedance of the op-amp is high, through C_1. Therefore,

$$V_{OUT} = -\frac{1}{C_1}\int I_{IN}\,dt = -\frac{1}{C_1}\int \frac{V_{cc}}{R_1}\,dt = -\frac{1}{C_1R_1}\int V_{cc}\,dt$$

i.e. the output voltage is proportional to the time integral of the input voltage. If a constant voltage is applied, as in the figure,

$$V_{OUT} = -\frac{1}{C_1R_1} \cdot V_{cc}t$$

and is a negative-going ramp function which returns to zero when the switch S is closed (Fig. 8.35*b*).

The operation of Fig. 8.35*a* can also be explained by considering the *Miller effect* discussed earlier in this book. The equivalent circuit of Fig. 8.35*a* is shown by Fig. 8.35*c*. Assuming, as in the previous paragraph, that the input impedance of the op-amp is very high, the time constant of the circuit is

$$R_1C_1(1+A_v)$$

and hence the capacitor voltage is given by

$$v_c = V_{cc}(1-e^{-t/R_1C_1(1+A_v)}) \tag{8.37}$$

and $V_{OUT} = A_v v_c = A_v V_{cc}(1-e^{-t/R_1C_1(1+A_v)})$ (8.38)

The improvement in ramp linearity can be shown by expanding equations (8.35) and (8.38), using the series $e^x = 1+x+\frac{1}{2}x^2+\text{etc}$. Thus for equation (8.35),

$$v_c = V_{OUT} = V_{cc}\left[1-\left(1-\frac{t}{C_1R_1}+\frac{t^2}{2C_1^2R_1^2}+\cdots\right)\right]$$

$$V_{OUT} = V_{cc}\left[\frac{t}{C_1R_1}-\frac{t^2}{2C_1^2R_1^2}\right] = \frac{V_{cc}t}{C_1R_1}\left[1-\frac{t}{2C_1R_1}\right] \tag{8.39}$$

and for equation (8.38)

$$V_{OUT} = A_vV_{cc}\left[1-\left(1-\frac{t}{C_1R_1(1+A_v)}+\frac{t^2}{2C_1^2R_1^2(1+A_v)^2}\right)\right]$$

$$= A_vV_{cc}\left[\frac{t}{C_1R_1(1+A_v)}-\frac{t^2}{2C_1^2R_1^2(1+A_v)^2}\right]$$

$$= \frac{A_vV_{cc}t}{C_1R_1(1+A_v)}\left[1-\frac{t}{2C_1R_1(1+A_v)}\right]$$

$$V_{OUT} \simeq \frac{V_{cc}t}{C_1R_1}\left[1-\frac{t}{2C_1R_1(1+A_v)}\right] \tag{8.40}$$

The second term in equations (8.39) and (8.40) represents the departure from linearity of the ramp waveform and can be shown to be *identical* with the results obtained by applying equation (8.34).

Example 8.9

An amplifier with an inverting gain of 400 and a very high input impedance is used as a ramp generator. The d.c. supply voltage is 12 V and the resistance and capacitance used are 12 kΩ and 100 nF respectively. Calculate i) the time taken to complete a 10 V ramp, ii) the percentage deviation from linearity at the end of the ramp.

Solution From the first term of equation (8.40),

$$10 = \frac{12t}{100 \times 10^{-9} \times 12 \times 10^{3}} \qquad t = 1\text{ ms} \quad (Ans)$$

From the second term, the percentage deviation from linearity in the ramp is

$$\frac{1 \times 10^{-3}}{2 \times 100 \times 10^{-9} \times 12 \times 10^{3} \times 401} \times 100\% = 0.105\% \quad (Ans)$$

To obtain a linear ramp, a capacitor should be charged by a constant *current.*

The charge on a capacitor is given by $Q = CV = It$ and hence, if the current is constant, the capacitor voltage is $v_c = It/C$ volts. When a capacitor is charged from a constant voltage supply, the current flowing is constant for only the beginning of the charge since, immediately the capacitor receives some charge, its voltage rises and the current flowing is reduced to $i = (V - v_c)/R$.

It is clear from the first term of equation (8.40) that the Miller integrator approximates to a constant current generator. The first term is $V_{cc}t/C_1R_1$ and, since V_{cc}/R_1 is the constant current, this can be written as It/C_1. Note that although a high gain is needed the actual *value* of the gain does not matter.

The switch S shown in the basic circuit is normally a bipolar or a field effect transistor that is turned ON and OFF to open or close the switch.

Fig. 8.36*a* shows the circuit of a bipolar transistor **Miller ramp generator** (see also Fig. 8.51). Before a trigger pulse has been applied, T_2 is OFF and the circuit can be redrawn as shown in Fig. 8.36*b*. Capacitor C_1 charges up to a voltage $V_c = V_{cc} - V_{be} \simeq V_{cc}$ volts with a time constant of C_1R_2 seconds. When a positive trigger voltage pulse is applied to the circuit to turn T_2 ON, R_2 is connected as the collector load resistor for T_1. The collector voltage for T_1 drops suddenly by an amount $A_v V_{be}$ and the negative-going voltage pulse is passed through C_1 to the base of T_1 and causes T_1 to almost turn OFF. The circuit can be redrawn to give Fig. 8.36*c*. The voltage applied across the series circuit R_1C_1 is the voltage across R_2 and this is equal to $V_{cc} - V_{OUT} = V_{cc} - V_{c(max)}$. The initial current i flowing in the circuit is

$$i = (V_{cc} - V_{c(max)})/R_1$$

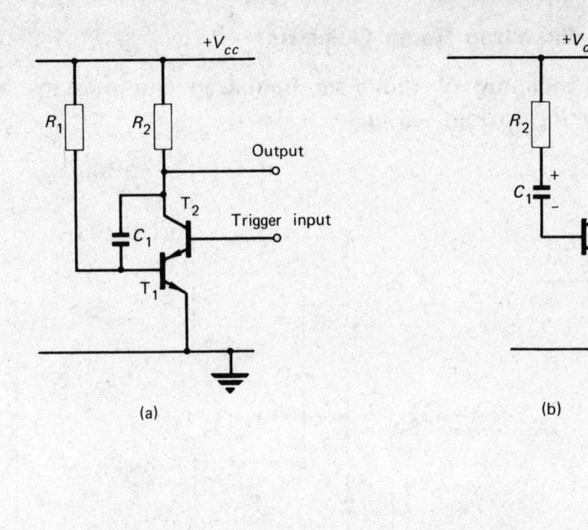

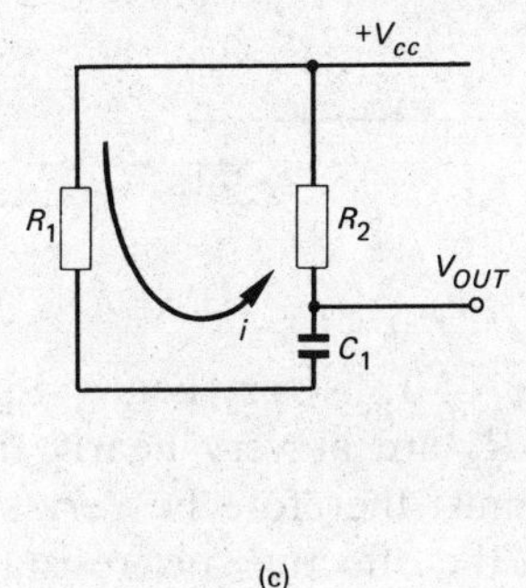

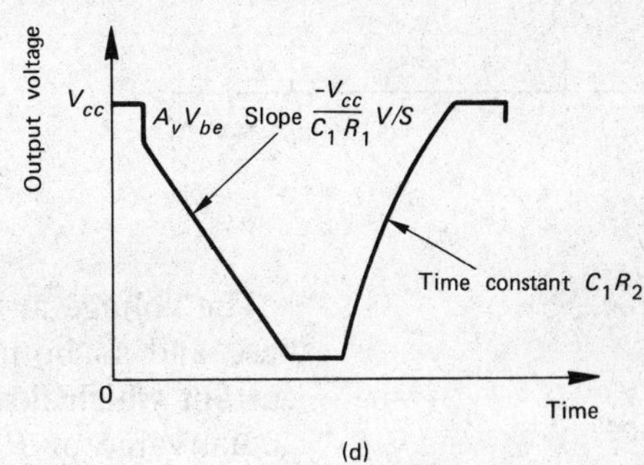

Fig. 8.36 Bipolar transistor Miller ramp generator.

and is the discharge current of the capacitor and as the capacitor voltage falls the effective voltage across the circuit $V_{cc} - v_c$ remains constant and so a constant current $i = V_{cc}/R_1$ flows. This means that V_{OUT} falls at the constant rate of

$$it/c = V_{cc}/R_1C_1 \text{ V/s}$$

as shown by Fig. 8.36*d*. The minimum voltage is determined by the p.d. across T_1 and T_2. The output voltage will remain at this value until the trigger voltage is removed and will then recover, with a time constant of C_1R_2, to the starting state of the ramp, i.e. $V_{cc} - V_{be}$ volts.

An alternative Miller circuit is given in Fig. 8.52. For either circuit the use of a fet as the amplifying device will result in a bigger initial drop in the output voltage.

The Bootstrap Ramp Generator

The meaning of the term **bootstrap** can perhaps best be illustrated by referring to Fig. 8.37*a*.

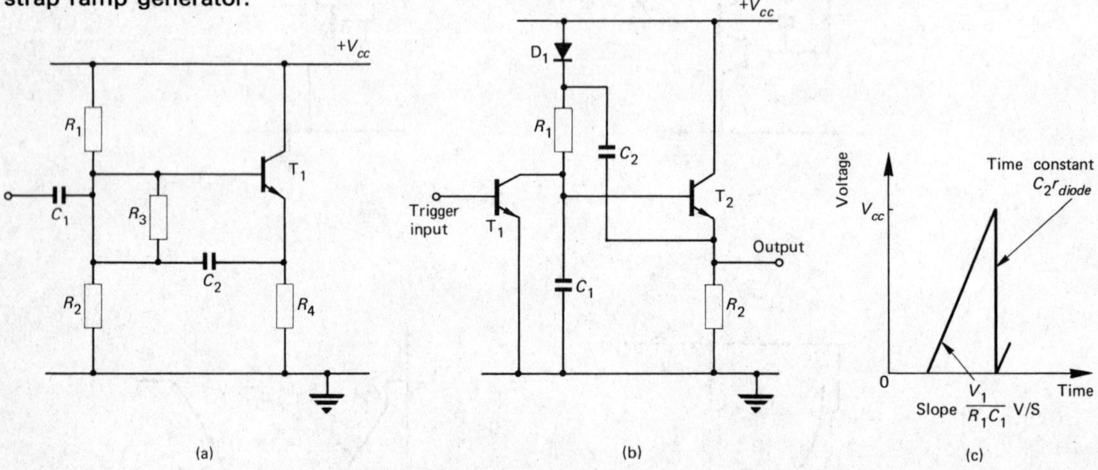

Fig. 8.37 The bootstrap ramp generator.

The voltage at the emitter of T_1 is very nearly equal to the voltage at its base and so both ends of R_3 are at very nearly the same potential. The current which flows in R_3 must therefore be very small—regardless of the actual value of R_3—and so the effective a.c. resistance of R_3 is very high. This principle is used in the bootstrap ramp generator which possesses the advantage of producing a ramp voltage which starts from 0 V.

The circuit of a **bootstrap ramp generator** is shown in Fig. 8.37*b*. For the voltage across C_1 and hence the output voltage (T_2 is connected as an emitter follower) to rise linearly it must be charged by a constant current. This can be achieved if a *constant* voltage can be maintained across resistor R_2. Then the constant charging current is equal to

$$\frac{\text{Voltage across } R_2}{R_2}$$

Suppose that a positive trigger voltage has been applied to the circuit to turn T_1 ON and so discharge capacitor C_1. The base voltage of T_2 is then 0 V and this transistor is OFF. The emitter voltage of T_2 is also 0 V and capacitor C_2 is charged to V_{cc} volts minus the small voltage drop across the diode D_1.

When the trigger terminal is taken to 0 V, T_1 is turned OFF and C_1 is able to commence charging towards V_{cc} volts. Because T_2 is connected as an emitter follower its emitter voltage *follows* the voltage across C_1 and this charge is transferred via C_2 to the junction of D_1 and R_1. D_1 turns OFF and the potential difference across R_1 is maintained at V_{cc} volts. R_1 acts like a constant current generator to supply a constant current V_{cc}/R_1 to capacitor C_1. Then

$$V_{cc}/R_1 = C_1\frac{dv_{c1}}{dt} = C_1\frac{dV_{OUT}}{dt}$$

$$\frac{dV_{OUT}}{dt} = \frac{V_{cc}}{C_1R_1}\ \text{V/s} \qquad (8.41)$$

Thus the output voltage rises from zero at the constant rate of V_{cc}/C_1R_1 volts/sec (Fig. 8.37*c*). The ramp will not be exactly linear because *a*) the emitter follower will have less than unity gain, *b*) some of the current "supplied" by R_1 will flow into the base of T_2, *c*) C_2 is not of infinite capacitance and will therefore discharge slightly during the sweep.

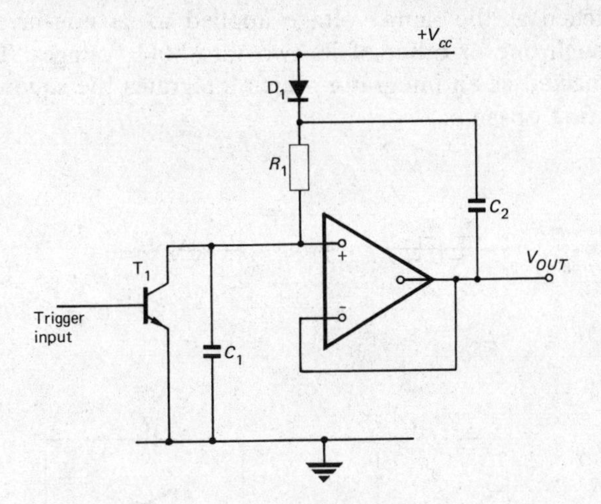

Fig. 8.38 Op-amp bootstrap ramp generator.

The maximum amplitude of the ramp is limited to V_{cc} volts since, immediately the ramp voltage exceeds this value, the collector/base junction of T_2 will become forward biased and put C_1 in parallel with the collector supply voltage. At the end of the sweep the trigger voltage turns T_1 ON again to discharge C_1. Recovery to the starting state of the ramp occupies the time needed for the voltage across C_2 to return to V_{cc} volts and this is determined by the time constant C_2r_{diode}.

The circuit of an op-amp bootstrap sawtooth generator is shown in Fig. 8.38. Comparison with Fig. 8.37*b* shows that an op-amp voltage follower has replaced the emitter follower T_1.

Example 8.10

A bootstrap ramp generator has the following component values: $R_2 = 82\ \text{k}\Omega$, $C_1 = 0.01\ \mu\text{F}$ and $C_2 = 10\ \mu\text{F}$. The supply voltage is 12 V and the trigger pulse has a duration of 60 μs. Calculate *a*) the amplitude of the ramp, *b*) the frequency of the ramp waveform.

Solution

a) From equation (8.41)

$$V_{OUT} = \frac{12\times 60\times 10^{-6}}{82\times 10^3\times 10^{-8}} = 0.88\ \text{V} \quad (\textit{Ans})$$

b) The ramp occurs in 60 μs, hence it has a rate of change of 0.88 V/60 μs or 14.67 kV/s. The frequency of the ramp is

$$f = 1/T = 10^6/60 = 16.67\ \text{kHz} \quad (Ans)$$

Triangular Waveform Generation

A **triangular waveform** is the time integral of a square waveform and so the easiest way of generating a triangular waveform is to use the method shown in Fig. 8.39.

An op-amp version of this is given in Fig. 8.40.

The first op-amp is connected as a Schmitt trigger whose output is switched as the ramp voltage applied to its non-inverting terminal passes through one or other of its two threshold voltages. The second op-amp is connected as an integrator which integrates the square waveform output of the first op-amp.

Fig. 8.39 Method of producing a triangular waveform.

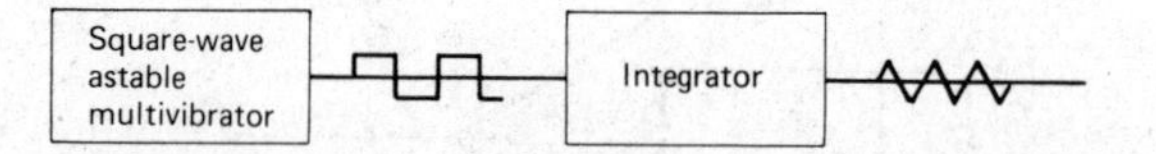

Fig. 8.40 Op-amp triangular waveform generator.

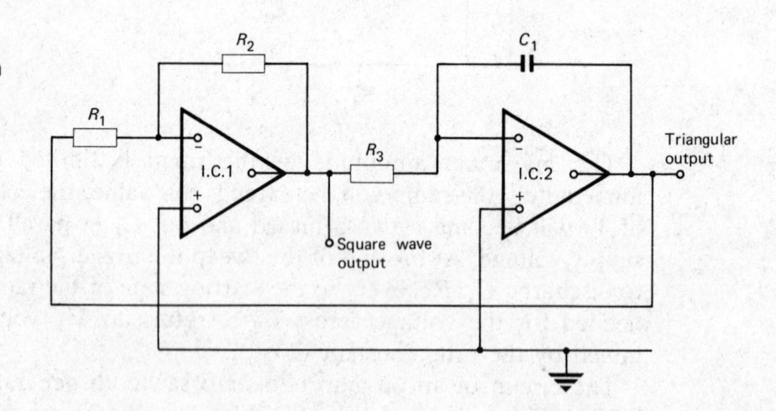

Suppose that the Schmitt trigger output is at its positive saturation value $V^+_{o(SAT)}$, then the output of the integrator is a falling ramp waveform V_{ramp}. Then the voltage V_+ at the non-inverting input to the trigger is

$$V_+ = \frac{V^+_{o(SAT)} R_1}{R_1 + R_2} + \frac{-V_{ramp} R_2}{R_1 + R_2}$$

The trigger switches when $V_+ = 0$ and so

$$V^-_{ramp} = \frac{-V^+_{o(SAT)} R_1}{R_2}$$

Similarly, on the rising half of the triangular output waveform

$$V^+_{ramp} = \frac{V^-_{o(SAT)} R_1}{R_2}$$

and the peak-peak triangular voltage is

$$V^+_{ramp} - V^-_{ramp} = 2V_{o(SAT)} R_1/R_2$$

assuming equal saturation voltages.

The current charging C_1 is

$$I = -\frac{V_{o(SAT)}}{R_3} = \frac{C_1 dv_c}{dt} = -\frac{C_1 dV_{ramp}}{dt}$$

Therefore,

$$\frac{+V_{o(SAT)}}{C_1 R_3} = \frac{dV_{ramp}}{dt} = \text{rate at which the ramp voltage increases (or decreases)}$$

The time for the ramp to be completed is

$$t_1 = \frac{\text{Peak-peak ramp voltage}}{\text{Ramp rate}}$$

$$= \frac{2V_{o(SAT)} R_1/R_2}{V_{o(SAT)}/C_1 R_3} = \frac{2R_1 R_3 C_1}{R_2}$$

The negative-going ramp voltage occupies an equal time t_2 and hence the periodic time $T = t_1 + t_2$ of the triangular waveform is

$$T = 4R_1 R_3 C_1/R_2$$

Therefore frequency $= 1/T = R_2/4R_1 R_3 C_1$ Hz (8.42)

In some instances, for example television tube deflection circuits, a linear ramp of *current* must be passed through an inductance. Since the back e.m.f. developed across an inductance is given by $e = -L\,di/dt$ the voltage required to drive this current will *not* have a ramp waveform.

Suppose that initially a pure inductance, i.e. with zero resistance, is considered. During the ramp the current in the inductance is changing at the rate of

$$I_{max}/(\text{time for ramp})$$

where I_{max} is the peak value of the ramp current. During the flyback period

$$\frac{di}{dt} = \frac{I_{max}}{\text{Time for flyback}}$$

With an inductance of L this will require an applied e.m.f. of

$$\frac{LI_{max}}{\text{Time for ramp}} \quad \text{and} \quad \frac{LI_{max}}{\text{Time for flyback}}$$

respectively, and the current and voltage waveforms are as shown by Fig. 8.41.

Any practical inductance possesses self-resistance which is effectively connected in series with the inductance and a voltage must be applied to overcome the voltage developed by the current ramp in this resistance. This means that the total voltage applied across the inductance must be the sum of that shown in Fig. 8.41*a* and a sawtooth voltage, and the required waveform is shown in Fig. 8.41*b*.

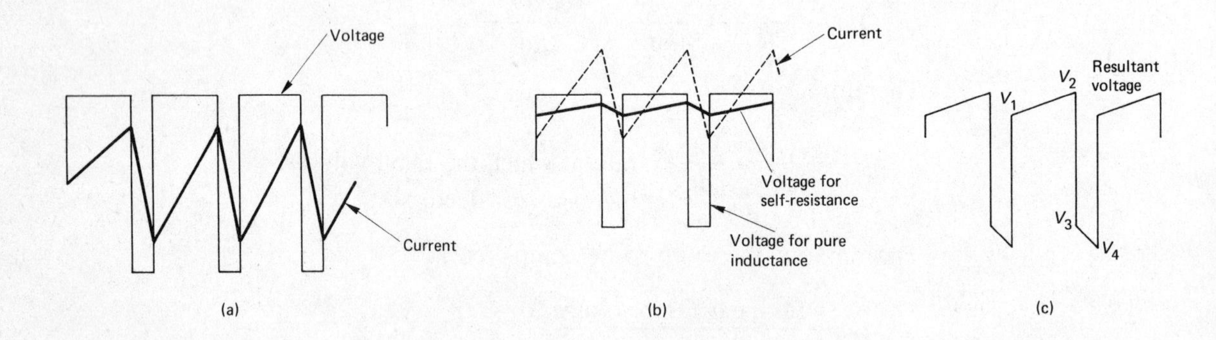

Fig. 8.41 Showing the voltage needed to drive a linear ramp of current through an inductance.

To determine the required ratio of step to slope for a linear current ramp the equation for the voltage applied across the inductance is written down;

$$V = L\frac{di}{dt} + Ri \qquad \text{and taking transforms}$$

$$V(s) = i(s)(sL + R)$$

It is required that the current flowing in the inductor has a sawtooth waveform, i.e. $i(s) = I/s^2$ where I is the rate of change of the current. Therefore

$$V(s) = \frac{I}{s^2}(sL + R) = \frac{IL}{s} + \frac{IR}{s^2}$$

and

$$V(t) = \underset{\text{(step)}}{IL} + \underset{\text{(ramp)}}{IRt} \qquad (8.43)$$

The required ratio of step to slope is

$$IL/IRt = L/Rt$$

where t is the time for which the ramp lasts.

Example 8.11

An inductor has an inductance of 100 mH and a resistance of 150 Ω. A sawtooth current of peak-to-peak amplitude 200 mA, frequency 15 625 Hz and flyback time 12 μs is to be passed through the inductor. Determine the waveform of the required voltage across the inductor.

Solution

$$f = 15\,625 \text{ Hz} \qquad T = 1/15\,625 = 64\ \mu\text{s}$$

Therefore, sweep time = 52 μs and flyback time = 12 μs.
During sweep

$$L\frac{di}{dt} = \frac{100\times10^{-3}\times0.2}{52\times10^{-6}} = 385 \text{ V}$$

During flyback

$$-L\frac{di}{dt} = \frac{100\times10^{-3}\times0.2}{12\times10^{-6}} = -1667 \text{ V}$$

Also, the sawtooth voltage component needed to overcome the coil resistance has a peak-peak value of $0.2\times150 = 30$ V.

Hence, referring to Fig. 8.41*c*

$$V_1 = +385 \text{ V} \qquad V_2 = 415 \text{ V} \qquad V_3 = -1667 \text{ V} \qquad V_4 = -1637 \text{ V}$$

The frequency quoted in Example 8.11 is the line frequency of a television receiver, the field frequency is only 50 Hz and the required field timebase voltage waveform is of almost sawtooth waveform.

Integrated Circuit Waveform Generators

A number of integrated circuits are on the market which are capable of waveform generation and two of these, the 555/6 timer and the Intersil 8038, are particularly useful and commonly employed.

The 555 Timer

The 555 timer was originally introduced by Signetics but it is now available from several other manufacturers. The device can be used for various timing purposes, producing accurate timing periods from a few microseconds to hundreds of seconds. It can also be connected to operate as a monostable or astable multivibrator or as a Schmitt trigger. The pin connections of the 8 and 14 pin d.i.l. packages are shown in Figs. 8.42*a* and *b* respectively while Fig. 8.42*c* shows internal block diagram of the timer.

Essentially the i.c. contains two op-amps both of which are connected as voltage comparators, one SR flip-flop, an output amplifier and a separate transistor. Also provided are three equal value resistors, labelled *R*, which are connected between the positive supply voltage (at pin 8 or 14) and earth. The upper comparator therefore has its inverting terminal held at $\frac{2}{3}V_{cc}$ volts while $\frac{1}{3}V_{cc}$ appears at the inverting terminal of the lower comparator. The outputs of the two comparators are connected, respectively, to the S and R inputs of the SR flip-flop. In turn, the output of the flip-flop is connected to both the base of the transistor and the input to the output amplifier. Lastly, the flip-flop can be reset by the application of the appropriate signal to pin 4 (or 6).

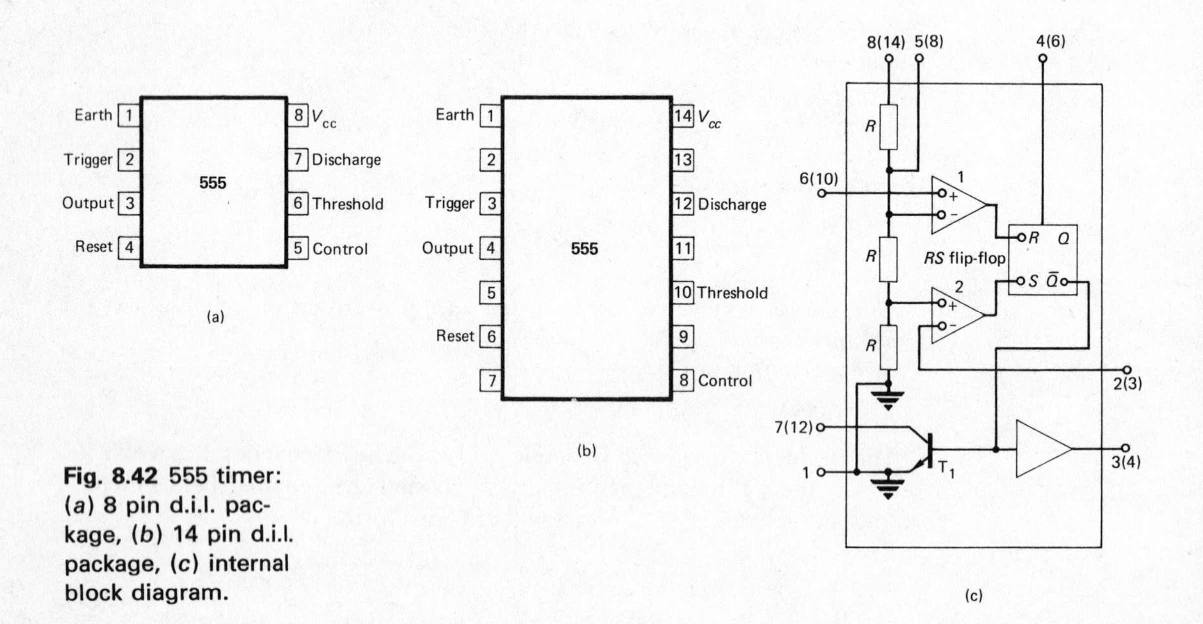

Fig. 8.42 555 timer: (*a*) 8 pin d.i.l. package, (*b*) 14 pin d.i.l. package, (*c*) internal block diagram.

The 555 as a Monostable Multivibrator

For the timer to operate as a **monostable multivibrator** it must be connected as shown by Fig. 8.43*a*. In the stable state, pin 2 is high, more positive than $V_{cc}/3$, so that the output of comparator 2 is low and resets the flip-flop. Then $\bar{Q}$ is at logical 1 and transistor T_1 is turned ON so that the output terminal 3 is low. How low depends upon the load current but is typically about 0.1 V. Pin 7 will be at very nearly 0 V and C_4 will be discharged. The output of comparator 1 is low since its inverting terminal is at $2V_{cc}/3$ volts. When a negative trigger pulse is applied to pin 2 to take the inverting terminal of comparator 2 below $V_{cc}/3$, the output of comparator 2 goes high and sets the flip-flop, so that its $\bar{Q}$ output goes low. Now transistor T_1 is turned OFF and the output voltage at pin 3 goes high. With the transistor OFF, C_4 is able to charge towards the supply voltage V_{cc} with a time constant of C_4R_2 seconds. When the capacitor voltage reaches $2V_{cc}/3$ volts comparator 1 has its + terminal more positive than its − terminal and switches to have a high output (1.5 to 2.5 V). This resets the flip-flop so that $\bar{Q} = 1$ and the transistor T_1 again turns ON and pin 3 goes low. Thus, the output of the circuit is a positive pulse whose duration is equal to the time interval required for the capacitor to charge to $2V_{cc}/3$ volts. Therefore,

$$2V_{cc}/3 = V_{cc}(1 - e^{-t/R_2C_4}) \qquad e^{-t/R_2C_4} = 1/3$$

$$t = C_4R_2 \log_e 3 = 1.0986C_4R_2 \simeq 1.1C_4R_2 \text{ sec} \tag{8.44}$$

Typical waveforms are shown in Fig. 8.43*b*.

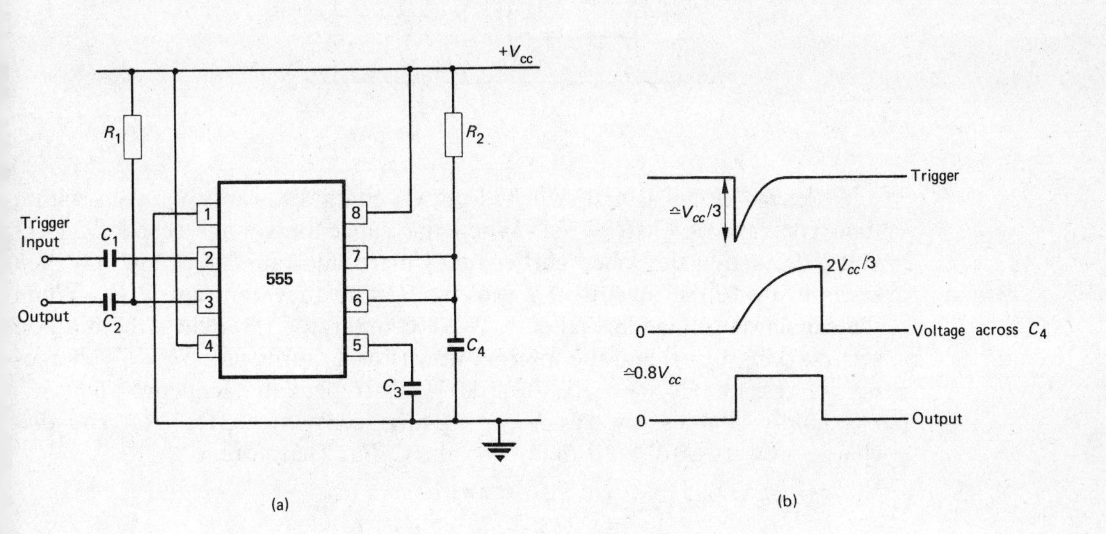

Fig. 8.43 555 monostable multivibrator.

Once the circuit has been triggered it will not respond to any other pulses that may appear at the trigger input terminal until the timing period, determined by C_4R_2, has been completed. The reset terminal allows C_4 to be discharged prematurely to interrupt the timing cycle and return the output to zero. As long as pin 4 is low, $\bar{Q} = 1$ and T_1 is ON so that C_4 cannot charge. When this facility is not wanted, pin 4 must be connected to the power supply line V_{cc}. The remaining pin, 5, is the control terminal and a voltage applied to this pin will vary the timing period and hence the pulse width by modifying the d.c. voltages set up by the three resistors. When this facility is not wanted pin 5 should be connected to earth via a capacitor of about 0.01 μF.

The 555 as an Astable Multivibrator

Fig. 8.44 shows how the 555 i.c. can be connected as an **astable multivibrator**. The connections differ from the monostable case only in that *a*) the trigger terminal 2 is now connected to the threshold terminal 6 and *b*) terminals 6 and 7 are no longer connected together.

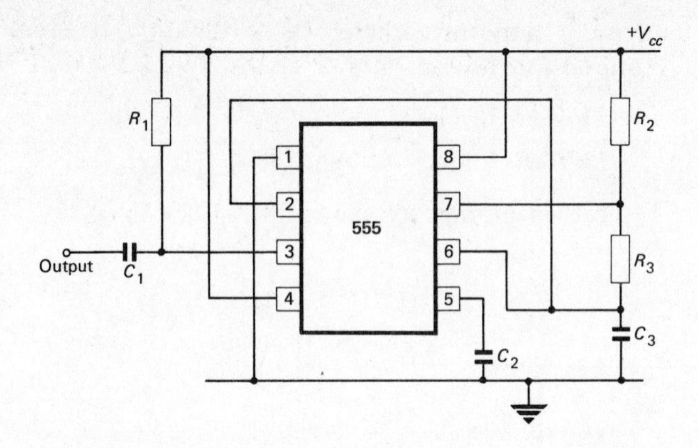

Fig. 8.44 555 astable multivibrator.

When the circuit is first switched on, C_3 charges towards V_{cc} volts with a time constant of $C_3(R_3+R_2)$. When the capacitor voltage reaches $2V_{cc}/3$ volts, the action described earlier takes place and pin 7 goes low. C_3 then starts to discharge towards 0 V into pin 7, with time constant C_3R_3. When the capacitor voltage has fallen to $V_{cc}/3$ comparator 1 switches, the flip-flop sets marking $\bar{Q}=0$ and the internal transistor T_1 turns OFF. Now C_3 charges up via resistors R_2 and R_3 towards V_{cc} to repeat the sequence. Thus C_3 alternately charges towards V_{cc} with time constant $C_3(R_2+R_3)$ and discharges towards 0 V with time constant C_3R_3. Therefore,

$$\tfrac{2}{3}V_{cc} = V_{cc} - (V_{cc} - V_{cc}/3)e^{-t_1/C_3(R_2+R_3)}$$
$$= V_{cc}(1 - \tfrac{2}{3}e^{-t_1/C_3(R_2+R_3)})$$
$$\tfrac{2}{3} = 1 - \tfrac{2}{3}e^{-t_1/C_3(R_2+R_3)}$$
$$\tfrac{1}{2} = e^{-t_1/C_3(R_2+R_3)}$$
$$t_1 = 0.69C_3(R_2+R_3)$$

Also, during the discharge period,

$$\tfrac{1}{3}V_{cc} = \tfrac{2}{3}V_{cc}e^{-t_2/C_3R_3} \qquad \tfrac{1}{2} = e^{-t_2/C_3R_3}$$
$$t_2 = 0.69C_3R_3$$

The periodic time T of the output waveform is

$$T = t_1 + t_2 = 0.69C_3(R_2+2R_3) \tag{8.45}$$

Hence the frequency f of the output waveform is

$$f = 1/T = 1/0.69C_3(R_2+2R_3) \tag{8.46}$$

If $R_3 \gg R_2$, then $f \simeq 1/1.4C_3R_3$ (8.47)

and the output waveform is very nearly square.

The 555 astable multivibrator has three particular advantages over most other types:

a) A wide frequency range can be covered with a single variable resistor control.

b) It can provide a large output current of up to 200 mA.
c) It can easily be modulated by applying a modulating signal to pin 5.

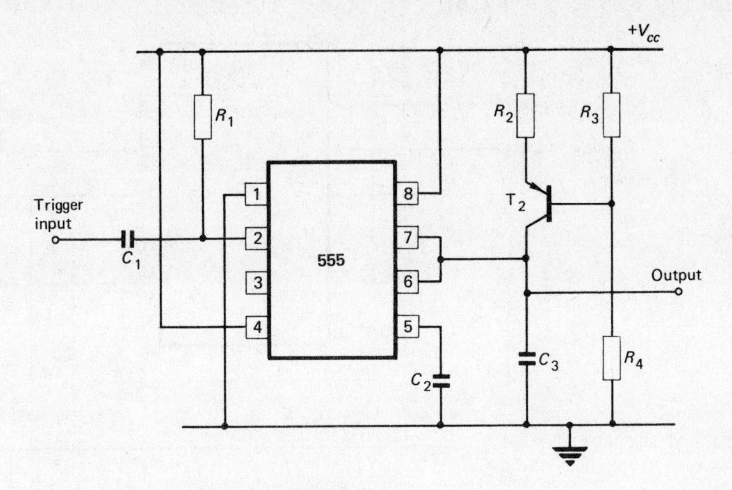

Fig. 8.45 555 ramp generator.

The 555 as a Ramp Generator

If the monostable multivibrator circuit (Fig. 8.43) is modified by adding an external resistor as shown by Fig. 8.45, a sawtooth or **ramp generator** can be produced. The operation of the circuit is similar to that of the monostable multivibrator except that capacitor C_3 is now charged by the *constant* current produced by the transistor T_2. This means that the capacitor voltage rises linearly with time until pin 7 goes low and flyback is initiated. C_3 still charges from 0 V to $2V_{cc}/3$ and the time for a ramp is $\frac{2}{3}V_{cc}/i$ where i is the constant current provided by T_2.

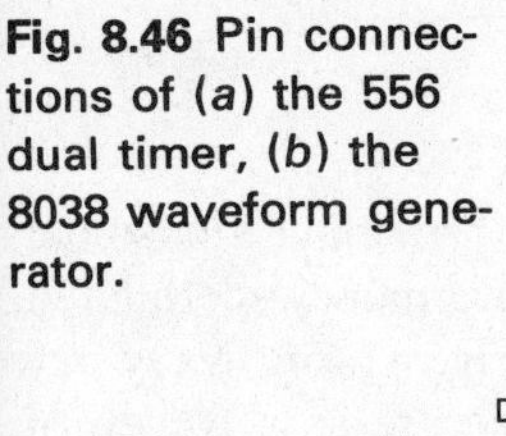

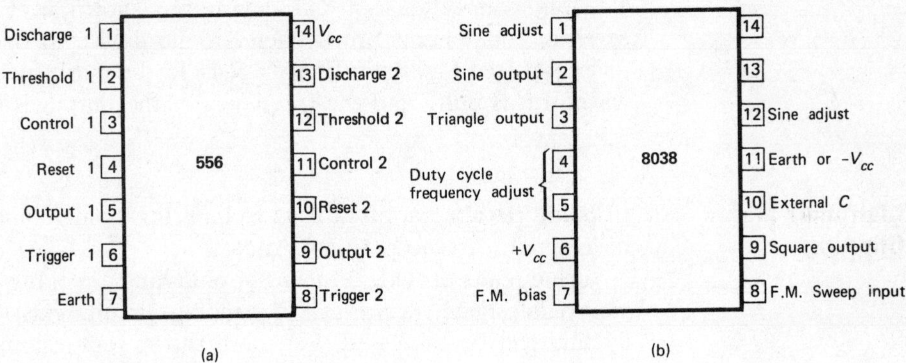

Fig. 8.46 Pin connections of (*a*) the 556 dual timer, (*b*) the 8038 waveform generator.

556 and CMOS 555

The 556 i.c. consists of two 555 timers in the one package and Fig. 8.46*a* gives its pin connections. Also available are a number of other timers but perhaps the most important is the cmos version of the 555 and the 556, labelled the 7 555/6. The cmos devices have identical pin connections to the

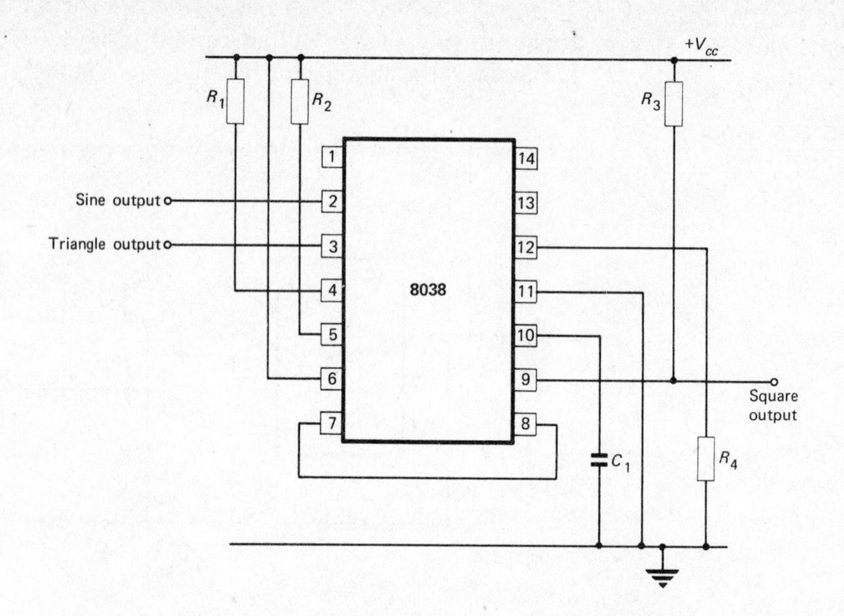

Fig. 8.47 8038 connected to deliver sinusoidal, square and triangular output waveforms.

bipolar versions and offer the following advantages: *a*) a much smaller current is taken from the power supply, *b*) a higher input impedance, and *c*) it is much less prone to interference resulting from transients in the power supply.

The 8 038 I.C.

The 8 038 is an integrated circuit that has outputs providing sinusoidal, square, and triangular waveforms and which can be frequency-modulated. The pin connections of the device are shown in Fig. 8.46*b*. Very few external components are needed for the device to operate as a waveform generator (see Fig. 8.47). If $R_1 = R_2 = R$, the mark/space ratio of the output waveform is unity and the frequency of the output is $0.15/RC_1$ Hz.

Clamping and Clipping

Clipping circuits are often used to limit the positive and/or negative peak(s) of a signal to some required value.

Such circuits provide a means of obtaining a rectangular waveform from a sinusoidal signal. A variety of clipping circuits exist but here only diode circuits will be discussed. Fig. 8.48 shows four examples of diode clipper circuits together with the output waveforms to be expected when the input is of sinusoidal waveform. Referring to circuits (*a*) and (*b*): the diode D_1 is held OFF by the d.c. bias voltage until the input voltage exceeds V_{dc} when the diode turns ON. Once the diode is ON, the output voltage remains constant at V_{dc} volts until such time as the input voltage falls below V_{dc} and allows the diode to turn OFF. In the case of circuit (*c*), the diode is always ON and the output voltage is constant at $-V_{dc}$ volts except when the input

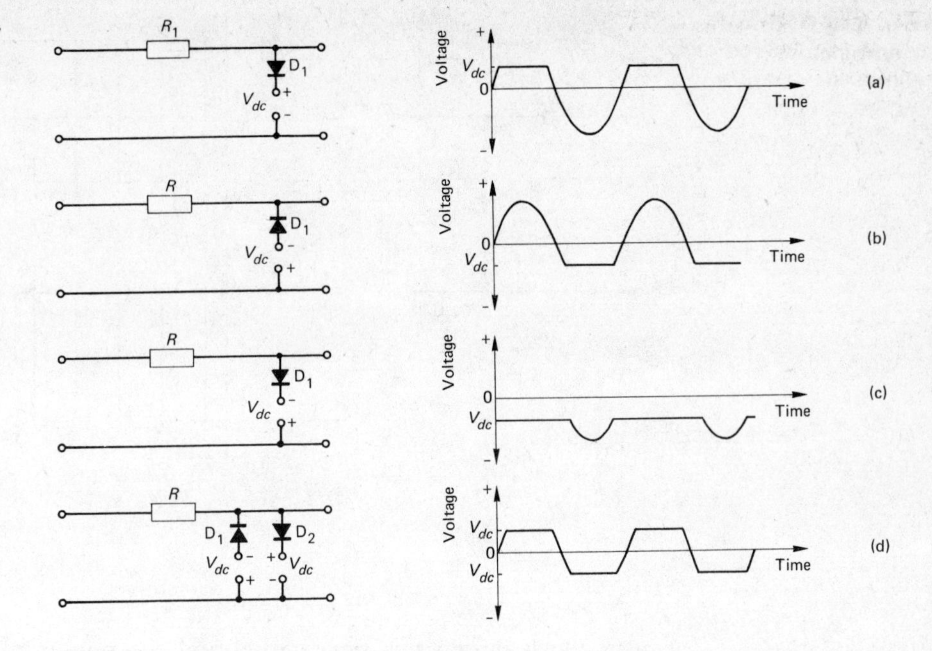

Fig. 8.48 Diode clipping circuit.

voltage is more negative. Finally, circuit (*d*) is a combination of circuits (*a*) and (*b*); it can be seen that the input sinusoidal voltage has been turned into an approximation of a square wave. Further amplification and clipping of this waveform would produce a much better approach to a square waveshape. Sometimes the diode is of the Zener type. Clipping circuits are used to remove noise voltages superimposed upon a wanted signal and for the improvement of pulse waveforms.

Whenever a waveform is passed through a capacitor or a transformer, its d.c. component is lost. For many applications the loss of the d.c. component cannot be tolerated and in such cases some means of establishing a d.c. level is necessary. A **clamping circuit** is shown in Fig. 8.49*a*. When a positive voltage is applied to the circuit, diode D_1 is turned ON and connects a very low resistance path across the output terminals. Conversely, when a negative input voltage is applied, D_1 is OFF and the input voltage appears at the output terminals. This means that any input alternating waveform will have its positive peaks *clamped* to earth potential.

Suppose that a rectangular waveform (Fig. 8.49*b*) is applied to the circuit. The first positive voltage pulse will turn D_1 ON and rapidly charge C_1 (time constant $C_1 r_{diode(ON)}$). The output voltage is approximately 0 V. At the end of this positive pulse, the input waveform suddenly changes to its negative voltage, a change of -2 V volts. The voltage across a capacitor cannot change instantaneously and so the right-hand terminal of C_1 must also experience a voltage change of -2 V volts. Capacitor C_1 now commences to discharge through the diode but, since the *off-resistance* of D_1 is high, the discharge time constant is long. Capacitor C_1 does not therefore discharge

Fig. 8.49 D. C. clamping.

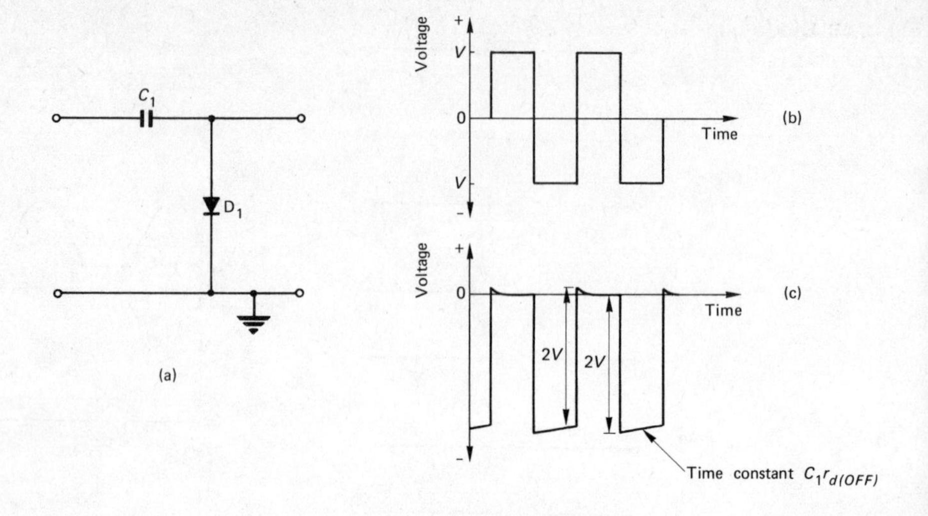

very much during this time period and the output voltage of the circuit remains at −2 V. When the next positive pulse arrives, the input voltage rises positively by 2 V and so the output voltage becomes 0 V and so on (see Fig. 8.49*c*).

A reversal in the polarity of the diode D_1 will result in the negative peaks of the waveform being clamped to 0 V.

Example 8.12

Draw the output waveform of the circuit shown in Fig. 8.50*a* if the diode has an ON resistance of 100 Ω and an OFF resistance of 200 kΩ.

Solution When the input is positive the circuit time constant is

$$0.1\times 10^{-6}\times 2\times 10^{5} = 20\times 10^{-3}\ \text{s}.$$

When the input is negative the time constant is

$$0.1\times 10^{-6}\times 100 = 1\ \mu\text{s}$$

Hence the output waveform is as shown in Fig. 8.50*b*.

Fig. 8.50

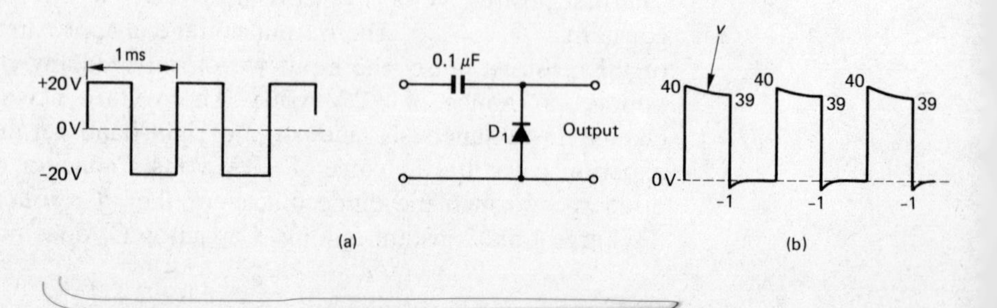

Exercises 8

8.1 A bistable multivibrator uses npn transistors having $V_{BE}=0.6$ V, $V_{CE(SAT)}=0.1$ V, $h_{FE}=100$ and $h_{ie}=1000\ \Omega$. The power supply voltages are $V_{cc}=+10$ V and $V_{BB}=-2$ V. The collector load resistors are each 2.2 kΩ and the four coupling resistors are all 56 kΩ. Calculate the stable-state bias conditions and the loop gain.

Fig. 8.51

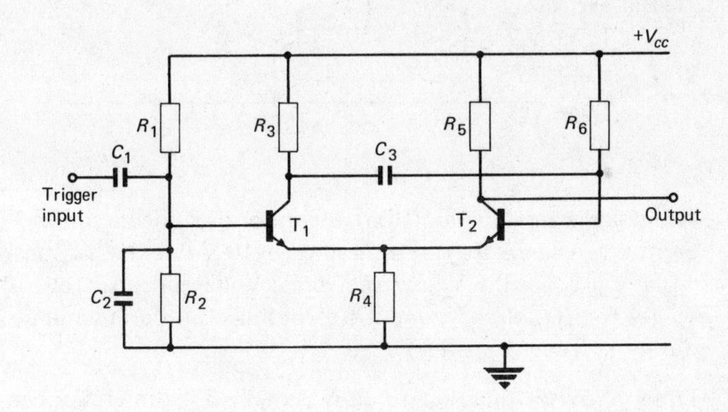

8.2 Fig. 8.51 shows the circuit of an emitter-coupled monostable multivibrator. Explain the operation of the circuit and derive an expression for the time the circuit will remain in its unstable condition.

8.3 An op-amp astable multivibrator has saturated output voltages of ±12 V. Calculate its frequency of oscillation if $R_1=100$ kΩ, $R_2=33$ kΩ, $R_3=47$ kΩ and $C_1=0.01\ \mu$F. Repeat the calculation if the saturated output voltages are +12 V and −10 V. Derive any expressions used.

8.4 Draw the circuit of an i.c. Schmitt trigger connected to operate as an astable multivibrator and show that the periodic time of its output waveform is given by

$$T=2C_1R_1\log_e[V_{dd}/(V_{dd}-V_1)]$$

8.5 Describe the operation of a Schmitt trigger circuit and draw the operating characteristics.

An op-amp of voltage gain 10^4 is connected as a Schmitt trigger with feedback resistors $R_1=100\ \Omega$ and $R_2=10$ kΩ. The output saturation voltages are $V^+_{o(SAT)}=12$ V and $V^-_{o(SAT)}=-11$ V. Determine the approximate voltage levels at which the input signal causes switching to be initiated.

8.6 Fig. 8.52 shows the circuit of a sawtooth generator. i) What kind is it? ii) Explain the operation of the circuit with the aid of the appropriate waveform diagrams. iii) Explain the effect of using a fet as T_2.

8.7 A symmetrical astable multivibrator is made using discrete components. Draw the circuit diagram and explain its operation. Derive an expression for the frequency of the output waveform. Enumerate, if the collector resistors are each 1000 Ω. The base resistors are each 30 kΩ, and the timing capacitors are each 0.02 μF and the collector supply voltage is 12 V.

8.8 Draw the circuit diagram of a sawtooth generator using an op-amp and explain the operation of the circuit. Calculate component values to give an output ramp voltage that changes by 10 V/ms. Derive any expression used.

Fig. 8.52

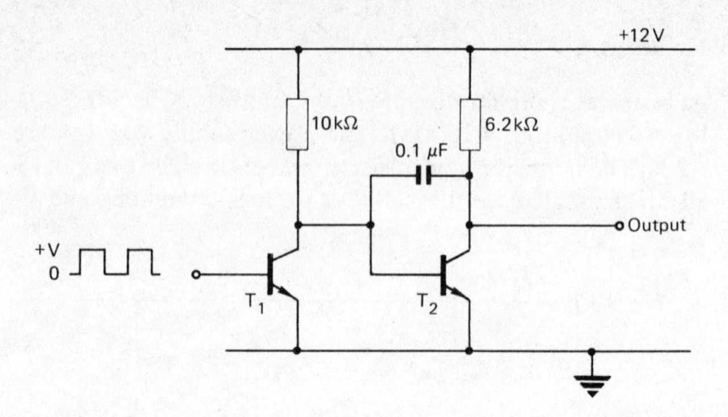

8.9 A bipolar transistor bootstrap generator uses the circuit of Fig. 8.37*b* except that its emitter resistance R_2 is returned to a -10 V line. If the transistors have $h_{fe} = 120$ and $V_{be} = V_{CE(SAT)} = 0.6$ V and the diode voltage drop is also 0.6 V, calculate suitable values for R_1, R_2, C_1, and C_2 to produce an output ramp of 10 V peak and 18 μs duration. Take $V_{cc} = 12$ V.

8.10 Draw a circuit diagram to show how a 555 timer i.c. can be used as an astable multivibrator. Derive an expression for the periodic of the output waveform and hence determine suitable values for the external components if the periodic time is to be 1 ms and the waveform is i) approximately square, ii) with a mark/space ratio of 1:2.

8.11 Plot to scale the waveform of the voltage across a coil of inductance 6 mH and resistance 10 Ω if the current flowing in the coil is a ramp of peak-peak value 200 mA and the ramp frequency is 12 kHz with 5% of the time used for flyback.

Short Exercises

8.12 Fig. 8.53 shows the pin connections of a 4 001 quad 2-input NOR gate i.c. Show how it can be connected to produce a bistable multivibrator circuit.

Fig. 8.53

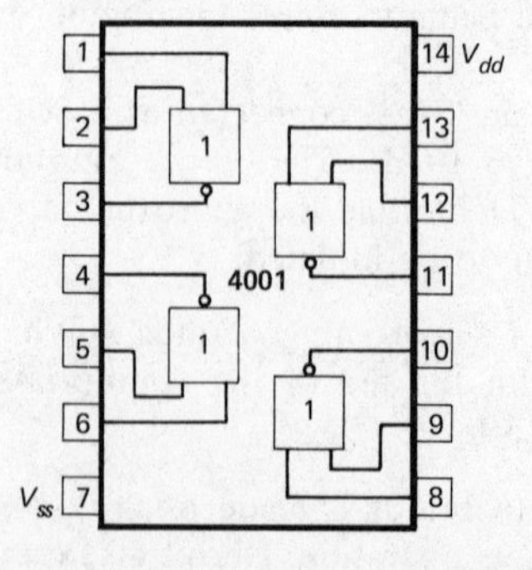

8.13 Draw the circuit of a non-inverting op-amp Schmitt trigger and give its transfer characteristic.

8.14 Quote the expression for the periodic time of a symmetrical discrete component astable multivibrator. If the base resistors are each 33 kΩ and the frequency of the output square waveform is 500 Hz, calculate the value of each coupling capacitor.

Fig. 8.54

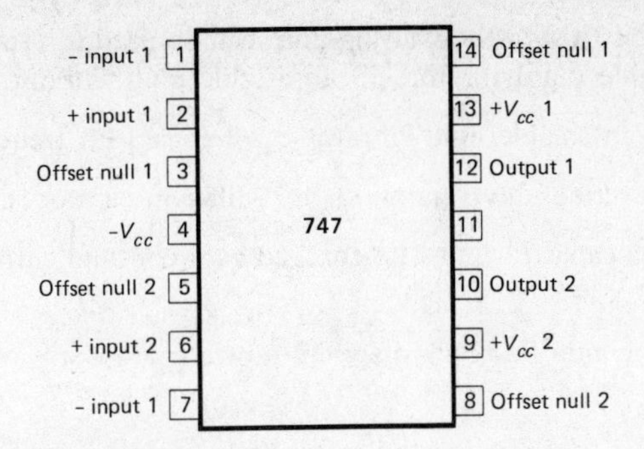

8.15 A 741 op-amp (slew rate 0.5 V/μs) is to be used as a ramp generator. If a 20 V ramp amplitude is wanted, determine the maximum possible frequency of operation.

8.16 Fig. 8.54 shows the pin connections of a 747 dual op-amp. Draw a diagram to show how this i.c. can be connected as a triangle/square wave generator.

8.17 Design an astable multivibrator using a 555 timer to have a mark/space ratio of 1.5 : 1 at a frequency of 2000 Hz. Use equation (8.45).

Fig. 8.55

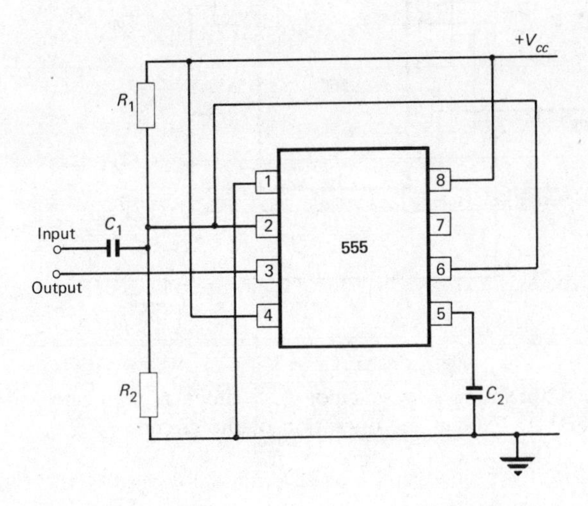

8.18 Explain the operation of the 555 Schmitt trigger circuit given in Fig. 8.55.

8.19 State the effects of circuit capacitance and delays on the trigger pulse width in a bistable multivibrator.

8.20 Describe the factors which determine the switching speed of a bistable multivibrator. What is the effect of a speed-up capacitor?

8.21 Describe how a Schmitt trigger can be used i) to a square a waveform, ii) for level detection.

8.22 List some applications of monostable multivibrators and show how one can be made from a 555 timer.

8.23 Draw diagrams to show how an op-amp can be used to provide i) a Schmitt trigger, ii) a monostable multivibrator, iii) an astable multivibrator.

8.24 Explain how an astable multivibrator can be used for frequency division.

8.25 Draw, and describe, any version of the Miller integrator ramp generator.

8.26 Explain why a capacitor must be charged by a constant current if a linear ramp is to be obtained.

8.27 Show, with the aid of a block diagram, how a triangular wave can be derived from a square wave.

8.28 A 7 4121 i.c. is to be connected as a bistable multivibrator with an output pulse of duration 1 ms. Draw a suitable circuit.

8.29 Repeat 8.28 using a 7 4122 i.c.

8.30 List some applications for a monostable multivibrator.

Fig. 8.56

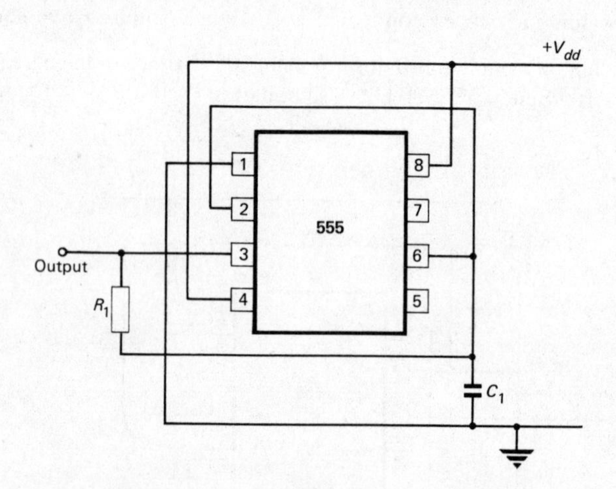

8.31 Fig. 8.56 shows how a cmos 555 timer is connected to operate as a sawtooth generator. Explain the operation of the circuit.

9 Noise

Introduction

The output waveform from any electronic circuit will always contain some unwanted voltages and currents in addition to the wanted signal. These unwanted components are known as **noise** and can originate from a variety of different sources. Some of these noise sources are situated within the circuit itself, while others will already be present when the signal first appears at the input terminals. This means that the amplification process will *always* degrade the output signal-to-noise ratio. Noise having a constant energy per unit bandwidth over a particular frequency band is said to be **white noise**.

Because noise is of random nature, instantaneous values are of little significance. If instantaneous values are squared, the resultant quantity is always positive and has a definite average value over a sufficiently long period of time. It is customary therefore to employ the mean square value as a measure of noise level. The average power dissipated in a resistance is proportional to the mean square voltage or current and hence the average power can also be employed as a measure of noise level. When two noise voltages or currents are applied in series, the resultant voltage is the arithmetic sum, instant by instant, of the individual mean square voltages. The sum waveform fluctuates randomly in the same manner as the waveforms of the component waves. The mean square voltage of the sum waveform is equal to the sum of the mean square voltages of the component noise voltages; similarly the power of the sum waveform is equal to the sum of the powers of the component waves.

The noise power per unit bandwidth is known as the **power density spectrum** (p.d.s.). The p.d.s. of a noise source tells how the noise power is distributed over the frequency spectrum. The p.d.s. of white noise is a horizontal line as shown by Fig. 9.1 and it has an equal amount of power in each hertz of the bandwidth. For white noise

$$\text{p.d.s.} = \frac{\text{Total noise power in band between } f_1 \text{ and } f_2}{\text{bandwidth } f_2 - f_1 \text{ (Hz)}} \text{ W} \qquad (9.1)$$

Example 9.1.

If the total noise power in the band 1–2 MHz is 80 μW determine the p.d.s. if the noise can be assumed to be white.

Solution From equation (9.1.)

$$\text{p.d.s.} = \frac{80 \times 10^{-6}}{1 \times 10^{-6}} = 8 \times 10^{-11} \text{ W/Hz} \quad (Ans)$$

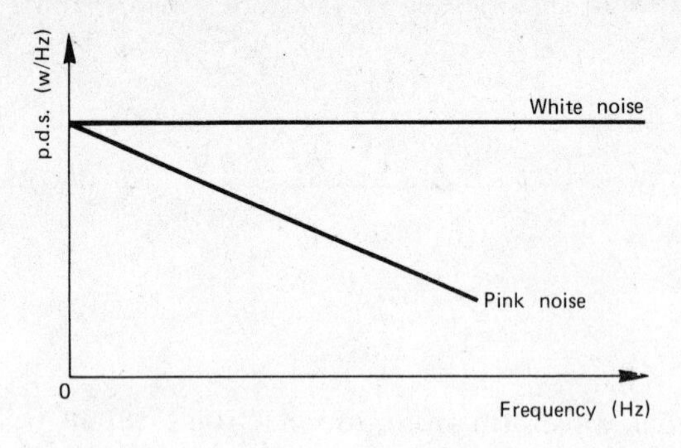

Fig. 9.1 Power density spectrum of white and pink noise.

The term **pink noise** is given to noise that has a p.d.s. which is inversely proportional to frequency (see Fig. 9.1). Pink noise has an equal amount of power in each decade of bandwidth, i.e. there is the same noise power in the frequency bands 0.1–1 Hz, 1–10 Hz, and 10–100 Hz.

Thermal Agitation Noise

The electrons in a conductor are in a state of thermal agitation and this results in minute currents flowing in the conductor which vary continuously and randomly in both magnitude and direction. Because of this a randomly varying voltage is developed across the conductor and this unwanted voltage is known as **thermal agitation noise** or *resistance noise.*

The r.m.s. value of this voltage is given by equation (9.2), i.e.

$$V_n = \sqrt{[4kTBR]} \tag{9.2}$$

where k = Boltzmann's constant = 1.38×10^{-23} J/°

T = *absolute* temperature of conductor in °. Note that ° = °C + 273.

B = bandwidth (Hz) over which the noise is measured or of the circuit at whose output the noise appears, whichever is the smaller.

R = resistance, or real part of the impedance, of the circuit in ohms.

It is the *bandwidth* and not the frequency of operation that is important with regard to thermal agitation noise. Thus a wideband amplifier is noisier than a narrowband amplifier whatever their operating frequencies may be. Thermal agitation noise is white.

Example 9.2

Calculate the noise voltage produced in a 100 kΩ resistor in a 4 MHz bandwidth at a temperature of 20°C.

Solution From equation (9.2)

$$V_n = \sqrt{[4 \times 1.38 \times 10^{-23} \times 293 \times 4 \times 10^6 \times 10^5]} = 80.4\ \mu\text{V} \quad (Ans)$$

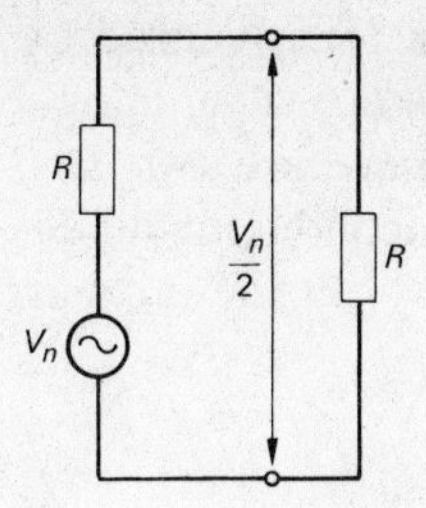

Fig. 9.2 Available noise power.

The thermal noise e.m.f. may be regarded as acting in series with the resistance producing it. Maximum power transfer from one resistor to another occurs when their resistances are equal. Consider Fig. 9.2, which shows a resistance connected across another resistance of the same value. The *maximum* noise power P_a delivered to the load resistance is

$$P_a = \frac{(V_n/2)^2}{R} = \frac{4kTBR}{4R} = kTB \quad \text{watts} \tag{9.3}$$

The maximum or *available* noise power that can be delivered by a resistance is proportional to *both* temperature *and* bandwidth.

If the two resistors are at equal temperatures they will each supply kTB watts noise power to the other and zero net transfer of energy will take place. If the resistors are at different temperatures there *is* a net power transfer that is proportional to the difference between their absolute temperatures.

Example 9.3

Calculate the available noise power from a resistance at 17°C in a 1 MHz bandwidth.

Solution

$$P_a = 1.38 \times 10^{-23} \times 290 \times 10^6 = 4 \times 10^{-15}\ \text{W} \quad (Ans)$$

Generally, resistors are at room temperature, nominally 290°, and this is denoted by the symbol T_0.

Noise in Semiconductors

Bipolar Transistors

There are several sources of noise in a bipolar transistor and the total noise output is likely to have some contribution from each of them.

1 Thermal agitation noise A thermal agitation noise voltage is produced in the base spreading resistance of a transistor and is given by

$$V_n = \sqrt{[4kTBr_b]}$$

Similar sources are also present in the emitter and collector regions of the device but these can usually be neglected.

2 Shot noise Shot noise in a transistor is caused by the random arrival and departure of charge carriers across a p-n junction. Since there are two p-n junctions in a bipolar transistor there are two sources of shot noise. Shot noise will also be present in a semiconductor diode. The magnitude of each source of shot noise depends upon both the *d.c.* current I flowing across the junction and the bandwidth B:

$$\text{Shot noise} = \sqrt{[2eIB]} \tag{9.4}$$

where e = electronic charge = 1.6×10^{-19} coulombs.

Shot noise has a mean square value which is directly proportional to the bandwidth and this means that the p.d.s. is constant and hence shot noise is white. Note particularly that shot noise does not depend upon the temperature. This is the predominant source of semiconductor noise other than at low frequencies or low currents.

Example 9.4

Calculate the shot noise produced by a semiconductor diode when it is passing a current of 2 mA and the signal bandwidth is 100 kHz.

Solution From equation (9.4)

$$\text{Shot noise} = \sqrt{(2 \times 1.6 \times 10^{-19} \times 2 \times 10^{-3} \times 10^{5})} = 8\ \text{nA} \quad (Ans)$$

3 Flicker noise Flicker noise is also often known as $1/f$ noise and is caused by fluctuations in the conductivity of the semiconductor material. Flicker noise is most significant at low frequencies and is generally negligible at frequencies in excess of 10 kHz or so. Flicker noise is pink.

4 Burst or popcorn noise Burst noise is another source of low-frequency noise that is exhibited by some devices and it is given this name because it consists of bursts of noise that characteristically sound like "popping". This source of noise is most prevalent in integrated circuit devices.

5 Interference noise The term *interference noise* is used here to include the various sources of noise which may be picked up by an electronic circuit. This will not include noise sources that are predominantly the interest of communication engineers be they switching, line or radio [EIII]. Amongst the large number of possible sources are included: power supply ripple, magnetic couplings, vibration, and poor switch contacts.

The various noise sources that are inherently present in a transistor can be represented in a low-frequency hybrid-π circuit as shown in Fig. 9.3. The meanings of the various noise generators shown are as follows:

$i_{n1} = \sqrt{[2eI_bB]}$ is the shot noise in the base current plus the flicker and burst noise.

$i_{n2} = \sqrt{[2eI_cB]}$ is the shot noise in the collector current.

$v_{n1} = \sqrt{[4kTBr_{bb'}]}$ is the thermal agitation noise in $r_{bb'}$.

Equations can be derived using Fig. 9.3 that will describe the noise performance of the transistor but in this book an alternative approach will be used (page 226).

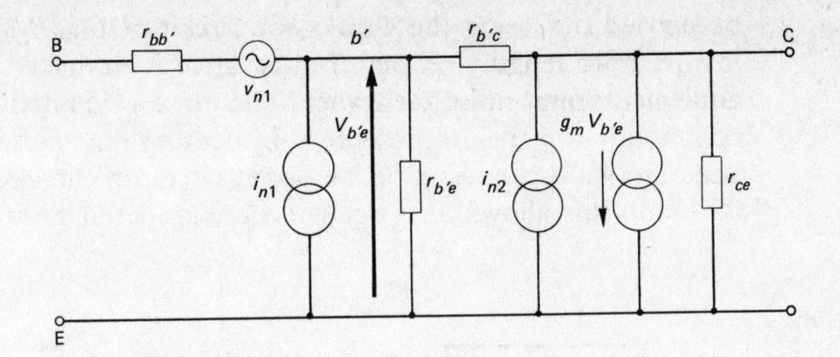

Fig. 9.3 Noise equivalent circuit of a transistor.

Field Effect Transistors

The major sources of noise in a fet are thermal agitation noise produced in the channel resistance and flicker noise. Some shot noise does appear in the gate leakage current of a jfet but this is normally very small. The thermal and flicker noise sources are generally lumped together and represented in the fet equivalent circuit (Fig. 1.25) by a current generator drawn in parallel with r_{ds}.

Inherently the fet is a lower-noise device than the bipolar transistor but the latter can give a comparable, or perhaps better, noise performance when it is operated with a very small collector current.

Noise Calculations

The internal noise sources of a transistor and of an electronic circuit either are the result of thermal agitation or are semiconductor noise. In either case they can be represented by either a current or a voltage generator. When calculating the noise performance of a transistor or a circuit, the total noise arising from several sources is determined using the **mean square value** of each noise generator. Consider Fig. 9.4 which shows two resistors R_1 and R_2 connected in series. The r.m.s. noise voltages produced by the two resistors are

Fig. 9.4 Noise generators in series.

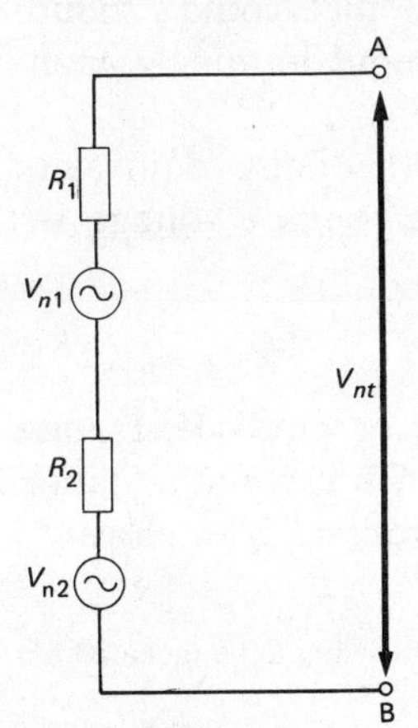

$$V_{n1} = \sqrt{[4kTBR_1]} \quad \text{and} \quad V_{n2} = \sqrt{[4kTBR_2]}$$

The total mean square noise voltage appearing between the terminals A and B is

$$V_{nt}^2 = V_{n1}^2 + V_{n2}^2 = 4kTB(R_1 + R_2)$$

and clearly this is the same result as would be obtained if the thermal agitation noise in the combined resistance $R_1 + R_2$ had been calculated. This would not be true however if the two resistors were at different temperatures.

A similar result is obtained for two resistors connected in parallel; the total mean square noise voltage is then

$$V_{nt}^2 = 4kTBR_1R_2/(R_1 + R_2)$$

The determination of the noise performance of a transistor or a circuit *can* be carried out using the equivalent circuit of Fig. 9.3 but the calculations involved are lengthy. A better alternative is to make use of the concept of **equivalent input noise resistance**. The noise generated within any transistor or circuit can be represented by equivalent noise generators. These generators are connected in the input circuit of the circuit as shown by Fig. 9.5*b* and this allows the circuit to be considered as noiseless.

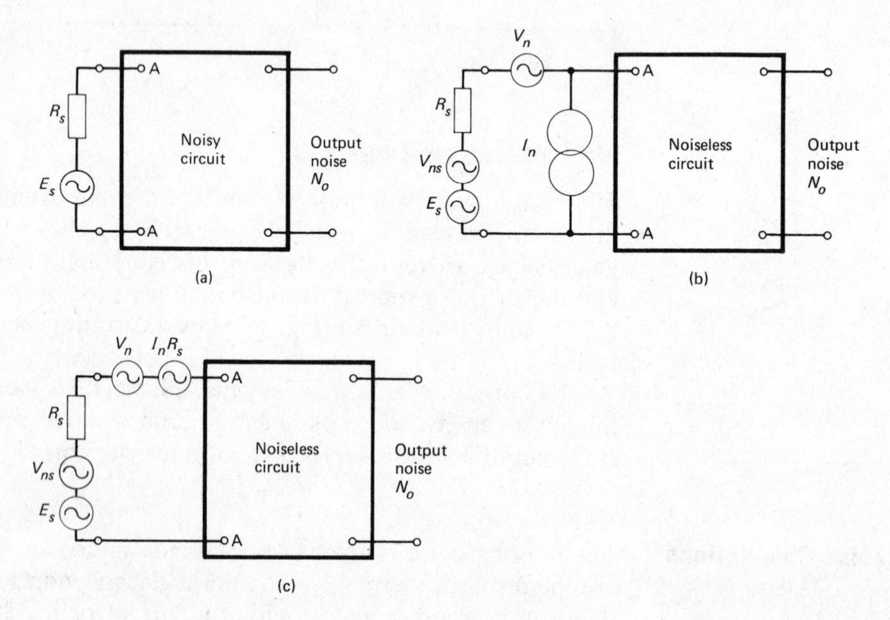

Fig. 9.5 (*a*) Amplifier fed by a source of e.m.f. E_s and resistance R_s, (*b*) noise equivalent circuit of (*a*), (*c*) Thevenin version of (*b*).

The values of the two generators can be fairly easily determined since

V_n = equivalent input r.m.s. noise voltage in volts that would produce the same noise output as the circuit with its input terminals short-circuited

I_n = equivalent input r.m.s. noise current in amperes that would produce the same noise output as the circuit with its input terminals open-circuited

Applying Thevenin's theorem to Fig. 9.5*b* gives the noise equivalent circuit shown in Fig. 9.5*c*. From this the total mean square noise voltage V_{nt}^2 at the input terminals of the "noiseless" circuit is

$$V_{nt}^2 = V_{ns}^2 + V_n^2 + I_n^2 R_s^2 \qquad (9.5)†$$

It is generally more convenient to work in terms of **equivalent noise resistances.** These are the (imaginary) resistances in which thermal agitation would generate the noise voltage V_n or the noise current I_n. Therefore

† Really a term representing any correlation between the two generators should be included but since it is usually small it will be neglected.

$$V_n^2 = 4kTBR_{nv} \quad \text{or} \quad R_{nv} = \frac{V_n^2}{4kTB} \tag{9.6}$$

$$I_n^2 = 4kTBG_{ni} = \frac{4kTB}{R_{ni}} \quad \text{or} \quad R_{ni} = \frac{4kTB}{I_n^2} \tag{9.7}$$

An analysis of the hybrid-π noise equivalent circuit given in Fig. 9.3 shows that for a bipolar transistor

$$R_{nv} \simeq r_{bb'} + 1/2g_m \quad \text{and} \quad R_{ni} \simeq 2h_{fe}/g_m \tag{9.8}$$

For a fet

$$R_{nv} \simeq 0.7/g_m \quad \text{and} \quad R_{ni} \simeq 0 \tag{9.9}$$

Using R_{nv} and R_{ni} equation (9.5) can be written as

$$V_{nt}^2 = 4kTB(R_s + R_{nv} + R_s^2/R_{ni}) \tag{9.10}$$

Noise Factor (or Figure)

The **noise factor** or noise figure F of a circuit is defined as

$$F = \frac{\text{total mean square noise voltage at output}}{\text{that part of the above which is due to the thermal noise at the source}} \tag{9.11}$$

The source is supposed to be at an absolute temperature of 290° and the circuit must be linear. The terms noise figure and noise factor are both commonly used but in this book **figure** will be preferred.

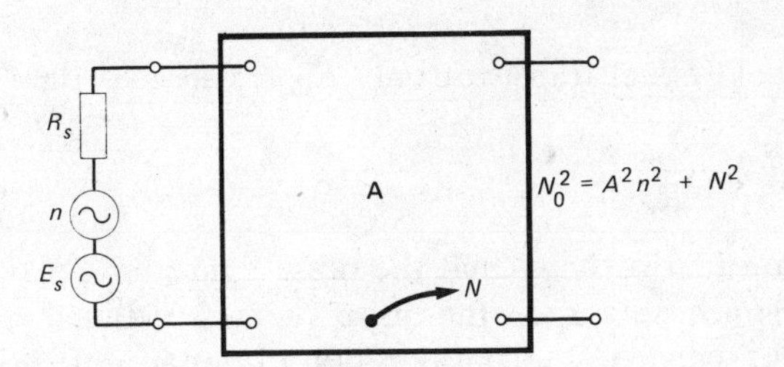

Fig. 9.6 Circuit for the definition of noise figure.

Fig. 9.6 shows a circuit of voltage gain A, that introduces a mean square noise voltage N^2, fed by a source of e.m.f. E_S and mean square noise voltage n^2.

From equation (9.11)

$$F = \frac{A^2n^2 + N^2}{A^2n^2} = \frac{1/n^2}{A^2/(A^2n^2 + N^2)} = \frac{E_s^2/n^2}{A^2E_s^2/(A^2n^2 + N^2)}$$

$$F = \frac{\text{Input signal-to-noise ratio}}{\text{Output signal-to-noise ratio}} \tag{9.12}$$

Noise figure is a measure of the degradation of the input signal-to-noise ratio caused by a circuit but it is *not* a measure of the output signal-to-noise ratio itself.

Example 9.5

If the signal-to-noise ratio at the input to an amplifier having a noise figure of 10 is 40 dB, calculate the output signal-to-noise ratio.

Solution Noise figure is essentially a power ratio and hence $F=10$ corresponds to $F=10$ dB. From equation (9.12)

Output signal-to-noise ratio $=40-10=30$ dB (*Ans*)

A circuit with a noise figure of 3 dB or less would be regarded as a low-noise circuit while a noise figure of 10 dB would be average.

The noise figure of a circuit can be determined in terms of the equivalent noise resistances R_{nv} and R_{ni} using Fig. 9.5*c*. The **noiseless circuit** will amplify both signal and noise to the same extent so that the output signal-to-noise ratio of the circuit will be the same as that at the input terminals AA of the amplifier.

The input signal-to-noise ratio is

$$E_s^2/V_{ns}^2 = E_s^2/4kTBR_s$$

The signal voltage at the terminals AA is E_s and the total noise at AA is given by equation (9.10). Therefore, the output signal-to-noise ratio is

$$E_s^2/[4kTB(R_s+R_{nv}+R_s^2/R_{ni})]$$

From equation (9.12) the noise figure of the circuit is

$$F=1+\frac{R_{nv}}{R_s}+\frac{R_s}{R_{ni}} \tag{9.13}$$

For a field effect transistor only R_{nv} exists and then

$$F=1+\frac{R_{nv}}{R_s} \tag{9.14}$$

Equation (9.13) shows that the noise figure obtained for a particular circuit depends not only upon the noise sources within the circuit but also on the source resistance R_s. Consideration of equation (9.13) shows that F will be very large *either* if R_s is large *or* if R_s is small and this is an indication that there must be some optimum value for R_s at which the minimum possible value for the noise figure is obtained. The optimum value for R_s can be determined by differentiating F with respect to R_s and equating the result to zero. Thus,

$$\frac{dF}{dR_s}=\frac{-R_{nv}}{R_s^2}+\frac{1}{R_{ni}}=0$$

Hence $R_{s(optimum)}=\sqrt{[R_{nv}R_{ni}]}$ (9.15)

Substituting equation (9.15) into equation (9.13) gives

$$F_{min}=1+\frac{R_{nv}}{\sqrt{[R_{nv}R_{ni}]}}+\frac{\sqrt{[R_{nv}R_{ni}]}}{R_{ni}}=1+2\sqrt{\frac{R_{nv}}{R_{ni}}} \tag{9.16}$$

Example 9.6

An amplifier has equivalent noise resistances of $R_{nv} = 1300\ \Omega$ and $R_{ni} = 600\ \Omega$. Calculate i) the optimum source resistance and ii) the minimum noise figure for the amplifier.

Solution From equation (9.15),

$$R_{s(optimum)} = \sqrt{(1300 \times 600)} = 883\ \Omega \quad (Ans)$$

and from equation (9.16)

$$F_{min} = 1 + 2\sqrt{\frac{1300}{600}} = 3.94 = 5.96\ \text{dB} \quad (Ans)$$

Noise Output of a Circuit

The available noise power into a circuit is kT_0B watts and the noise factor can be written as

$$F = \frac{P_{IN}/kT_0B}{GP_{IN}/N_0}$$

where N_0 is the output noise power, P_{IN} is the input signal power, and G is the power gain of the circuit. Therefore

$$F = N_0/GkT_0B \quad \text{and} \quad N_0 = FGkT_0B \tag{9.17}$$

GkT_0B is the amplified input noise power and so the noise power due to sources within the amplifier appearing at the output is given by

$$N_0' = (F-1)GkT_0B \tag{9.18}$$

Variation of Noise Figure with Frequency

Since the gain, or loss, of a circuit and the noise generated per unit bandwidth are often a function of frequency, noise figure may be frequency-dependent. Thus a distinction between single-frequency noise figure and integrated or full-band noise figure must be made. If the bandwidth B is narrow enough for variations in gain and generated noise to be ignored, the **spot noise figure** is obtained. The spot noise figure can be measured at a number of different points in the overall bandwidth of the amplifier and then plotted against frequency. If the bandwith is wide, the average noise figure is obtained. If the spot noise figure is constant over a specified bandwidth, it will be equal to the average noise figure in that bandwidth.

Overall Noise Figure of Several Circuits in Cascade

Fig. 9.7 shows two circuits connected in cascade. The circuits have noise figures F_1 and F_2 and power gains G_1 and G_2. It is assumed that the bandwidths of the two circuits are equal to one another and are also equal to the overall bandwidth of the cascade. The available output noise power is

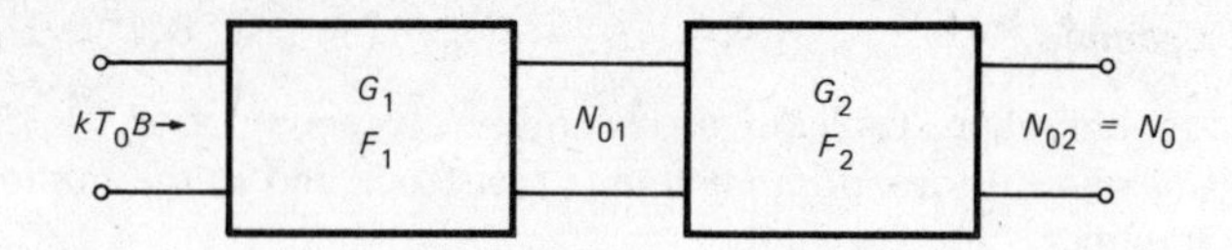

Fig. 9.7 Noise figures in cascade.

$$N_0 = F_{12}G_1G_2kT_0B \tag{9.19}$$

where F_{12} is the overall noise figure. For the first circuit

$$N_{01} = F_1G_1kT_0B$$

and for the second circuit

$$N_{02} = N_0 = G_2F_1G_1kT_0B + (F_2 - 1)G_2kT_0B \tag{9.20}$$

Comparing equations (9.19) and (9.20)

$$F_{12} = F_1 + \frac{F_2 - 1}{G_1} \tag{9.21}$$

Similarly it can be shown that for n circuits connected in cascade

$$F_{in} = F_1 + \frac{F_2 - 1}{G_1} + \frac{F_3 - 1}{G_1G_2} + \cdots + \frac{F_n - 1}{G_1G_2 \cdots G_{n-1}} \tag{9.22}$$

Equations (9.21) and (9.22) show that the effect of noise sources in stages other than the first is reduced, unless the first stage introduces loss, because of the division by the gains of all the preceding circuits. If the first stage introduces loss, the noise introduced by the second stage will be accentuated; this means that any loss in the first circuit must always be minimized.

Example 9.7

An amplifier of power gain 20 dB and noise figure 5 dB is connected in cascade with a 6 dB attenuator to provide an overall gain of 14 dB. Determine the overall noise figure obtained with each of the two possible cascade arrangements.

Solution Equation (9.21) has been derived assuming the power gains and noise figures of the amplifier to be quoted as power ratios. Hence the first step is to convert the given dB values into the corresponding power ratios. Thus

5 dB = 3.16 : 1 power ratio

6 dB = 3.98 : 1 power ratio

14 dB = 25.12 : 1 power ratio

20 dB = 100 : 1 power ratio.

The noise figure of an attenuator is equal to its attenuation, i.e. F = 6 dB.

i) With the amplifier followed by the attenuator

$$F_{12} = 3.16 + \frac{(3.98 - 1)}{100} = 3.19 = 5\text{ dB} \quad (Ans)$$

ii) With the attenuator followed by the amplifier

$$F_{12} = 3.98 + \frac{(3.16-1)}{1/3.98} = 12.58 = 11\ \text{dB} \quad (Ans)$$

Effective Noise Figure of a Circuit Fed by a Source at $T_s \neq T_0$

The noise figure of a circuit is defined with reference to a temperature of 290°. If the source is not at 290° but some other temperature T_s the degradation of the input signal-to-noise ratio is different from that indicated by the noise figure. For example, if the source is at a temperature below 290° the degradation is greater but if the source temperature is above 290° the degradation in signal-to-noise ratio is reduced. This can be taken account of by the concept of the **effective noise figure** F_{eff}.

From equation (9.17) the noise figure of a circuit referred to temperature $T_0 = 290°$ is

$$F = N_0/GkT_0B$$

and hence the available output noise power due to the network alone is

$$(F-1)GkT_0B \text{ watts}$$

If, now, a source at temperature T_s is connected to the input terminals of the circuit, the available output noise power will be

$$N_0 = GkT_sB + (F-1)GkT_0B$$

The effective noise figure F_{eff} is

$$F = N_0/GkT_sB$$

and substituting for N_0

$$F_{eff} = \frac{GkT_sB + (F-1)GkT_0B}{GkT_sB} = 1 + \frac{T_0}{T_s}(F-1) \qquad (9.23)$$

Noise Temperature

The available noise output power N_0 generated within a circuit may be considered as originating from its output resistance which is at a temperature T where

$$T = \frac{N_0}{GkB} \qquad (9.24)$$

T is the **noise temperature** of the circuit and it is related to the standard temperature T_0 by $T = tT_0$ where t may be greater than, or smaller than, unity.

There is a simple relationship between the noise figure and the noise temperature of a circuit and it can be obtained in the following manner. The noise produced within a circuit is

$$GkTB = (F-1)GkT_0B$$

Therefore $T=(F-1)T_0$ or $t=F-1$ (9.25)

The *total* noise output from a circuit (see equation 9.17) can now be written as

$$N_0=GkB(T_s+T) \quad (9.26)$$

where T_s is the temperature of the source which may, of course, be equal to T_0.

Example 9.8

An amplifier has a bandwidth of 100 kHz, a power gain of 30 dB, and a noise factor of 6 dB, and it is fed by a source at temperature T_0. Calculate the output noise power using i) equation (9.17), ii) using equation (9.26).

Solution

i) $N_0=4\times10^3\times1.38\times10^{-23}\times290\times10^5=1.6\times10^{-12}$ W (*Ans*)

ii) $t=F-1=3$ and $T=tT_0=3\times290°$. Therefore

$$N_0=10^3\times1.38\times10^{-23}\times10^5(3\times290+290)=1.6\times10^{-12}\text{ W} \quad (Ans)$$

This example makes it clear that either noise figure or noise temperature can be used to determine output power or output signal-to-noise ratio. Usually noise temperatures are used for aerial and microwave calculations and very often for low-noise circuits where more convenient numbers then result.

Noise Temperatures in Cascade

If 1 is subtracted from both sides of equation (9.22),

$$F_{in}-1=F_1-1+\frac{F_2-1}{G_1}+\frac{F_3-1}{G_1G_2}+\text{etc}$$

$$t_{in}=t_1+\frac{t_2}{G_1}+\frac{t_3}{G_1G_2}+\text{etc.}$$

$$T_{in}=T_1+\frac{T_2}{G_1}+\frac{T_3}{G_1G_2}+\text{etc.} \quad (9.27)$$

Measurement of Noise Figure and Noise Temperature

The measurement of the noise figure of a circuit is usually carried out using some kind of **noise generator**. For frequencies from 0 Hz up to some hundreds of megahertz, the noise generator usually consists of either a temperature-limited thermionic diode or a reverse-biased semiconductor diode. The shot noise produced by a diode is given by equation (9.4), i.e.

$$i_n=\sqrt{[2eIB]}$$

Fig. 9.8 shows the arrangement used to measure the noise figure of a circuit. The r.m.s. noise current produced by the noise generator can be set to any desired value by varying the current passed by the diode. The noise

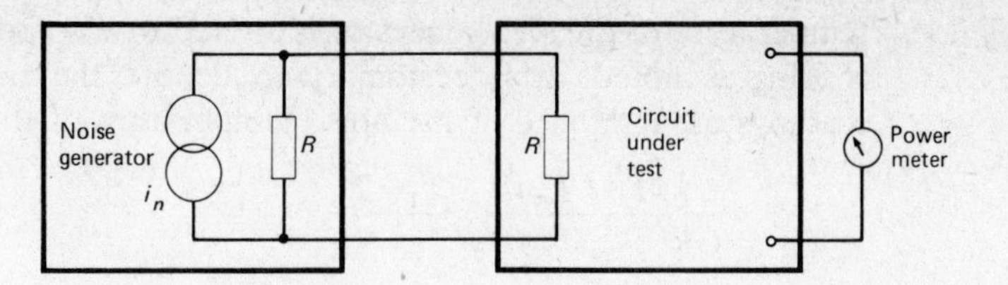

Fig. 9.8 Measurement of noise figure.

generator has an output resistance R, which is matched to the input resistance of the circuit under test. The total noise output power developed by the noise generator is

$$N_0 = kT_0B + (i_n/2)^2R$$

or $N_0 = kT_0B + 2eIBR/4 = kT_0B + \frac{1}{2}eIBR$

This is also the input noise to the circuit under test. With the noise generator switched off, the reading of the power meter, or true r.m.s. ammeter, at the output of the circuit is noted. This noise power is equal to $N_0 = FGkT_0B$ watts. The noise generator is then switched on and its noise output is increased until the reading of the power meter is doubled. Then

$$2FGkT_0B = FGkT_0B + GeI_aBR/2$$

or $FGkT_0B = GeI_aBR/2$ Therefore,

$$F = \frac{eIR}{2kT_0} = 20IR \qquad (9.28)$$

Noise figure measurements can be employed to determine the values of the equivalent noise generators R_{nv} and R_{ni}. If the noise figure is measured with a very low value of R_s the result will enable R_{nv} to be calculated; a second measurement of F with R_s at a very high value will then give R_{ni}.

The standard method of measuring the noise temperature of a circuit is to apply two noise sources of known temperatures T_1 and T_2 in turn to the circuit and then to obtain the ratio Y of the output powers thus obtained. Two separate standard noise sources are used only if precise measurements are being carried out; commonly a single noise generator is used and this is switched between two known temperatures.

The block diagram of the arrangement used to measure noise temperature is shown by Fig. 9.9. With the noise source at effective noise temperature T_1 connected to the circuit the reading of the output meter is noted. The circuit is then connected to the other noise generator at temperature T_2 (or the

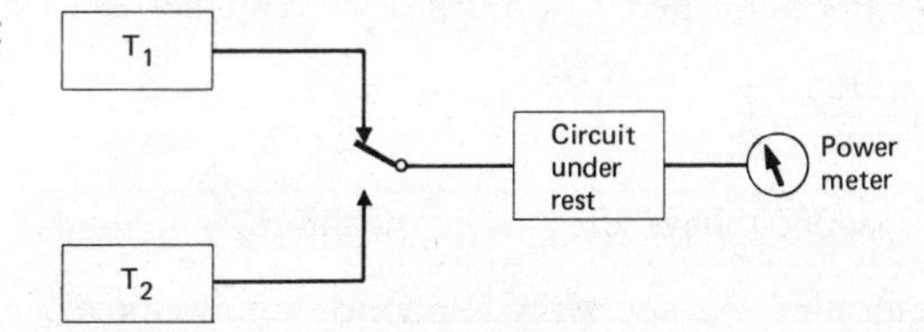

Fig. 9.9 Measurement of noise temperature.

temperature of the sole generator is altered to T_2) and the new output meter reading is noted. It is customary to denote the ratio of the two meter readings as Y. Hence, if the noise temperature of the circuit is T_A,

$$Y = \frac{Gk(T_1 + T_A)B}{Gk(T_2 + T_A)B} = \frac{T_1 + T_A}{T_2 + T_A}$$

$$T_A = \frac{T_1 - YT_2}{Y - 1} \qquad (9.29)$$

Exercises 9

9.1 Define the term noise figure and explain its importance when applied to amplifiers.

A bipolar transistor has $r_{bb'} = 50\ \Omega$ and operates with a collector current of 1 mA and $h_{fe} = 100$. Calculate its optimum source resistance and the minimum noise figure for this device.

9.2 List the sources of noise in a field effect transistor, and explain the meaning of the term equivalent noise resistance. A generator of e.m.f. 2 mV r.m.s. and internal resistance 100 Ω is applied to the 1:4 step-up input transformer of a common-source fet amplifier. The equivalent noise resistance of the fet is 600 Ω. Calculate the output signal-to-noise ratio of the amplifier if the bandwidth is 60 kHz.

Fig. 9.10

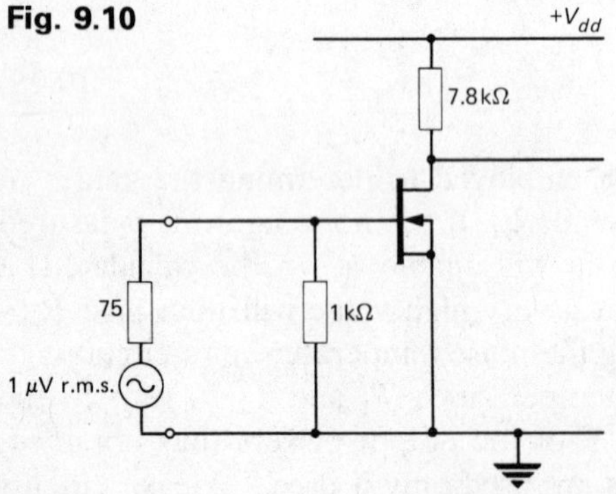

+V_{dd}
7.8kΩ
75
1kΩ
1 μV r.m.s.

Table 9.1

	Power gain (dB)	Noise figure (dB)
First stage	6	3
Second stage	20	8
Third stage	20	5

9.3 Define the term noise figure as applied to an amplifier. Why is it necessary for the first stage in a multi-stage amplifier to have a low noise figure?

Fig. 9.10 shows a fet amplifier. If the signal-to-noise ratio at the output of the amplifier is 10 dB in a bandwidth of 40 kHz calculate the noise figure of the amplifier.

9.4 Prove that the noise figure of a transistor amplifier can be written as

$$F = 1 + \frac{R_{nv}}{R_s} + \frac{R_s}{R_{ni}}$$

where the symbols have their usual meanings.

Hence determine the necessary relationship between R_{nv} and R_{ni} for $F = 1$ dB.

9.5 Show that the output signal-to-noise ratio of an amplifier is less than the input signal-to-noise ratio by the ratio $T_s/(T+T_s)$ where T_s is the noise temperature of the source and T is the noise temperature of the amplifier.

9.6 List the various sources of noise that are generated within an amplifier. State how each of these sources vary with frequency. Explain with the aid of power density spectrum graphs the meanings of the terms white noise and pink noise.

9.7 An amplifier has a bandwidth of 1 MHz and a noise figure of 10 dB measured with the source at 290°. A source of noise temperature 580° is applied to the input terminals of the amplifier when the output signal-to-noise ratio is found to be 20 dB. Calculate the input signal-to-noise ratio of the amplifier.

Short Exercises

9.8 Draw the hybrid-π equivalent circuit of a bipolar transistor and include the noise generators v_n and i_n. State the relationships between v_{ni} and i_{ni} and i) R_{nv} and R_{ni} and ii) the transistor parameters.

9.9 A bipolar transistor has noise resistances $R_{nv} = 400\ \Omega$ and $R_{ni} = 300\ \text{k}\Omega$. Determine the values of the equivalent noise generators v_n and i_n in units of V/$\sqrt{}$Hz and A/$\sqrt{}$Hz respectively.

9.10 A bipolar transistor has noise generators $v_n = 3$ nV/$\sqrt{}$Hz and $i_n = 0.8$ pA/$\sqrt{}$Hz. Determine its equivalent resistances R_{nv} and R_{ni} in a 1 MHz bandwidth.

9.11 Determine the r.m.s. voltage generated in a 1 MHz bandwidth by two 10 kΩ resistors connected i) in series, ii) in parallel.

9.12 The r.m.s. noise voltage generated in a resistor is 12 μV. What will be the r.m.s. noise voltage if i) the temperature is doubled, ii) the bandwidth is halved, iii) i) and ii) occur together?

9.13 The signal-to-noise ratio at the output of an amplifier is 10 dB. Discuss briefly whether this indicates an amplifier with a good or a poor noise performance.

9.14 An amplifier has a bandwidth of 200 kHz and a voltage gain of 100. Calculate its noise output voltage due to a 100 kΩ resistor at temperature T_0 connected across the input terminals. Is this the total noise output voltage?

9.15 The input noise to an amplifier is −100 dBm. What power gain is required to give an output signal power of 10 mV if the input signal-to-noise ratio is 40 dB?

9.16 An amplifier has a noise figure of 4.5 dB referred to 290°. What will be the output signal-to-noise ratio if the input signal-to-noise ratio is 36 dB and the source temperature is i) 290°, ii) 580°?

9.17 Calculate the noise output from a circuit which has a noise temperature of 300°, a power gain of 100, and a bandwidth of 2 MHz if the source temperature is i) 290°, ii) 580°?

9.18 Calculate the d.c. current flowing in a diode if the r.m.s. noise current generated is 12 nA/MHz.

9.19 An amplifier has three stages, the power gains and noise figures of which are given in Table 9.1. Calculate in dB the overall gain and the overall noise figure of the amplifier.

10 Controlled Rectification

Introduction

The amount of power delivered to a load can be controlled using a semiconductor device known as a *thyristor.* The power is controlled in one of two ways; one of these is known as *phase control* and consists of controlling the time for which a supply current is allowed to flow in the load per cycle of the mains supply (see Fig. 10.1*a*). The other method is known as either *integral cycling* or as *burst firing* and this entails allowing the supply current to flow in the load for an integral number of cycles in each given time interval (see Fig. 10.1*b*).

The applications of controlled rectification are many and include heater and lamp controls and the control of the speed of motors.

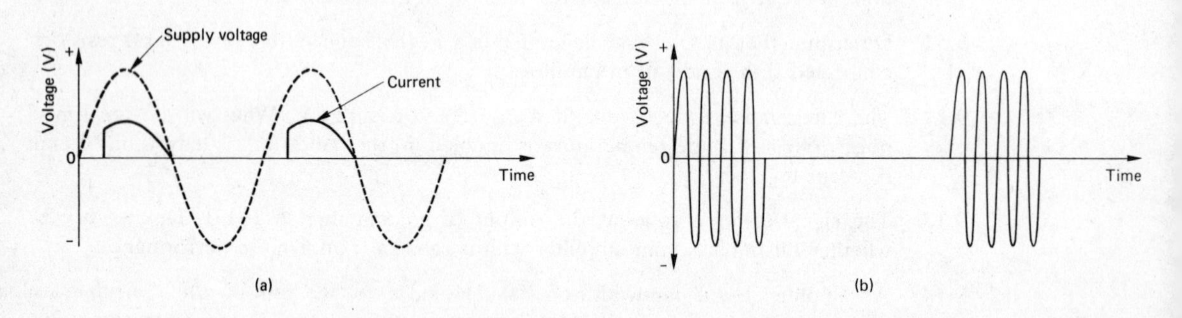

Fig. 10.1 Principle of (*a*) phase control, (*b*) integral cycling.

The Thyristor

The **thyristor** (sometimes known as the silicon controlled rectifier or s.c.r.) is a semiconductor device that consists of two n-type and two p-type layers formed as shown by Fig. 10.2*a*. The upper p-type layer is connected to a terminal known as the *anode*, while the lower n-type layer is in contact with the *cathode*. A third terminal, known as the *gate*, is connected to the lower p-type region. Often, the device is encapsulated within a case with a projecting screw thread and this acts as the anode connection (Fig. 10.2*b*). The symbol for a thyristor is shown in Fig. 10.2*c*.

When a voltage is applied across a thyristor with a polarity such that its anode is held negative with respect to its cathode, p-n junctions J_1 and J_3 are reverse-biased and the device only conducts a small leakage current. If the reverse voltage across the thyristor is increased, the leakage current will also increase, slowly at first and then, at the point where **avalanche breakdown** occurs, rapidly to a large value which may well destroy the device.

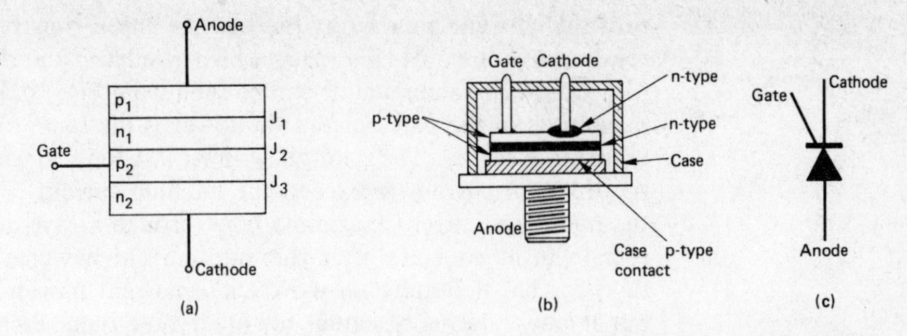

Fig. 10.2 The thyristor: (*a*) arrangement of p-type and n-type layers, (*b*) construction, (*c*) symbol.

If a forward voltage is applied across the thyristor, so that the anode is positive with respect to the cathode, the p-n junctions J_1 and J_3 will become forward-biased but the junction J_2 will be reverse-biased. Once again only a small leakage current will flow and the thyristor is said to be in its **forward blocking state,** or more simply OFF. This forward leakage current also increases slowly with increase in the forward voltage until the breakdown voltage V_{BO} is reached. At this point the current passed by the thyristor increases rapidly and the voltage across the device falls to a much lower value, V_T. This is shown by the static characteristics of a thyristor (see Fig. 10.3*a*).

A thyristor can also be made to conduct a large forward current when the forward anode-cathode voltage is smaller than the breakdown value V_{BO} by triggering, or firing, the device. Triggering is achieved by injecting a current pulse into the gate terminal. The gate current is amplified by the n-p-n transistor and then appears at the collector as $h_{fe2}I_G$. This current then flows in the base of the p-n-p transistor to appear at its collector as $h_{fe1}h_{fe2}I_G$. In

Fig. 10.3 (*a*) Static characteristics of a thyristor, (*b*) showing the effect of varying the gate current.

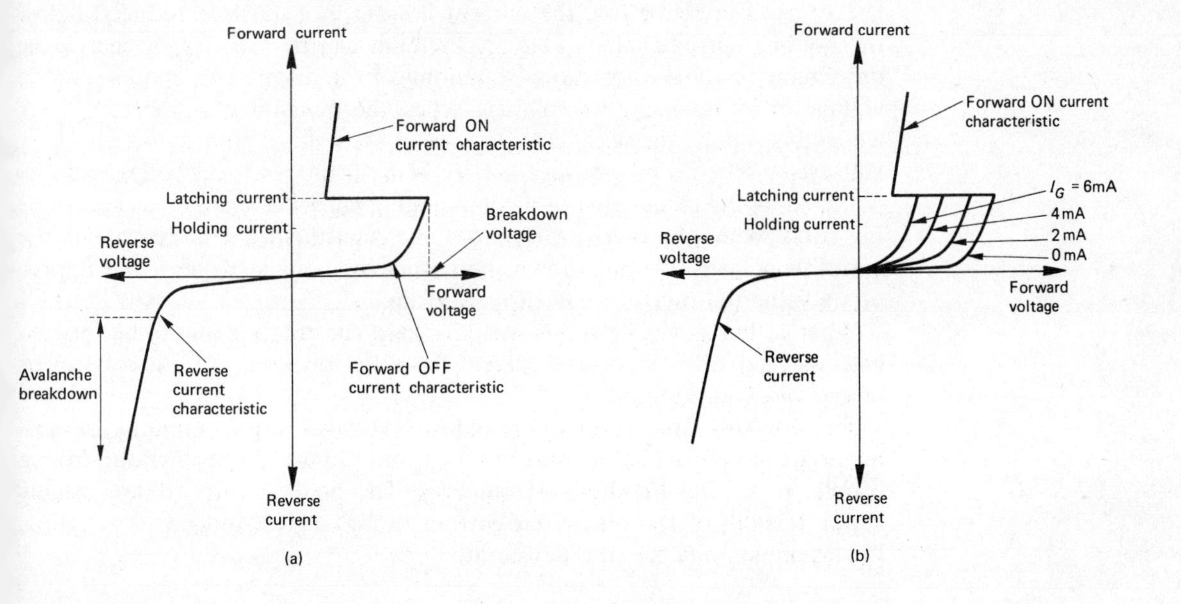

turn this current appears at the base of the n-p-n transistor and so on. A regenerative action takes place which results in the device switching to its forward ON characteristic. It can be seen from Fig. 10.3*b* that the greater the magnitude of the gate current the lower is the forward voltage at which the thyristor turns ON. The gate pulse must last for a long enough time to allow the thyristor current to exceed the latching current. The **latching current** is the minimum current that must flow through a thyristor before it is able to remain in its ON state after the gate current has ceased to flow. Once the thyristor has definitely turned ON, the current flowing in it can be reduced but it must always be equal to, or greater than, the holding current. The **holding current** is the minimum current that can keep the thyristor in its ON condition *after* it has been turned ON. The gate has no further control over the thyristor current after it has turned ON.

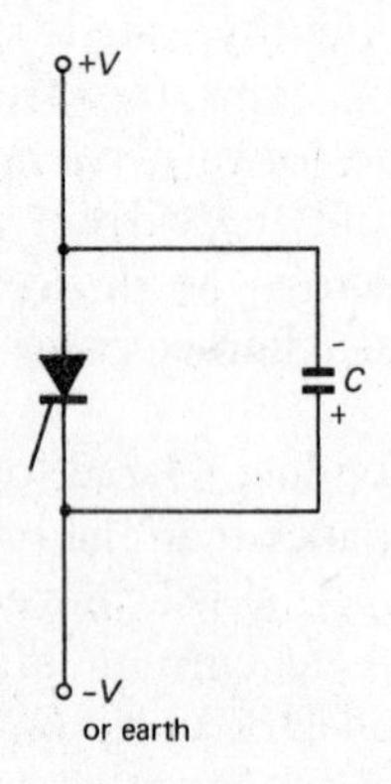

Fig. 10.4 Use of a commutating capacitor.

To turn a thyristor OFF, the current flowing in it must be reduced below the holding current value. This requirement can be satisfied by increasing the circuit resistance, or more commonly, by reducing the anode-cathode voltage or by reversing its polarity. When the thyristor is connected to the a.c. mains supply, the necessary reversal of the voltage polarity occurs every half-cycle. When a d.c. voltage supply is used, the reversed voltage can be obtained by the connection of a *commutating* capacitor charged as shown by Fig. 10.4. When the thyristor turns OFF the capacitor will charge up with the opposite polarity to that shown, and when its voltage reaches the appropriate value the thyristor will turn ON again.

When a thyristor is passing a small forward current, it is said to be *forward blocking* and when a reverse current flows the thyristor is said to be in its *reverse blocking* mode.

The forward and reverse breakdown voltages are commonly several hundreds of volts while the maximum current ratings can be anything from a couple, to several hundreds of amperes. The holding currents are usually about 1/100th of the maximum current rating, e.g. 20 mA for 2 A rating. For example, data for one device are

$$V_{BO(forward)} = V_{BO(reverse)} = 400\ \text{V} \qquad I_{max} = 3\ \text{A}$$
$$I_G = 20\ \text{mA for } 1.5\ \mu\text{s minimum} \qquad I_{holding} = 10\ \text{mA}$$

A thyristor can also be turned ON if the *rate* at which the forward voltage is increased is in excess of some critical value and manufacturer's data normally quote a maximum dv/dt value for a device.

An expression for the current flowing in a thyristor can be derived by considering Fig. 10.5. Hence,

$$I_{c2} = h_{fb2} I_{e2} + I_{CBO2} \tag{10.1}$$

$$I_{b1} = I_{e1} - I_{c1} = I_A - h_{fb1} I_A - I_{CBO1}$$
$$= I_A(1 - h_{fb1}) - I_{CBO1} \tag{10.2}$$

But $I_{c2} = I_{b1}$, therefore,

$$h_{fb2} I_{e2} + I_{CBO2} = I_A(1 - h_{fb1}) - I_{CBO1} \tag{10.3}$$

Now $I_{e2} = I_A + I_G$, hence

$$h_{fb2}(I_A + I_G) + I_{CBO2} = I_A(1 - h_{fb1}) - I_{CBO1}$$

or
$$I_A = \frac{h_{fb2} I_G + I_{CBO1} + I_{CBO2}}{1 - h_{fb1} - h_{fb2}} \tag{10.4}$$

This expression is known as the **characteristic equation**. Typically, $h_{fb1} + h_{fb2}$ is equal to about 0.9 and with $I_G = 0$ the thyristor is normally non-conducting. For the thyristor to turn ON, $h_{fb1} + h_{fb2}$ must be made to increase to very nearly unity since then the numerator of equation (10.4) will be very nearly zero and so I_A will be large. The value of $h_{fb1} + h_{fb2}$ can be increased by passing a gate current I_G into the gate terminal because the current gain of a transistor is a function of its emitter current.

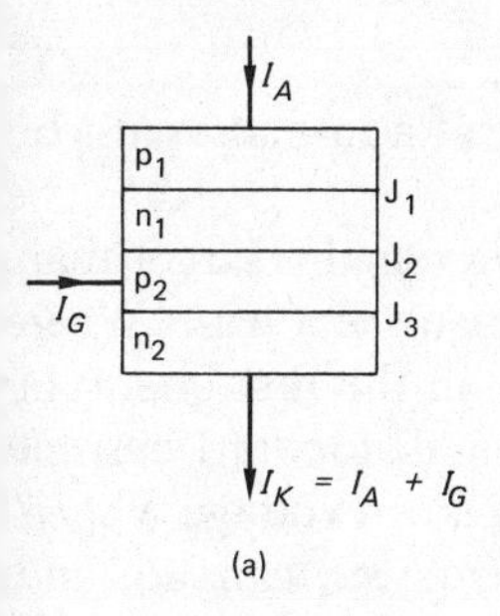

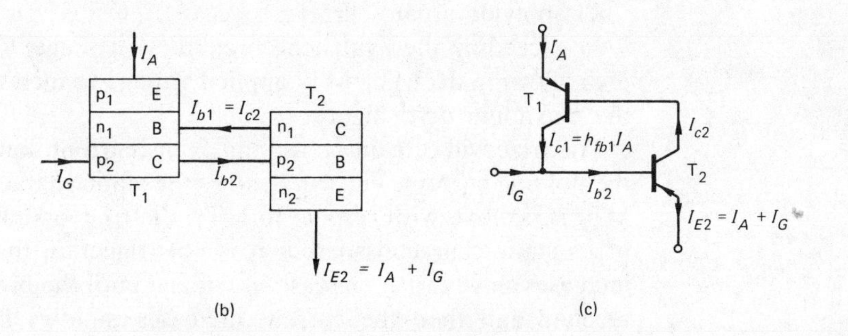

Fig. 10.5 "2-transistor" analogy of a thyristor.

The **junction temperature** of a thyristor must be kept to within the manufacturer's recommended limits because the leakage currents I_{CBO1} and I_{CBO2} are temperature-dependent and, if the junction temperature is allowed to increase too much, I_{CBO1} and I_{CBO2} could attain such values that the thyristor would falsely trigger ON. Very often a thyristor is mounted on a heat sink to assist in the removal of heat from the device and thus keep the temperature down.

The Triac

The **triac** is a semiconductor device whose operation is very similar to that of two thyristors connected in reverse parallel. A triac is able to conduct a large current in *both* directions, being triggered ON in one direction or the other by a gate pulse of the appropriate polarity.

The basic construction of a triac is shown by Fig. 10.6*a* and *b* and its symbol is given by Fig. 10.6*c*. The gate terminal is connected to both p_2 and n_3 so that the triac can be turned ON by either a positive-going or a negative-going gate current pulse. The input and output *main* terminals, labelled as MT_1 and MT_2 respectively, are also connected to both an n-type and a p-type region, MT_1 to n_4 and p_2, MT_2 to n_1 and p_1. These connections mean that it is possible for current to flow either

a) with MT_2 positive with respect to MT_1: path p_1, n_2, p_2, n_4.
b) with MT_1 positive with respect to MT_2: path p_2, n_2, p_1, n_1.

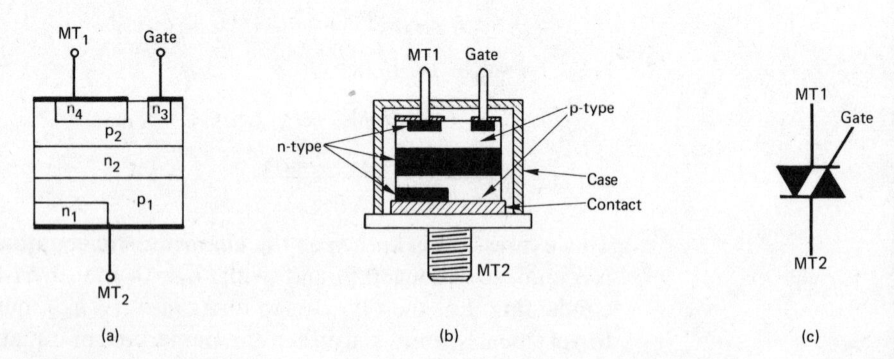

Fig. 10.6 The triac: (*a*) arrangement of p-type and n-type layers, (*b*) construction, (*c*) symbol.

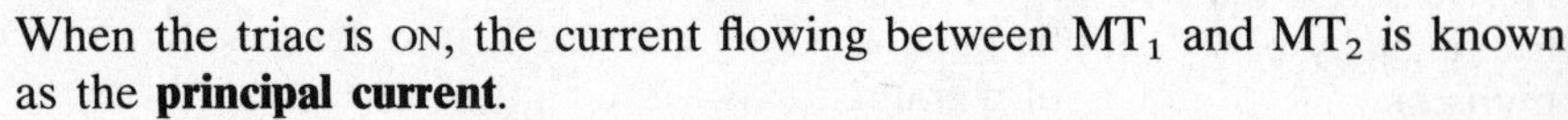

When the triac is ON, the current flowing between MT_1 and MT_2 is known as the **principal current**.

As with the thyristor a triac can be turned ON by

a) applying a gate current,
b) exceeding the avalanche breakdown voltage V_{BO},
c) allowing the MT_1–MT_2 applied voltage to increase at a rate in excess of the maximum dv/dt figure.

The triac will remain ON as long as the current flowing remains larger than the holding current. Fig. 10.7 shows the static characteristic of a triac. When MT_2 is positive with respect to MT_1, the triac operates in the first quadrant of its static characteristics; if it is not triggered, the small forward current increases slowly with increase in voltage until the breakdown voltage V_{BO} is reached and then the current increases rapidly. The device can be, and usually is, turned ON at a smaller forward current by injecting a suitable gate current, and the characteristics show the effect of increasing the gate current from zero to 4 mA. The gate current must be maintained until the main current is at least equal to the latching current.

When terminal MT_1 is positive with respect to MT_2, the triac operates in the third quadrant and the current flows in the opposite direction.

The triac can be triggered to operate in either quadrant by the application of *either* a positive *or* a negative gate current pulse (see Fig. 10.8). The

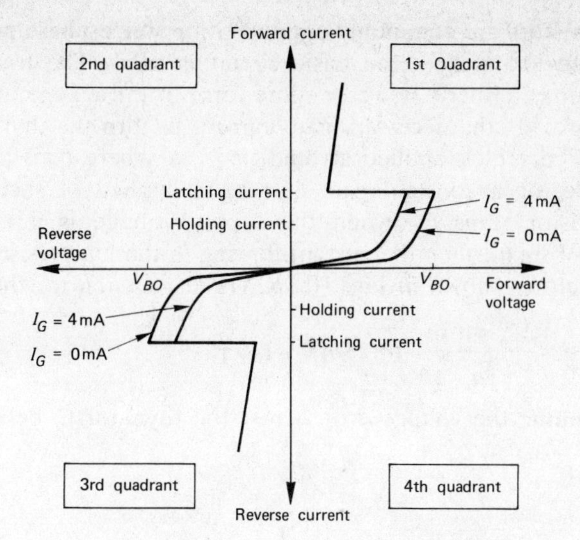

Fig. 10.7 Triac static characteristic.

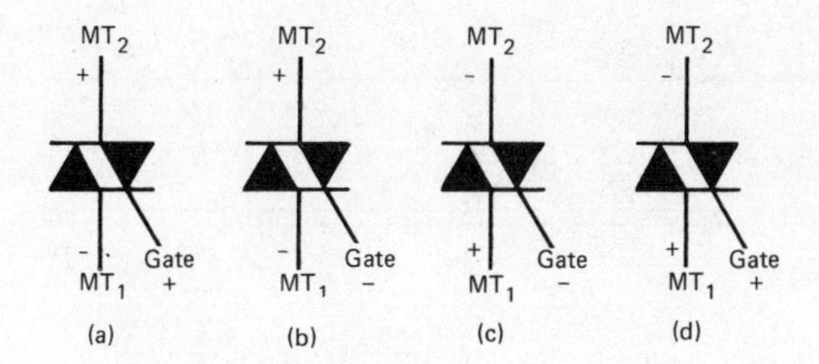

Fig. 10.8 Different modes of triac operation.

arrangements shown in Fig. 10.8*a* and *c* are the most often employed since the use of the same polarity voltage for both the gate and the MT_2 terminal allows the power supplies to be derived from the same source, and this considerably simplifies the circuitry of the trigger module.

The **relative merits** of thyristors and triacs are as follows:

1 One triac can perform the same function as two thyristors connected in reverse parallel, and hence occupies less space and is cheaper.

2 A large reverse voltage could well damage a thyristor but will merely turn a triac ON so that no harm would result provided the current was limited to a safe value.

3 A thyristor has at least a complete half-cycle of the supply voltage in which to turn OFF, whereas a triac has only the short period of time during which the supply voltage passes through zero. This means that the use of the triac is normally limited to a frequency of 50 Hz.

4 The maximum dv/dt a triac can withstand without false triggering taking place is less than for a thyristor.

Single-phase Power Control

Thyristors are commonly employed for **single-phase power control** and Fig. 10.9*a* shows how the basic circuit is used. The trigger module contains circuitry, either passive or some form of pulse generator, whose function is to provide the necessary gate current to turn the thyristor ON. If a pulse of gate current is applied at time $t = \theta_1/\omega$, where θ_1 is the **firing or triggering angle**, or a constant gate current is applied of such magnitude that the thyristor turns ON when the supply voltage is instantaneously equal to $v = V \sin \theta_1$, then the current flowing in the load resistance R_L will have the waveform shown in Fig. 10.9*b*. The expression for the current is given by

$$i = \frac{V_m \sin \omega t}{R_L} \quad \text{for} \quad \theta_1 \leqslant \omega t \leqslant \pi \tag{10.5}$$

assuming the voltage drop across the thyristor to be negligibly small.

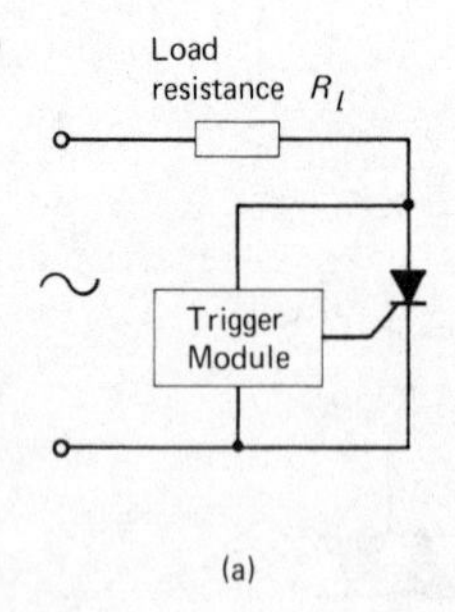

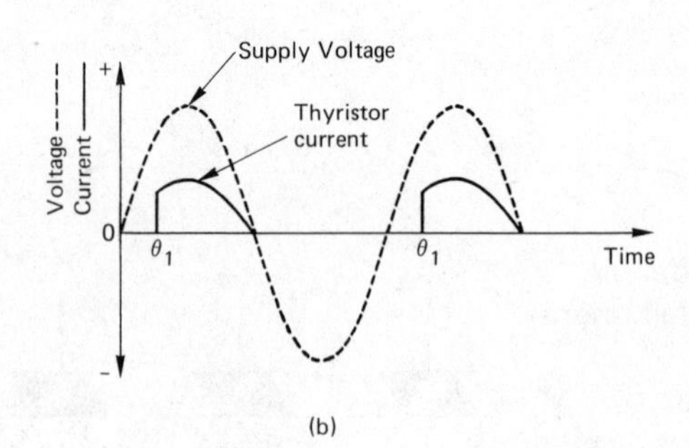

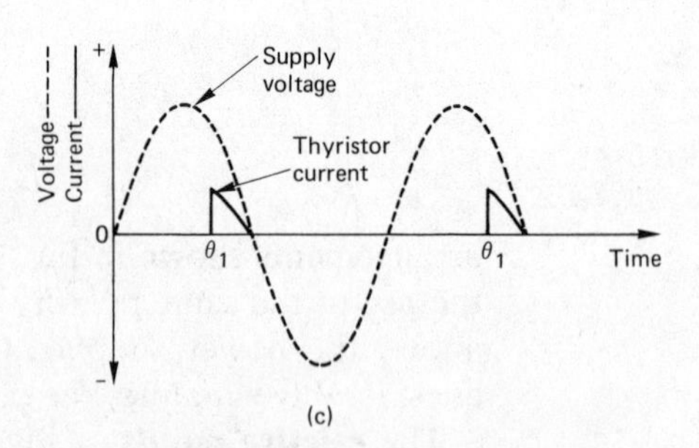

Fig. 10.9 Single-phase thyristor power control.

The d.c. component of the load current is its mean value over a complete cycle and it is clear from the waveforms given in Figs. 10.9*b* and *c* that the mean value is a function of the firing angle θ_1. The later in the positive half-cycles of the supply voltage that the thyristor is fired the smaller will be the mean value of the thyristor current. This means that the system shown in Fig. 10.9*a* is essentially one of power control and rectification. The amount of d.c. power dissipated in the load resistance is controlled by the relatively small power that is supplied to the gate during each positive half-cycle of the supply voltage. The load power is determined by the relative time positions of the supply voltage and the gate pulses.

The points of automatic commutation are $\omega t = \theta_1$ and $\omega t = \pi$ and these should be compared with the commutation method used in a d.c. generator to convert the generated d.c. voltage into a pulsating d.c. voltage.

From equation (10.5) the d.c. component of the thyristor current is

$$I_{dc} = \frac{1}{2\pi}\int_{\theta_1}^{\pi} i\, d\omega t = \frac{1}{2\pi}\int_{\theta_1}^{\pi} \frac{V_m \sin \omega t}{R_L}\, d\omega t$$

$$= \frac{V_m}{2\pi R_L}(1+\cos\theta_1) \tag{10.6}$$

The d.c. load power is

$$P_{dc} = I_{dc}^2 R_L = \frac{V_m^2(1+\cos\theta_1)^2}{4\pi^2 R_L} \tag{10.7}$$

Equation (10.7) shows that the d.c. load power depends upon the supply voltage and the firing angle but *not* upon the gate power.

Example 10.1

In the circuit of Fig. 10.9*a* the supply voltage has a peak value of 340 V and a load resistance of 50 Ω. Calculate the d.c. component of the load current if the triggering (firing) angle is 60°. Also calculate the d.c. load power.

Solution From equation (10.6),

$$I_{dc} = \frac{340}{2\pi \times 50}(1+\cos 60°) = 1.623\text{A} \quad (Ans)$$

From equation (10.7)

$$P_{dc} = \frac{340^2(1+\cos 60°)^2}{4\pi^2 \times 50} = 131.71 \text{ W} \quad (Ans)$$

The **r.m.s. current** delivered to the load is

$$I = \sqrt{\left\{\frac{1}{2\pi}\int_{\theta_1}^{\pi} \frac{V_m^2 \sin^2 \omega t}{R_L^2}\, d\omega t\right\}} = \frac{V_m}{R_L}\sqrt{\left\{\frac{1}{2\pi}\int_{\theta_1}^{\pi} \frac{(1-\cos 2\omega t)}{2}\, d\omega t\right\}}$$

$$= \frac{V_m}{R_L}\sqrt{\left\{\frac{1}{2\pi}\left[\frac{\omega t}{2} - \frac{\sin 2\omega t}{4}\right]_{\theta_1}^{\pi}\right\}} = \frac{V_m}{R_L}\sqrt{\left\{\frac{1}{2\pi}\left[\frac{\pi}{2} - \frac{\theta_1}{2} + \frac{\sin 2\theta_1}{4}\right]\right\}}$$

$$I = \frac{V_m'}{R_L}\sqrt{\left\{\frac{1}{2} - \frac{\theta_1}{2\pi} + \frac{\sin 2\theta_1}{4\pi}\right\}} \tag{10.8}$$

where V_m' is the r.m.s. supply voltage.

Half-wave phase-controlled rectification is relatively simple, cheap and reliable but it does possess a disadvantage in that a mean (i.e. direct) current is taken from the mains supply.

Example 10.2

Calculate the total load power in the system of example 10.1.

Solution From equation (10.8),

$$I = \frac{240}{50}\sqrt{\left[\frac{1}{2} - \frac{\pi}{3 \times 2\pi} + \frac{0.866}{4\pi}\right]} = 3.04 \text{ A}$$

Therefore, the total load power $= 3.04^2 \times 50 = 463.4$ W (*Ans*)

Inductive Loads

When the load on a thyristor is an **inductive load,** the current flowing in the load does *not* fall to zero immediately the supply voltage goes negative. As the load current starts to fall, a back e.m.f. is generated across the inductance and this, from Lenz's Law, has such a polarity that it tends to maintain the flow of the current. The current therefore falls at a slower rate until it becomes less than the holding current value of the thyristor at which point the thyristor turns OFF.

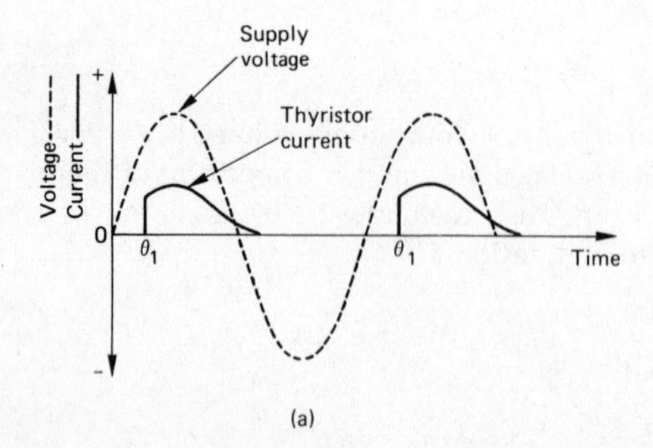

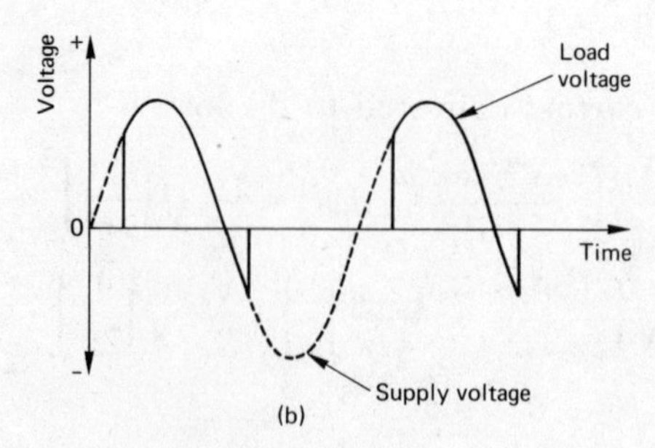

Fig. 10.10 Waveforms in a single-phase thyristor controlled circuit with an inductive load.

The waveforms of supply voltage and load current for an inductive load are shown in Fig. 10.10*a*. It is clear that the thyristor continues to conduct current for a portion of the negative half-cycle of the supply voltage. This effect *always* occurs when load inductance is present and it becomes more pronounced as the ratio inductance/resistance is increased. Because of this effect there is less time available per supply cycle for the thyristor to recover from conduction than for a purely resistive load. While the thyristor is conducting, the load voltage is equal to the supply voltage (remember that the thyristor volts drop is assumed to be zero), and so the load voltage exhibits negative blips as shown by Fig. 10.10*b*.

Another effect of an inductive load is that it will reduce the rate at which the load current can rise, and the possibility exists that the current may not reach the latching value before the gate current pulse ends. If this should happen the thyristor will fail to turn ON. This point is illustrated by Fig. 10.11.

Fig. 10.11

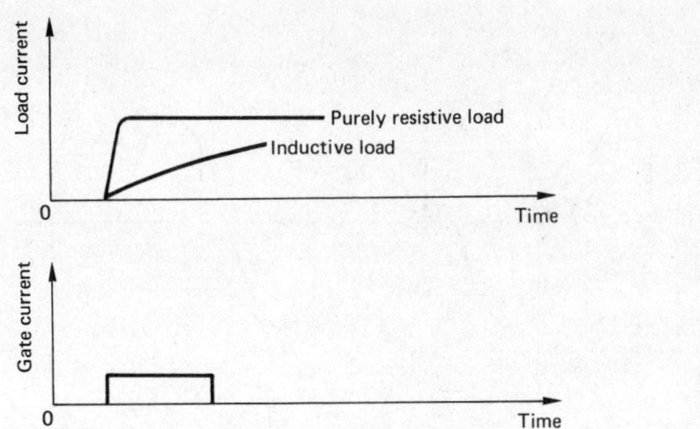

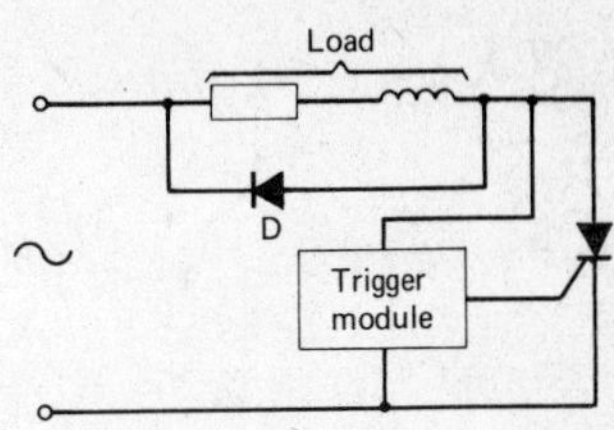

Fig. 10.12 Use of a flywheel diode.

Current can be prevented from flowing during the negative half-cycles of the supply if a *commutating* or *flywheel* diode D is connected in parallel with the inductive load as shown by Fig. 10.12. When the supply voltage goes negative, the diode conducts and provides a low-resistance path for the inductive load current and the thyristor is able to turn OFF.

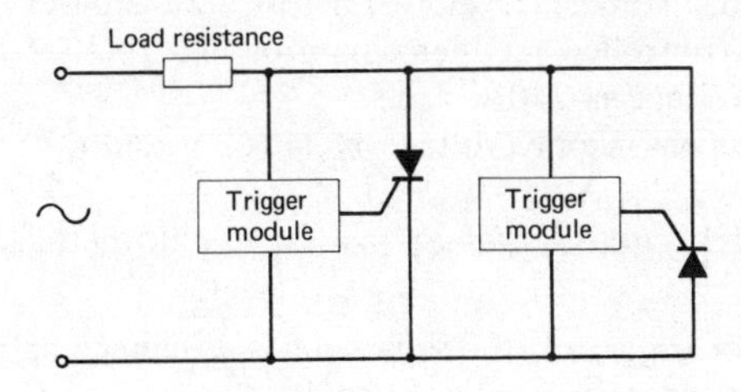

Fig. 10.13 Full-wave power control.

Full-Wave Power Control

1 Fig. 10.13 shows the circuit of a **full-wave power control** circuit which uses two thyristors connected in reverse parallel. Each of the thyristors is turned ON during alternative half-cycles of the supply voltage, the firing angles being determined by the two trigger modules. The operation of the circuit is shown by the waveforms given in Fig. 10.14. Assuming the thyristors have zero voltage dropped across them when conducting, the load voltage is the same as the supply voltage whenever one or other of the

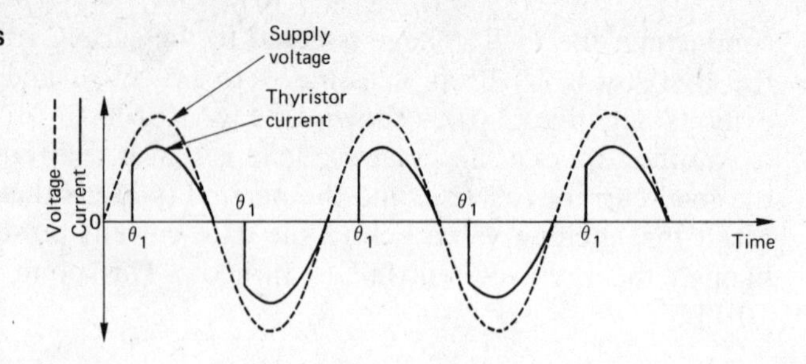

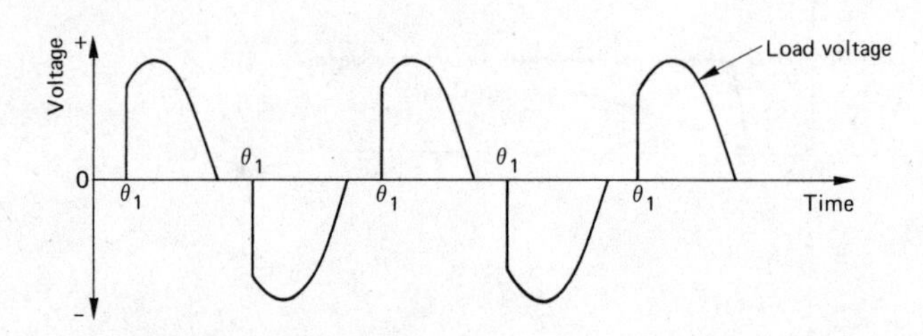

Fig. 10.14 Waveforms in a full-wave power control circuit.

thyristors is ON. It can be seen that the waveforms of the load current and voltage are symmetrical about the 0V axis and so have a mean value of zero. The r.m.s. value of the load current, or voltage, depends upon the firing angle θ_1 and the supply voltage.

Example 10.3

Calculate the r.m.s. voltage developed across a 25 Ω purely resistive load connected in the full-wave controlled rectifier circuit of Fig. 10.13 if the firing angle θ_1 is 60° and the supply voltage is 240 V r.m.s.

Solution The peak supply voltage is $240\sqrt{2}$ V and $60° = \pi/3$ radians. Therefore,

$$V = \sqrt{\left\{\frac{1}{2\pi}\left(\int_{\pi/3}^{\pi} (240\sqrt{2}\sin\omega t)^2\,d\omega t + \int_{4\pi/3}^{2\pi} (240\sqrt{2}\sin\omega t)^2\,d\omega t\right)\right\}}$$

Since the thyristor triggers at the same angles θ_1 in each half cycle, the r.m.s. voltage expression can be re-written as

$$V = \sqrt{\left\{\frac{1}{\pi}\int_{\pi/3}^{\pi} (240\sqrt{2}\sin\omega t)^2\,d\omega t\right\}} = 240\sqrt{2}\sqrt{\left\{\frac{1}{2\pi}\int_{\pi/3}^{\pi} (1-\cos 2\omega t)\,d\omega t\right\}}$$

$$= 240\sqrt{\left\{\frac{1}{\pi}\left[\omega t - \frac{\sin 2\omega t}{2}\right]_{\pi/3}^{\pi}\right\}} = 240\sqrt{\left\{\frac{1}{\pi}\left[\pi - \frac{\pi}{3} + 0.433\right]\right\}} = 215\text{ V} \quad (Ans)$$

2 The above circuit does not take a direct current from the supply but it has a disadvantage in that it needs two trigger modules. Another circuit which needs only one trigger module is shown in Fig. 10.15*a*. The circuit uses one thyristor and four diodes arranged in a full-wave bridge circuit. The

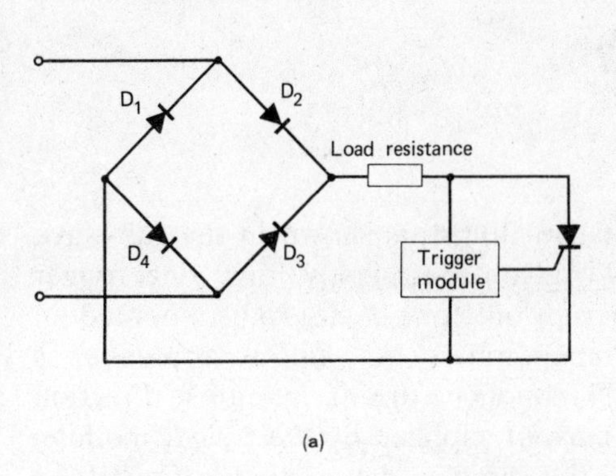

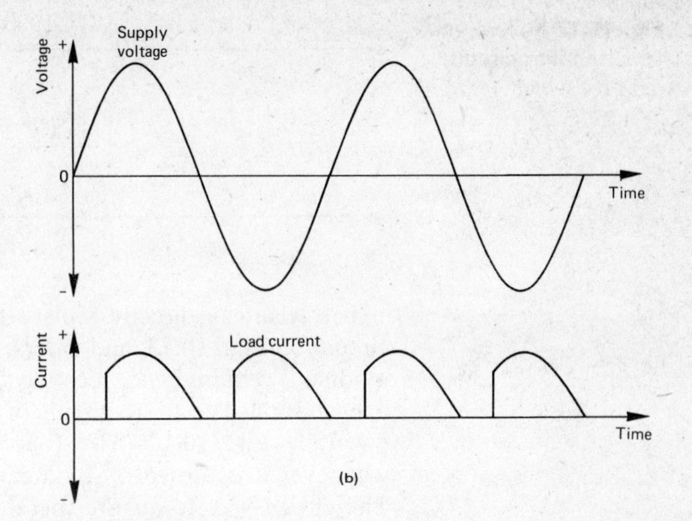

Fig. 10.15 Full-wave bridge power control.

voltage applied to the thyristor always makes the anode positive with respect to the cathode and so the thyristor conducts current twice per supply cycle (Fig. 10.15*b*). This arrangement possesses the advantages that *a*) the reverse blocking requirements of the thyristor are reduced, *b*) no mean (d.c.) current is taken from the supply provided the bridge is accurately balanced.

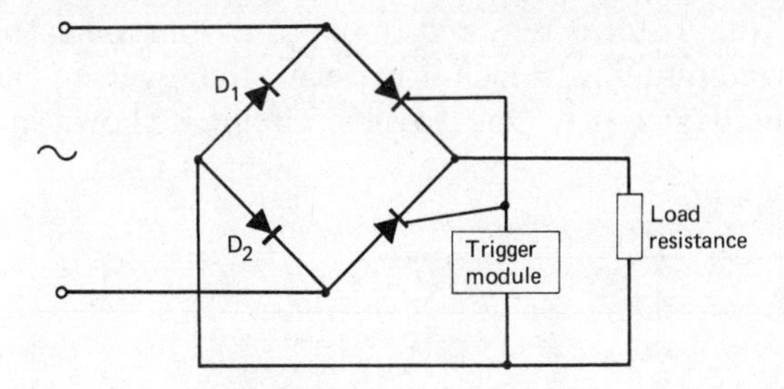

Fig. 10.16 Half-controlled bridge.

3 Another full-wave thyristor circuit is known as the half-controlled bridge because two of the four rectifying elements are thyristors (see Fig. 10.16). The upper thyristor is turned ON during the positive half-cycles of the supply voltage and the circuit is completed by diode D_2. During the negative half cycles, the lower thyristor is fired and now the circuit is completed by the diode D_1. The circuit waveforms are shown by Fig. 10.15*b*.

Comparing the waveforms given in Fig. 10.10 and 10.15*b* for half-wave and full-wave controlled rectifiers respectively, it is evident that the latter provides larger d.c. and r.m.s. currents to the load. This means that the load voltage and power are greater for the same load resistance, supply voltage and firing angle. In all cases, increasing the firing angle means that the thyristor is turned ON later in the supply cycle and this results in a reduction in the r.m.s. load current.

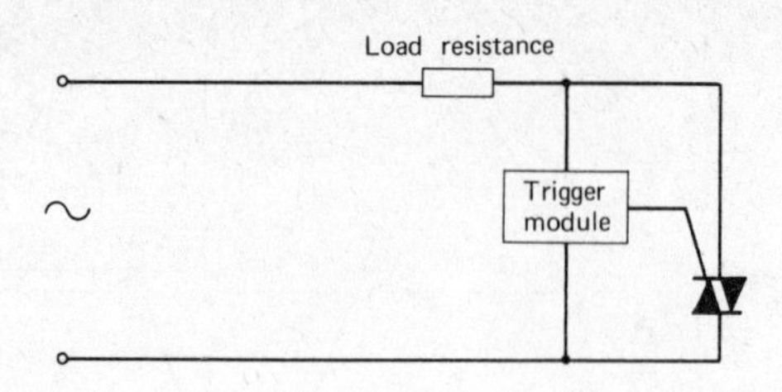

Fig. 10.17 Full-wave triac rectifier circuit.

The **triac** can directly replace the two thyristors shown in the full-wave circuit of Fig. 10.13 and offers two advantages: firstly, only one trigger module is required and, secondly, there is only one device to be mounted on a heat sink. The basic circuit of a single-phase triac full-wave rectifier is given by Fig. 10.17. The triac will conduct current in either direction whenever it is turned ON by a gate current supplied by the trigger module. The circuit waveforms are therefore the same as those given in Fig. 10.14 for the two reverse-paralleled thyristors.

When a full-wave controlled rectifier feeds an inductive load, the same effects as were previously described for a half-wave circuit will occur. As before the effect can be eliminated by the use of a flywheel diode connected in parallel with the load.

Operation from a D.C. Supply

When a thyristor is required to operate from a **d.c. supply,** it is necessary to use a commutating capacitor to derive the reverse polarity voltage needed to turn the device OFF. One possible circuit is shown in Fig. 10.18.

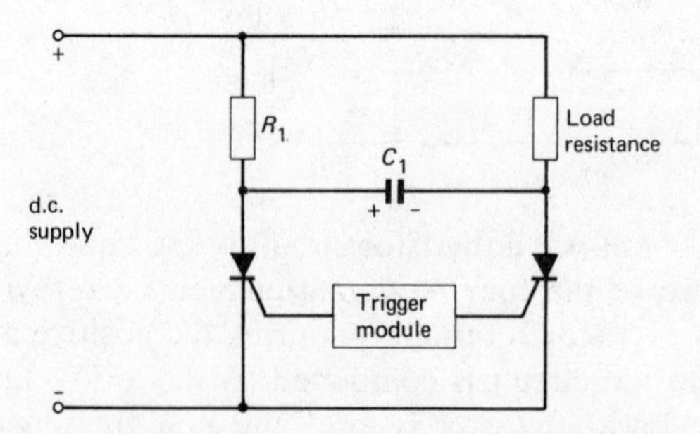

Fig. 10.18 Operation from a d.c. supply.

The right-hand thyristor controls the supply of power to the load and when it turns ON the d.c. supply voltage appears across the load. The commutating capacitor C_1 charges up to V volts, with the polarity shown, with a time constant of C_1R_1 seconds. When the left-hand thyristor is fired, the voltage at its anode falls to very nearly zero taking the left-hand plate of C_1 to 0V. The capacitor now places a reverse voltage of V volts across the right-hand thyristor to turn it OFF.

Single-phase controlled rectification is used in a variety of applications amongst which can be numbered electric ovens, cooker rings, a.c. and d.c. electric motors, lamps, heaters, television receiver power supplies.

Trigger Modules

The thyristor or triac in a phase-controlled rectifier circuit must be turned ON at regular intervals of time by a gate current supplied by a circuit which has hitherto been shown as a block marked **trigger module**. Trigger modules can be purchased from a number of sources but often it may prove to be more economical to use a simpler circuit. For example, Fig. 10.19*a* shows a circuit in which the choice of firing angle is determined by adjustment of the variable resistor R_1.

Very often a device known as a **unijunction transistor** is used to form a simple pulse generator and Fig. 10.19*b* shows a lamp dimmer circuit. Capacitor C_1 is initially discharged and the u.j.t. is non-conducting. When the power is first applied, C_1 charges, with a time constant C_1R_2 seconds, and when its voltage reaches a certain value the u.j.t. suddenly turns ON. T_1 now places a low resistance across C_1 and the capacitor discharges rapidly. T_1 then turns OFF again and the cycle is repeated. The frequency of operation of the circuit depends upon the time constant and this is adjusted by means of the variable resistor R_1.

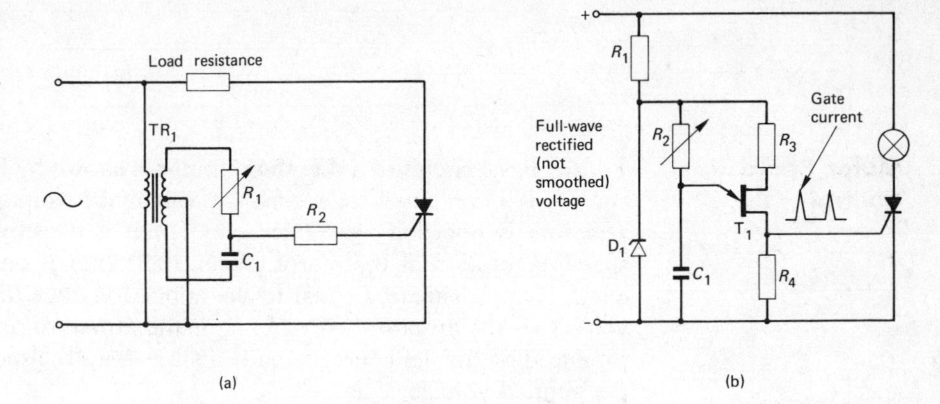

Fig. 10.19 (*a*) Use of a single trigger circuit, (*b*) a lamp dimmer circuit.

Another device that is frequently found in trigger circuits is the **diac** since it is a kind of low-cost bi-directional diode. Its symbol is shown in Fig. 10.20*a*. The diac is able to rectify a.c. voltages during both half-cycles and so it provides gate current for the triac in Fig. 10.20*b*, or for the thyristor in Fig. 10.20*c*, whenever the voltage applied to it is large enough to turn it ON. In both circuits this voltage is the voltage across the capacitor C_1 and this reaches the "turn-on" voltage after a time determined by the time constant R_1C_1 of the circuit. The waveforms at various points in the thyristor circuit are given by Fig. 10.20*d*.

Fig. 10.20 (*a*) Diac symbol, (*b*) diac control of a triac, (*c*) diac control of a thyristor, (*d*) waveforms in circuit (*c*).

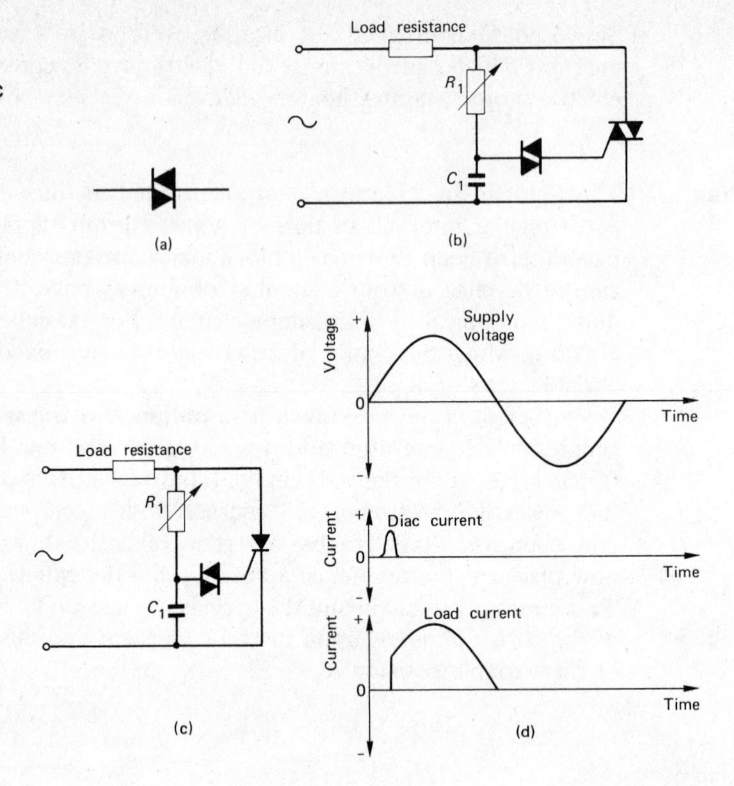

Motor Speed Control

1 The basic circuit of a **d.c. shunt motor** is shown by Fig. 10.21. If the field current is maintained at a constant value, the voltage applied across the armature is opposed by a back e.m.f. that is directly proportional to the speed of rotation of the motor. When the motor is on "no-load", the back e.m.f. is approximately equal to the applied voltage. If the back e.m.f. is E volts, I_a is the armature current, R_a is the armature resistance, Φ is the flux produced by the field current, and ω is 2π times the speed of the motor then the applied voltage V is

$$V = E + I_a R_a \simeq E$$

If the armature resistance is ignored the basic e.m.f. on no-load is equal to the applied voltage. Therefore,

$$E = k\Phi\omega \quad \text{or} \quad \omega = E/k\Phi = V/k\Phi$$

Since Φ is maintained constant and k is a constant, the motor speed is directly proportional to the applied voltage.

This means that the speed of a d.c. shunt motor can be controlled by varying the average value of the voltage applied across its armature. The thyristor is suitable for motor speed control because the average value of the

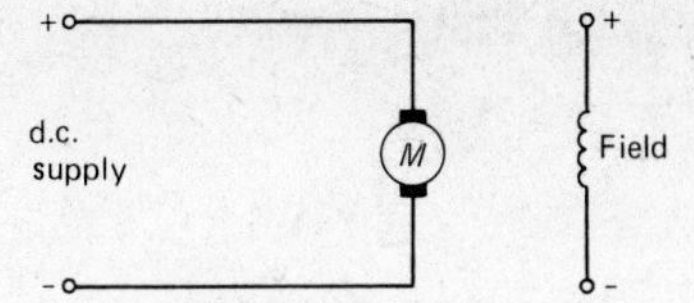

Fig. 10.21 Basic circuit of a d.c. shunt motor.

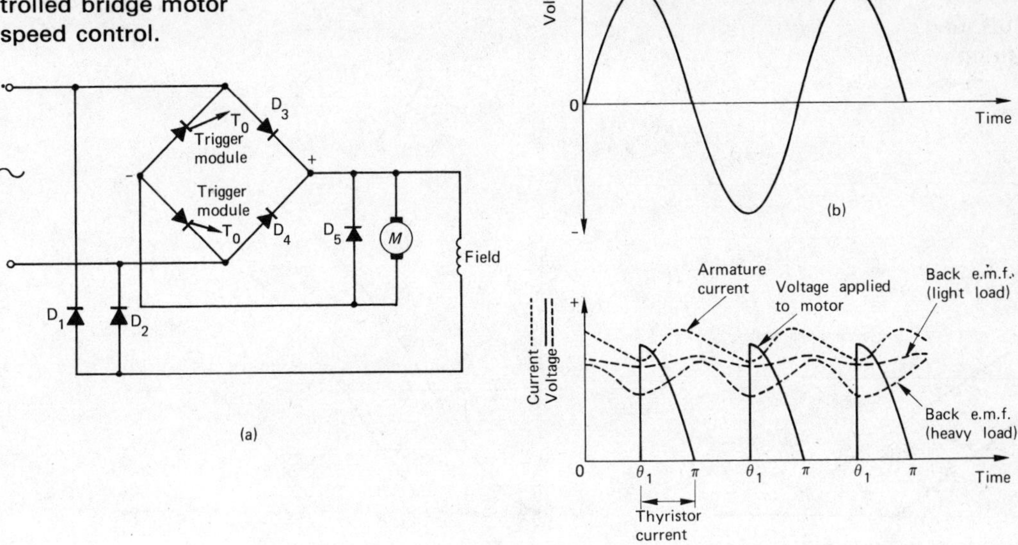

Fig. 10.22 Half-controlled bridge motor speed control.

rectified voltage applied to a motor can be varied by using phase control. Fig. 10.22*a* shows one possible circuit which uses a half-controlled bridge. Diodes D_1 and D_2 are provided with diodes D_3 and D_4 to complete a bridge rectifier to provide the constant field current. The circuit waveforms are given in Fig. 10.22*b* and *c*. The thyristor current flows in pulses between $\omega t = \theta_1$ and $\omega t = \pi$, and so the voltage applied across the motor has the waveform given in (*c*). The armature current is continuous and flows either through the thyristor or through the flywheel diode D_5.

The motor is supplied with two pulses of thyristor current per supply cycle; each time current is supplied the motor speed increases (see the curve labelled back e.m.f. since this is proportional to speed). When the thyristor current ceases to flow, the armature current falls exponentially at a rate determined by the time constant of the armature circuit and the back e.m.f. (hence the motor speed). The motor therefore slows down until the next conduction period occurs. The *average* motor speed depends upon the *average* thyristor current and hence upon the *average* value of the voltage applied across the motor. The back e.m.f. provides a reverse bias voltage

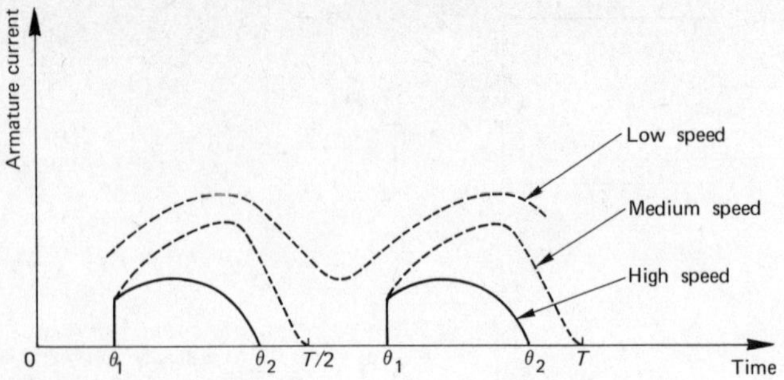

Fig. 10.23 Variation of armature current with motor speed.

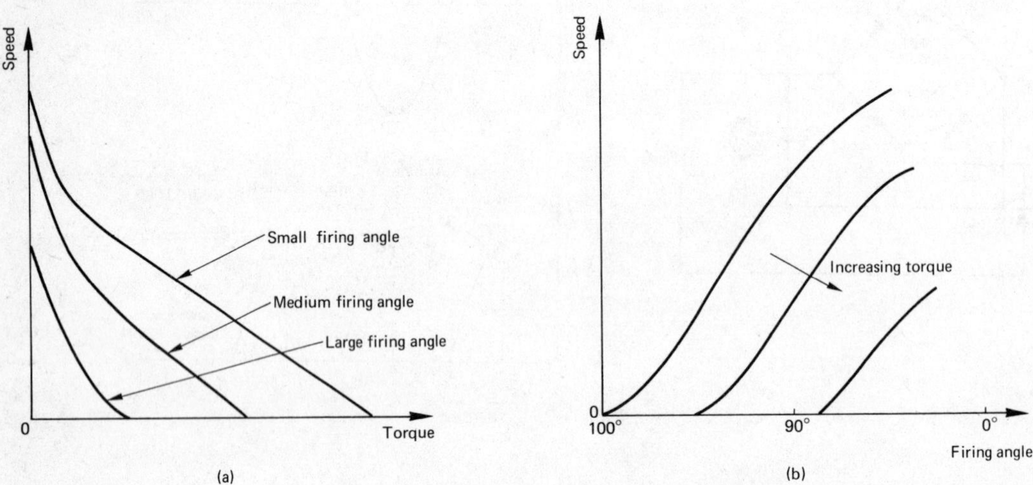

Fig. 10.24 Showing how motor speed varies with (*a*) load torque, (*b*) firing angle.

for the thyristors and this means that the conduction angle is not determined solely by the trigger module but also by the speed of the motor.

If the motor speed is high the back e.m.f. will be large and the thyristor will turn OFF at some angle θ_2 less than π, and this will mean that the flywheel diode will not conduct. The armature current is now discontinuous. For medium speeds, the thyristor will turn OFF at approximately $\theta_2 = \pi$ and the flywheel diode will conduct. The armature current now flows for a larger conduction angle but it is still not continuous. Lastly, when the motor speed is low, the thyristor turns OFF at $\theta_2 = \pi$ but the armature current will be still flowing when the thyristor fires again (see Fig. 10.23).

Graphs showing how the speed of a d.c. shunt motor varies with load torque for various firing angles, and how speed varies with firing angle for various load torques, are given in Fig. 10.24*a* and *b* respectively. Clearly, the greater the motor torque the lower the speed of the motor; but for a given torque, reducing the firing angle will increase the motor speed. This is, of course, to be expected since a reduction in the firing angle results in an increase in the mean thyristor current supplied to the armature of the motor.

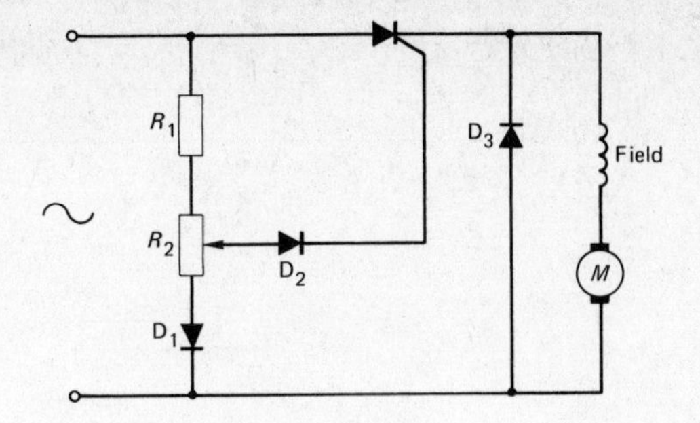

Fig. 10.25 Control of a small d.c. series motor.

2 A simple circuit for the control of a small **d.c. series motor** is shown in Fig. 10.25. During the negative half-cycles of the supply voltage the thyristor is unable to conduct and so no current is supplied to the motor. During the positive half-cycles diode D_1 is forward-biased and current flows in the R_1, R_2, D_1 chain, and when the voltage at R_2 becomes sufficiently positive D_2 conducts and provides a gate current to turn the thyristor ON. The supply voltage is supplied across the motor and this voltage reverse-biases diode D_2 so that the gate current is stopped. The thyristor will continue to conduct until the supply voltage passes through zero. The thyristor current will then cease but the motor current continues to flow via D_3. When the next positive half-cycle of the supply voltage occurs, the thyristor will turn ON again but the firing angle will be somewhat larger since the voltage at R_2 must now overcome the back e.m.f. of the motor as well as turn D_2 ON.

Integral Cycling (Burst Firing)

Phase-controlled rectification suffers from the disadvantage that the supply voltage waveform is interrupted at times when the voltage may be quite large and this process can lead to the generation of considerable radio-frequency interference. It is, of course, possible to suppress such interference but an alternative technique, known as **integral cycling** or **burst firing,** can sometimes be used which does not suffer from this defect.

With integral cycling, the thyristor or triac is turned ON or OFF only when the supply voltage is instantaneously at, or very near to, 0 V. The thyristor is then left ON or OFF, for several complete or *integral* cycles. The average power delivered to the load is determined by the ratio of the ON to the OFF cycles in a given period of time.

The principle of integral cycling is illustrated by Fig. 10.26. The thyristor(s) or triac(s) is turned ON at the instant the supply voltage passes through zero and starts its third cycle, and it is then kept ON for a period of time during which three cycles of the supply voltage are passed to the load. At the end of the third cycle, the supply voltage is once again instantaneously equal to zero and the thyristor is turned OFF. The thyristor then remains OFF for a time period equivalent to two supply voltage cycles and is then turned ON again and so on.

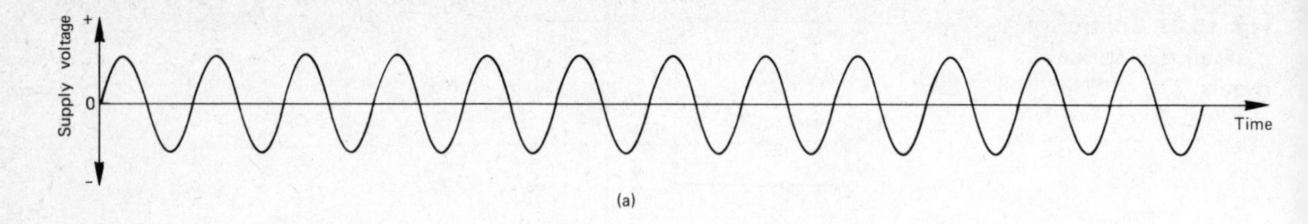

(a)

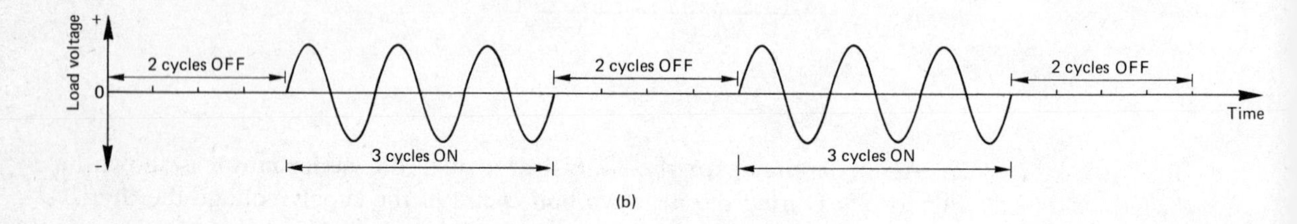

(b)

Fig. 10.26 Integral cycling.

The r.m.s. voltage developed across the load is equal to the supply voltage times the square root of the ratio of the ON to the ON plus OFF times, i.e.

$$V_{load} = V_{supply}\sqrt{\frac{\text{Time ON}}{\text{Time ON} + \text{Time OFF}}} \qquad (10.9)$$

Thus, referring to Fig. 10.26, the r.m.s. load voltage would be equal to $\sqrt{(3/5)}$ or 0.775 times the r.m.s. supply voltage. The load power is V_L^2/R_L and hence

$$P_L = \frac{V_{supply}^2}{R_L}\left(\frac{\text{Time ON}}{\text{Time ON} + \text{Time OFF}}\right) \qquad (10.10)$$

Example 10.4

An integral cycling system has a p.v. power transfer value of 0.4. Determine some possible load voltage waveforms.

Solution Since Time ON/(Time ON + Time OFF) = 0.4, there are several possible solutions. For example

a) 4 cycles ON, 6 cycles OFF
b) 2 cycles ON, 3 cycles OFF
c) 20 cycles ON, 30 cycles OFF (*Ans*)

Integral cycling is only suited for use with load having a slow reaction time. The **reaction time** of a load is the rate at which the condition of the load (i.e. its temperature or its speed or its light emission, for example) changes as the power is abruptly applied or removed. Examples of slow reaction loads are large electric furnaces and electric motors with large inertia loads. Loads having a fast reaction time such as filament lamps, motors driving low-inertia loads, and electromagnets are not suited to integral cycling control. If, for example, a filament lamp were to be operated from an integral cycling circuit, an annoying flicker would be evident.

Whether or not a particular load can be said to possess a fast or a slow reaction time depends upon its **thermal time constant**. This quantity is analagous to electrical time constant and can be defined as the time taken for the temperature of a load to reach 63% of its final value. (Alternatively it can be stated as the time in which the temperature would reach its final value if the original rate of increase were to be maintained.)

Thus, if T_f is the final temperature of the load the temperature T at time t after switching on a continuous power supply is

$$T = T_f(1 - e^{-t/\tau}) \tag{10.11}$$

where τ is the thermal time constant.

Electric furnaces are used for a wide variety of different processes ranging from sintering ceramic materials at temperatures of up to 1800°C to nitriding components at 550°C. The power ratings involved vary from a few kilowatts to a few megawatts. However, the majority of electric furnaces are used for heat treatment and similar processes up to about 1100°C. Generally the time taken for the temperature of the furnace to rise from the ambient temperature, say 17°C, to the required maximum temperature is between one and four hours, but for a large furnace could well be 16 hours or more. If, without much error, the initial temperature is taken as being 0°C, the time required for the final working temperature to be reached is about five thermal time constants. This means that the time constant of a furnace may vary considerably, but it is typically between 20 minutes and one hour.

Polyphase Controlled Rectification

Controlled rectification is not restricted to single-phase power supplies but can also be applied to any number of phases. Fig. 10.27*a* shows a **half-wave three-phase controlled rectifier** circuit and Fig. 10.27*b* gives the waveforms at various points in the circuit. In the diagram it has been assumed that one thyristor is triggered at the instant when the preceding line voltage is just passing through 0V. If the thyristors are triggered at a later point in the cycle, the load current will not flow continuously. Only the thyristor that has instantaneously the most positive voltage applied to it is able to conduct

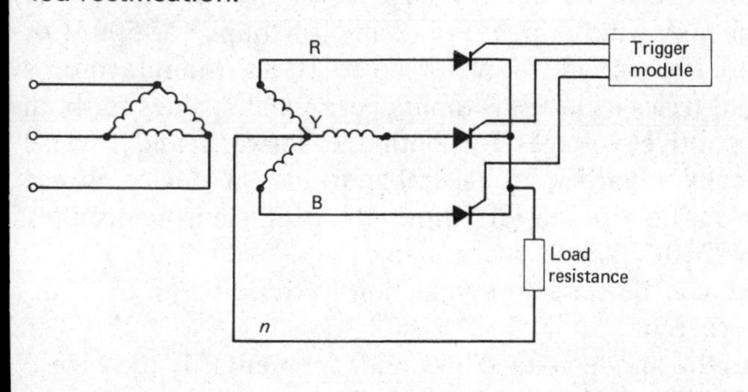

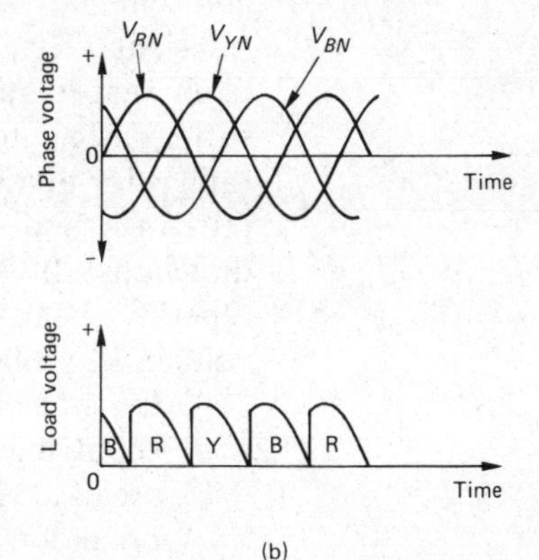

Fig. 10.27 Half-wave three-phase controlled rectification.

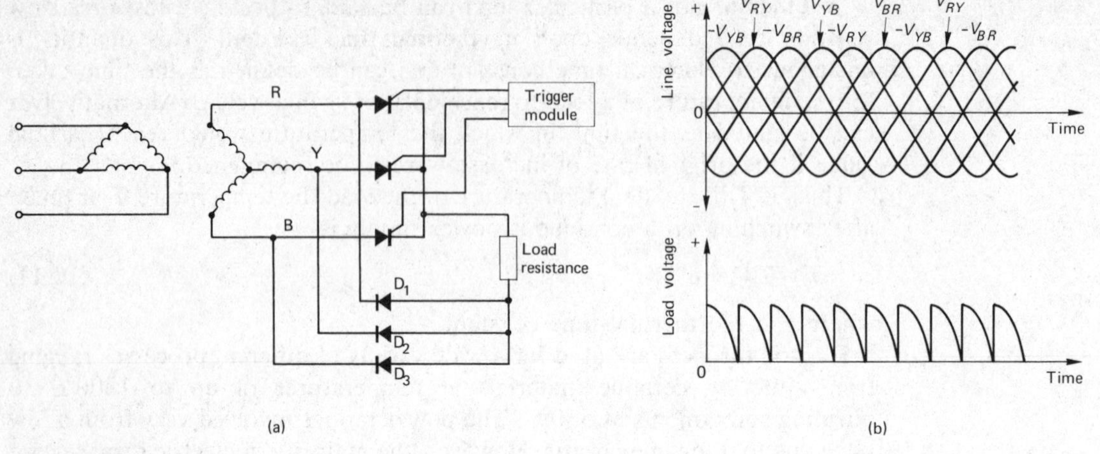

Fig. 10.28 Half-controlled full-wave three-phase rectification.

and, of course, it must be fired by the trigger module during this time interval.

For **full-wave** operation of a three-phase controlled rectifier a half-controlled bridge can be used (Fig. 10.28*a*). As with the half-wave circuit, the waveforms (Fig. 10.28*b*) have been drawn on the assumption that each thyristor is fired as the preceding line voltage (+or−) reaches 0V. If the thyristors are turned ON later than this, a discontinuous load current will be obtained. The firing angles are measured by taking the natural commutation points as the reference, i.e. as 0°.

In both the half-wave and the full-wave circuits, triggering the thyristors earlier than shown in the figures will result in the load voltage not falling to 0V periodically.

Protection of Thyristors and Triacs

The supply voltage applied to a thyristor/triac circuit is usually derived from the a.c. mains and often has a number of **transient voltages,** some periodic and some not, superimposed upon the sinusoidal waveform (Fig. 10.29*a*). A transient voltage may not itself be of very large amplitude but it is most likely that its rate of change will be high. For example, "blips" of 500 V or so may occur for up to 100 μs, or 750 V for up to 10 μs. Manufacturer's ratings for thyristors and triacs include maximum permissible values for both repetitive and non-repetitive voltages for both the forward and reverse directions. It is generally regarded as desirable to use a device whose repetitive peak voltage rating is at least 1.5 times the peak sinusoidal supply voltage, e.g. 600 V for 240 V single-phase mains.

A thyristor or a triac can be set into conduction in three ways:

a) supplying a gate current
b) breakdown due to the anode voltage becoming momentarily too high
c) breakdown due to the anode voltage increasing at too high a rate.

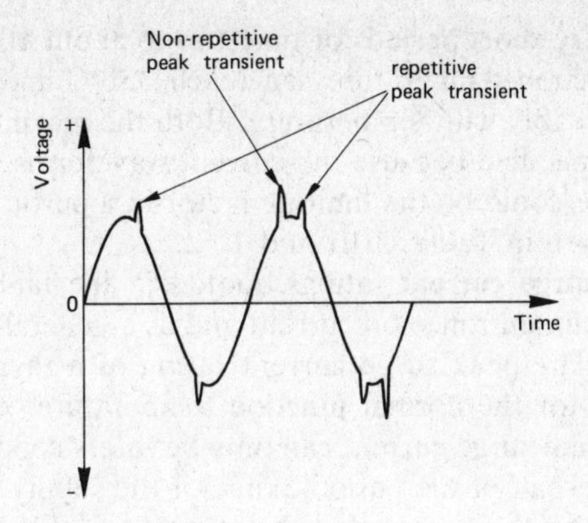

Fig. 10.29 Transients on a mains supply voltage.

Referring to Fig. 10.29 it should be evident that supply transients could cause false triggering of a thyristor or a triac by virtue of either *b*) or *c*) above. Also, a reverse voltage spike applied to a thyristor could cause a large current to flow in the reverse direction and this could destroy the device.

To avoid false triggering some kind of protection circuitry is needed which will *a*) reduce the magnitude of the transients and *b*) slow down their rate of increase.

Two methods of protecting thyristors are shown in Fig. 10.30*a* and *b* and a combination of the two can also be employed. Typically $R = 50\ \Omega$ and $C = 0.1\ \mu F$.

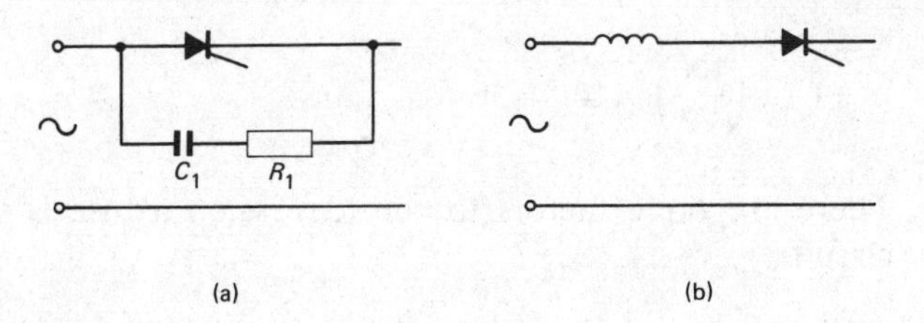

Fig. 10.30 Thyristor protection.

Fuses in Thyristor and Triac Circuits

The published data on thyristors and triacs also includes various current ratings. These are the repetitive and non-repetitive peak forward or *surge* currents, the maximum rate of increase of the ON current allowable, the maximum average and r.m.s. ON currents, and what is known as the I^2t rating. The **I^2t rating** is a measure of the maximum forward non-recurring current for very short periods of time (up to about 10 ms). During a current surge the junction temperature can reach 350°C and so the I^2t value does not vary noticeably with temperature. Both the maximum average and r.m.s. currents are specified because the current waveforms are far from sinusoidal and either one could be the limiting factor in a particular case. Some typical figures are given in Tables 10.1 and 10.2.

The peak **surge current** ratings quoted in the table range from 8 to 12 times the maximum r.m.s. ON current and as a general rule a factor of 10 can be assumed. The peak surge current rating of a thyristor is quoted by the manufacturer for the normal junction temperature of 75°C.

The maximum surge current can only be safely conducted by a thyristor or a triac for one-half of the periodic time of the supply voltage waveform. For 50 Hz a.c. mains this means that the maximum period of time is only 10 ms. If the surge current is maintained for longer than 10 ms, the thyristor/triac will be damaged and it must therefore be protected by a fuse. This fuse must operate to disconnect the supply voltage during the first half-cycle of the surge current.

The correct **fuse** for a particular application is chosen on the basis of the I^2t rating of the thyristor. The fuse must have an I^2t rating which is less than that of the thyristor so that it will afford protection under the worst fault conditions. Suppose $t = 10$ ms, then

$$I^2t = \int i^2\,dt = \left(\frac{\text{Peak surge current}}{\sqrt{2}}\right)^2 \times 10^{-2}\ \text{A}^2\text{s} \qquad (10.12)$$

Example 10.5

A thyristor has a peak surge current rating of 100 A. Calculate its I^2t rating.

Solution From equation (10.12)

$$I^2t = \left(\frac{100}{\sqrt{2}}\right)^2 \times 10^{-2} = 50\ \text{A}^2\text{s} \quad (Ans)$$

There are three factors to consider when a fuse is selected for a thyristor circuit:

a) The fuse must have an r.m.s. current rating less than that of the thyristor.

b) The fuse must have an I^2t rating less than that of the thyristor with t being the time in which the fuse is expected to operate.

c) The arc voltage of the fuse must be less than the peak surge voltage of the thyristor otherwise a prolonged arc may may exist and the fuse will not break the circuit rapidly enough. This means that the fuse should have a peak voltage rating somewhere in between the peak supply voltage and the peak transient voltage rating of the thyristor.

Table 10.1

Maximum average (A)	*ON current* r.m.s. (A)	*Holding current* (mA)	*Peak surge current* (A)	*Maximum gate current* (mA)
0.8	0.5	5	4.0	0.2
1.0	0.64	25	4.5	10
5.0	3.2	20	35	15
8.0	5.0	30	62	25

Table 10.2

Normal thyristor current rating (A)	*Peak surge current* (A)	I^2t (A^2s)
1.6	15	0.5
2.0	15	0.5
7.0	60	4.9
8.0	60	4.9
16.0	125	40
25	250	260

Exercises 10

10.1 Fig. 10.12 shows the circuit of a half-wave thyristor power supply. Explain its operation and mention the function of each component shown.

10.2 Draw the circuit of a full-wave half-controlled rectifier. Assuming a purely resistive load draw waveforms to show *a*) the supply voltage, *b*) the gate current, *c*) the load current, showing clearly what happens in the region of the holding current.

10.3 A fully-controlled bridge employs four rectifiers connected in a bridge circuit. Draw the circuit and describe the operation of such a rectifier. Assume a purely resistive load.

10.4 Fig. 10.31 shows a method of controlling the speed of a d.c. series motor from a d.c. supply. Explain the operation of the circuit.

Fig. 10.31

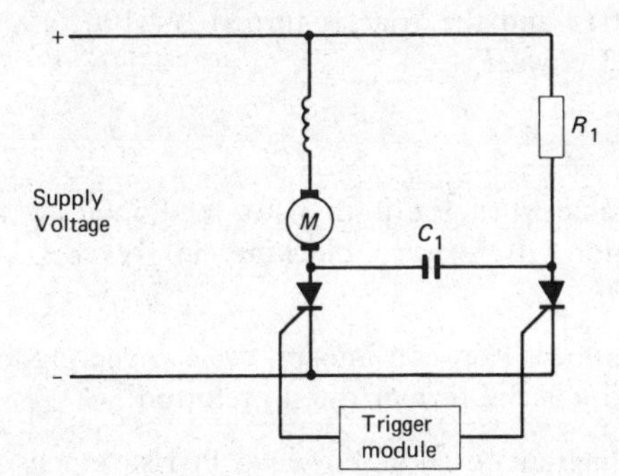

10.5 Draw the circuit of a half-wave three-phase controlled rectifier and explain, with the aid of waveform diagrams, its operation. Assume that the firing angles are such that the flow of load current is discontinuous.

10.6 An electric furnace reaches its maximum temperature of 1000°C from cold in 3 hours. Sketch the temperature/time graph for this furnace and use it to estimate the thermal time constant of the furnace.

Explain why integral cycling will probably be used and estimate the drop in temperature that occurs when the supply is switched off for 20 cycles. Comment on your result.

10.7 Sketch the circuit of a bi-phase half-wave d.c. power supply using thyristors. Show how the load power can be varied.

An electric furnace has a d.c. power consumption over its working temperature range of between 1200 W and 4 W. If the furnace is supplied by a bi-phase thyristor unit, calculate the range of firing angles required. The supply voltage is 240 V, 50 Hz and the load resistance is 30 Ω.

10.8 Draw and explain the operation of a thyristor power supply if *a*) the supply is alternating at 50 Hz, *b*) the supply is direct.

If the load resistance is 150 Ω and the control is such that the thyristor is fired 50° after the supply voltage passes through 0 V in the positive direction, calculate the mean and r.m.s. values of the load current. The supply voltage is 240 V r.m.s.

10.9 Describe the construction and operation of a thyristor. Draw a typical set of thyristor characteristics.

Fig. 10.32

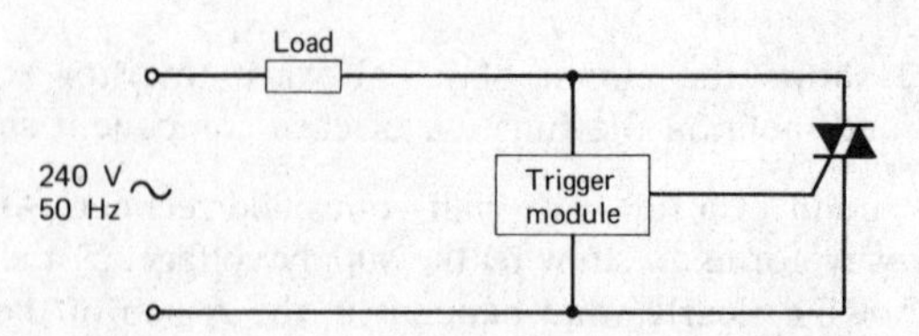

A 25 Ω resistive load is supplied by a thyristor circuit operating from a 240 V r.m.s. supply. Calculate the firing angle necessary for the load power to be 375 W.

10.10 Describe the operation of the circuit given in Fig. 10.32. If the load is a pure resistance of 100 Ω and the triac is turned ON 100° after the start of each cycle calculate the load power.

Short Exercises

10.11 Sketch the characteristics for a thyristor and indicate the following regions: i) forward conduction, ii) forward blocking, iii) reverse blocking, and iv) reverse breakdown.

10.12 Explain the differences between integral cycle triggering and phase angle triggering of a thyristor. Why is the former often preferred?

10.13 Draw a circuit diagram to show how two thyristors may be connected to provide half-wave rectification using i) a reverse parallel circuit, ii) a bridge circuit.

10.14 A thyristor is used to vary the power dissipated in a load. Calculate the percentage of the maximum possible power delivered to the load if the firing angle is 50°.

10.15 Sketch a diagram to show how a thyristor can control the current in an inductive load.

10.16 A single-phase half-wave thyristor circuit is used to control the current in a resistive load of 295 Ω from a sinusoidal supply voltage of 240 V r.m.s. Calculate the average current in the load, if $\theta_1 = 45°$.

10.17 For the circuit in **10.16** calculate the r.m.s. current.

10.18 A single-phase half-wave thyristor circuit controls the current in an inductive load, Draw typical waveforms of current and voltage for the circuit.

10.19 A thyristor has a firing or triggering angle of 40°. Calculate its conduction angle.

10.20 Draw the waveform of the load current of a full-wave half-controlled thyristor bridge if the load is inductive.

10.21 Draw the circuit of a fully-controlled full-wave three-phase rectifier. Outline its operation.

10.22 Determine the peak repetitive voltage rating for a thyristor that is to operate on a 415 V three-phase mains supply.

Numerical Answers to Exercises

1.1 31°C/W
1.2 5.43°C/W, 10.77 W
1.4 9.36 MΩ
1.5 575, −10.3%, −4.35%
1.6 2 kΩ, 20 kΩ, 140
1.8 44.72, 19.61, 9.95
1.9 6.8 MHz
1.10 1940 Ω
1.11 417 Ω
1.13 −3.77 V, 0.212 mA, 6.37 V
1.15 $6\underline{/-35^\circ}$
1.19 5.7 mA
1.20 5, 10
1.21 8 pF
1.23 705, 176
1.26 9800, 37.3 kΩ, 263

2.1 0.1155 cm × 0.866 cm
2.8 1500 Ω

3.1 114.59×10^8, 2.865 MHz
3.2 0.534
3.3 331 Ω, 60, 21.22 MHz
3.4 0.47 μF, 0.447
3.5 2 kΩ, 25 kΩ, 185, 59.6×10^6
3.6 7.273, 218.83 kHz, 0.316
3.7 −1.43 V, 0.96 mA, 5.86 V
3.8 234.74 kHz
3.9 275.74, 4800 Ω, 230, 45.45 kΩ
3.10 41 μF, 82 μF
3.11 119 pF, 9.5 pF
3.12 8 kΩ
3.13 $g_m/150$, 50 kΩ, $1200 - g_m/150$

4.1 101.7, 222 Ω, 802 Ω
4.2 117 kΩ, 5.06, 4.98, 137.9 kΩ
4.6 0.0197, 3333.3
4.7 19
4.8 66, 9.43

4.9 82.14, 115
4.10 0.4 dB, 1.2 MΩ
4.11 23 Hz
4.12 0.0032, 2.265
4.18 negative
4.20 4.8 kΩ
4.21 0.825 dB

5.3 1 kΩ, 100 kΩ, 1017 Ω, 2.48 Ω
5.5 418.3 kΩ
5.7 46.59
5.12 62.5 μV
5.14 1.25 V/μs
5.15 5.56 mV, 55.6 nV
5.16 0.8 V, 0.265 V

6.1 8 W, 10.4 W
6.2 7.68 W, 50.3%
6.4 1.84 W, 55%
6.5 9 W
6.12 12 Ω, 12 Ω
6.13 12.65 V, 1.58 A
6.14 5 W
6.17 5 W

7.1 205.47 kHz, 206.49 kHz
7.2 6.65 μH, 8.5
7.3 1592 Hz, 3
7.4 20.2 pF, 1010 pF
7.7 0.05
7.9 2441 Ω, 47.3
7.10 35 pF, 4.22 nF
7.11 100.62%

8.1 84.62 μA, 4.5 mA, −0.95 V, 7.81 V, 2.88
8.3 371 Hz, 370 Hz
8.5 0.1188 V, −0.1089 V
8.7 1208 Hz
8.8 If $V_{cc} = 15$ V, 100 kΩ, 15 μF
8.14 44 nF
8.15 25 kHz.

9.1 577 Ω, 0.86 dB
9.2 74.8 dB
9.4 $R_{ni} = 59R_{nv}$
9.7 27.4 dB
9.9 2.53×10^{-9} V/√Hz, 2.31×10^{-13}I/√Hz
9.11 17.89 μV, 8.95 μV
9.12 16.97 μV, 8.485 μV, 12 μV
9.14 1.79 mV, No
9.15 70 dB

9.16 31.5 dB, 33.6 dB
9.17 1.63×10^{-12} W, 2.4×10^{-12} W
9.18 0.45 mA
9.19 46 dB, 5.24 dB
10.7 40.87°–89.2°
10.8 0.59 A, 0.88 A
10.9 37.6°
10.10 39.85 W
10.14 64%
10.16 0.313 A

Electronics IV Learning Objectives (TEC)

(A) Small Signal Analysis of Transistors

page 17 (1) *Evaluates the use of h-parameter equivalent circuits of transistors.*

18 1.1 Defines the h-parameters of a bipolar transistor.

23 1.2 Relates the h-parameters to the static characteristics of the transistor.

21 1.3 Identifies the common emitter transistor amplifier's h-parameters, and obtains A_V, A_p, A_I, R_{IN} and R_{OUT}.

22 1.4 Justifies that $A_I \simeq h_{fe}$ and $R_{IN} \simeq h_{ie}$.

23 1.5 Designs and uses a simple circuit for measuring the h-parameters of a transistor.

23 1.6 Compares critically the values of h-parameters obtained in 1.2 with the values obtained in 1.5.

24 (2) *Evaluates the use of hybrid-π equivalent circuits of transistors.*

24 2.1 Draws the hybrid-π circuit of a transistor.

24 2.2 Relates the hybrid-π conductances to the parameters of the transistor.

24 2.3 Relates the hybrid-π parameters to the h-parameters.

26 2.4 Draws the high-frequency equivalent circuit of a CE amplifier using the hybrid-π circuit.

27 2.5 Derives an expression for the short-circuit current gain of the amplifier in 2.4.

3 2.6 Defines the parameter f_T.

27 2.7 Derives an expression for f_T.

25 2.8 Derives Miller's theorem.

25, 75, 77 2.9 Identifies the parameters of the amplifier in 2.4 by means of Miller's theorem.

28 2.10 Identifies the bandwidth and gain-bandwidth product from the analysis of 2.9.

30 (3) *Evaluates the equivalent circuits of fets.*

62, 68 3.1 Draws the circuit diagram of i) a common-source, and ii) a common-drain fet amplifier.

31 3.2 Uses the parameters r_{ds} and g_m to explain the action of the amplifier in 3.1.

30 3.3 Draws the low-frequency equivalent circuit of a fet.

30 3.4 Draws the low-frequency equivalent circuit of a common source (CS) amplifier.

31, 63 3.5 Derives an expression for the voltage gain of a CS amplifier.

78	3.6	Draws the high-frequency equivalent circuit of a fet.
79	3.7	Identifies the CS amplifier performance characteristics from the model in 3.6.
	(B)	**Electronic Noise**
221	(4)	*Appreciates the effects of noise on the amplification process.*
221	4.1	Defines the term power density spectrum (p.d.s.).
222, 223	4.2	Identifies the different forms of electronic noise.
222, 224	4.3	States that the thermal noise voltage in a resistor is $\sqrt{(4kTBR)}$ and that noise in a diode is $\sqrt{(2eIB)}$.
226	4.4	Draws the noise equivalent circuit of an amplifier in terms of its series (R_{nv}) and parallel (R_{ni}) noise resistors.
227	4.5	Defines noise figure.
231	4.6	Defines noise temperature.
228, 232	4.7	Derives expressions for the noise figure and noise temperature in terms of the resistors R_{nv} and R_{ni}.
229	4.8	Derives an expression for the noise figure of a multi-stage amplifier.
227	4.9	States the relationship between R_{nv} and R_{ni} and transistor characteristics.
225	4.10	Compares the noise properties of bipolar and field effect transistors.
232	(5)	*Demonstrates how electronic noise can be measured.*
221, 222	5.1	Explains the terms: i) white noise, and ii) pink noise.
222	5.2	Draws the p.d.s. of white and pink noise.
232	5.3	Uses a block diagram to explain how a noise generator can be used to obtain the noise characteristics of active and passive networks.
232	5.4	Uses a noise generator to obtain the noise characteristics of active and passive networks.
	(C)	**Large-signal Amplifiers**
134	(6)	*Explains the operation of power amplifiers.*
135	6.1	Draws the circuit diagram for a single-ended Class A transistor power amplifier and describes its operation.
136	6.2	States disadvantages of single-ended operation.
136, 137	6.3	Draws circuit diagrams for Class A push-pull amplifiers, with and without an output transformer.
136	6.4	Explains the operation of circuit of 6.3.
136, 138	6.5	Explains differences between Class A and Class B push-pull amplifiers.
136	6.6	Draws circuit diagram of a Class B push-pull amplifier employing input and output transformers and describes its operation.
136, 140	6.7	States the maximum theoretical efficiencies of Class A and Class B push-pull amplifiers.
136, 140	6.8	States practical efficiencies realized for Class A and Class B push-pull amplifiers.
139, 142	6.9	Draws circuit diagram of a Class B push-pull amplifier using a complementary pair.

138	6.10	Explains operation of circuit of 6.9.
138	6.11	Explains cross-over distortion.
138	6.12	Describes how cross-over distortion can be minimized.
145	6.13	Demonstrates experimentally the power output and efficiency of a single-stage power amplifier.
140	(7)	*Analyses large signal (power) amplifiers.*
140	7.1	Analyses push-pull transformer and complementary pair circuits to obtain their efficiencies.
137, 142	7.2	Shows, by using a power series, that harmonic distortion will be produced in the outputs of 7.1 for a sinusoidal input.
138	7.3	Obtains a composite dynamic characteristic from composite static transistor output characteristics.
147	7.4	Uses 7.3 to show the production of output distortion for a sinusoidal input.
145	7.5	Tests i) transformer and ii) complementary pair types of push-pull amplifiers for their power output, efficiency and harmonic distortion.

(D) Feedback

84	(8)	*Understands the concept of feedback.*
85, 91, 94, 97	8.1	Draws a block diagram of a feedback amplifier.
84	8.2	Defines positive and negative feedback in an amplifier.
86, 94	8.3	Derives the general expression for the stage gain of a feedback amplifier.
99	8.4	Lists the effects of applying negative feedback to an amplifier.
89, 92, 96, 99	8.5	Draws some practical negative feedback amplifier circuits.
85, 91, 94, 97	8.6	Shows the meaning of the following types of negative feedback (block diagram form): i) voltage-series, ii) voltage-shunt, iii) current-series, iv) current-shunt.
86, 92, 95, 97	8.7	Derives expressions for the gain, input and output resistance of the circuits of 8.6.
66, 68, 88	8.8	Derives expressions for the voltage gain, input and output resistances of the emitter and source followers.
102	(9)	*Understands and analyses problems concerning stability in feedback amplifiers.*
85, 86, 103	9.1	States that the loop gain of a feedback amplifier is a complex quantity.
103, 106	9.2	Uses Nyquist and Bode plots to predict whether a feedback amplifier is stable or unstable.
105	9.3	Determines gain margin and phase margin of a feedback amplifier.
103, 104	9.4	Predicts the frequency of oscillation of an unstable feedback amplifier.
110, 128	9.5	Discusses and analyses methods of improving the stability of a feedback amplifier.

(E) Single-phase Power Control

242 (10) *Describes the behaviour of a thyristor circuit with single phase supply, a resistive/inductive load.*

242 10.1 Sketches and labels a circuit diagram showing a thyristor load, trigger module (block) and supply.

242 10.2 Marks on a supply voltage waveform sketch those sections in which a thyristor can pass current to a load.

242 10.3 Examines the load voltage waveform oscillogram for a resistive load with varying conduction angle and identifies those sections representing load current conduction.

242 10.4 Infers that since there is a net d.c. component the system is one of control and recification.

243 10.5 Shows that the a.c. component varies with conduction angle.

243 10.6 Identitifies the points of automatic commutation and relates these to the method of commutation used in d.c. systems.

242 10.7 States that load power is controlled by the injection of a small amount of gate energy regularly in each positive half cycle of supply, and the amount of load power depends on the relative time position of gate pulse and supply waveform.

243 10.8 Deduces that load power is not dependent on gate power, only on timing.

243, 244 10.9 Calculates load power from a given formula, for a specified load resistance and conduction angle.

244 10.10 Obtains load voltage oscillograms for a resistor/inductor load with a fixed conduction angle.

244 10.11 Sketches the waveforms obtained in 10.10 overlaid on a supply voltage waveform sketch to the same scale.

244 10.12 Examines 10.11 and concludes that the thyristor continues to conduct for a portion of the other half cycle.

244 10.13 Predicts that the behaviour described in 10.11 and 10.12 will occur with any resistive/inductive load, and that the effect is more marked as the inductance/resistance ratio is increased.

244 10.14 Discusses the mechanism of the conduction in the negative half cycle in terms of the tendency for current in an inductor to continue to flow, and induced e.m.f.

244 10.15 Deduces that under these circumstances the time available for the thyristor to recover from conduction in each supply cycle is less than for a pure resistive load.

245 10.16 Obtains oscillograms for the L/R load arrangement with a commutating diode connected across the load.

245 10.17 Discusses the action of the diode in 10.16 and concludes that it allows the thyristor to turn off when the supply goes negative by conducting the inductive decay current.

247 (11) *Appreciates the major characteristics of operation of a half-controlled bridge with resistive and resistive/inductive loads.*

247 11.1 Sketches a circuit diagram of a half-controlled bridge with a resistive load and single-phase supply.

247	11.2	Describes the operation of the circuit in 11.1 with reference to load and supply voltage waveforms, and a gate pulse waveform.
247	11.3	Compares load voltage waveforms for the same conduction angle with a single thyristor arrangement.
247	11.4	Infers that load voltage, current and power are greater than for the single thyristor arrangement for the same load, supply and conduction angle.
247	11.5	Confirms 11.4 by measurement.
248	11.6	Obtains oscillograms of load voltage for a half-controlled bridge feeding a resistive/inductive load with a commutating diode connected.
248	11.7	Examines the oscillograms in 11.6 and concludes that the commutating diode permits each thyristor in turn to commutate at the point of zero supply voltage.
248	11.8	Sketches the waveforms obtained in 11.7 and labels the conduction sequence.
249	11.9	Lists some engineering situations which use single phase controlled rectification.

250	(12)	*Discusses the control of a d.c. shunt motor with constant field current and armature controlled by a single phase half-controlled bridge.*
250	12.1	States that for a fixed field strength, an applied armature voltage is opposed by an induced e.m.f. which varies linearly with speed of rotation.
250	12.2	States that, ignoring armature resistance, the induced e.m.f. on no load is equal to the average applied armature voltage.
250	12.3	Infers that for a controlled rectified armature supply the maximum (running light) speed is controlled by the average bridge output voltage.
251	12.4	Sketches on a waveform diagram of a half-controlled bridge output voltage a line representing the induced e.m.f. of the armature when running light.
251	12.5	States that when the motor is put on load, the line drawn in 12.4 is shifted downwards.
251	12.6	Explains why the shift in 12.5 occurs.
251	12.7	Sketches a circuit diagram of a half-controlled bridge supplying a d.c. shunt motor armature and includes a commutating diode.
251	12.8	Sketches for a fixed firing angle two armature voltage waveforms, one for light load and one for heavy load on the motor, and labels the areas of armature current flow.
251	12.9	Explains the armature and thyristor conduction sequence, and the action of the commutating diode for the waveforms of 12.8.
251	12.10	Obtains oscillograms for the armature voltage of the controlled motor.
251	12.11	Compares the oscillogram of 12.10 with the sketches of 12.4.
252	12.12	Performs a load test on a thyristor controlled shunt motor for:

a) a fixed firing angle and variable load torque,
b) a fixed load torque and variable firing angle.

252 12.13 Graphs for 12.12.
a) speed vs load torque,
b) speed vs firing angle.

252 12.14 Interprets the graphs of 12.13 in terms of general trends

248 (13) *Appreciates some single-phase control situations using triacs.*
248 13.1 Sketches a circuit diagram showing supply, triac, resistor load and trigger module.
248 13.2 Sketches waveforms of supply, gate and load voltage for a given conduction angle.
248 13.3 Distinguishes between the load waveform of 13.2 and that of the half-controlled bridge with the same load.
248 13.4 States that the load voltage waveform of 13.2 is symmetrical (has no d.c. component).
248 13.5 Infers that increasing the firing angle (i.e. reducing the conduction angle) results in reduced r.m.s. load voltage.
249 13.6 Lists some loads suitable for control by a triac or single-phase supply: ovens, cooker rings, series a.c. motors (drills etc.), lamps, bar heaters.

(F) Polyphase Controlled Rectification

255 (14) *Appreciates the concepts of half and full wave three-phase controlled rectification with resistive load.*
255 14.1 Sketches a circuit diagram of a three-phase half-wave controlled rectifier showing transformer, thyristor stack, load and trigger module.
255 14.2 Sketches a load voltage waveform for a given firing angle and labels this in terms of conducting phase and thyristor.
255 14.3 Compares the waveforms of 14.2 with those of a single-phase half-wave controlled rectifier and draws conclusions about the continuity of output voltage.
255 14.4 Obtains oscillograms for the system in 14.1.
256 14.5 Sketches the circuit diagram for a three-phase half-controlled bridge showing transformer, thyristor and diode stack, trigger module and load.
256 14.6 Explains by reference to labelled load voltage waveforms the conduction sequence of the system in 14.5.
256 14.7 Identifies on a 3-phase waveform sketch the points of natural commutation.
256 14.8 States that measurements of firing angle use the natural commutation points as zero reference points.

(G) Integral Cycling (Burst Firing)

253 (15) *Judges if an electrical control situation is suitable for integral cycling.*
254 15.1 Defines P.V. power transfer into a resistive load in terms of the number of ON and OFF cycles in a sequence.

254 15.2 Calculates P.V. power transfer values for a sequence given in waveform diagram form.
254 15.3 Constructs waveform diagrams representing given P.V. values.
254 15.4 Explores several possible waveform interpretations of a given P.V. value.
15.5 Tests an integral cycling system on single phase with a filament lamp load.
254 15.6 Makes subjective judgement as to whether integral cycling is suitable for lighting control.
254 15.7 Defines the "reaction time" of a load as the rate at which the load condition changes with abrupt delivery or removal.
254 15.8 Classes a filament lamp among the "fast reaction" loads.
254 15.9 Classes a large electric furnace among the "slow reaction" loads.
255, 260 15.10 Sketches a typical temperature/time graph for a furnace switched direct on line from cold, to a marked time scale.
255, 260 15.11 Estimates the thermal time constant from 15.10.
260 15.12 Estimates the temperature change from the working temperature when the supply is switched off for 5, 20 and 50 complete cycles.
255, 260 15.13 Infers that integral cycling control of a furnace can hold a desired temperature fairly closely.
255 15.14 Infers that large electric furnaces are suitable loads for integral cycling control.
254, 255 15.15 Predicts the suitability of integral cycling control for the following loads:
a) Fluorescent lamp.
b) Motor with a light (low inertia) load, e.g. electric drill, electrical saw.
c) Motor with a large inertia load, e.g. traction motor for a crane, or heavy vehicle.
d) Electromagnet.

(H) Thyristors

256 (16) *Evaluates the effect on thyristor/triac circuits of suppply transients, and some standard protection circuits.*
256 16.1 States that a thyristor or triac can be set into conduction by three methods.
a) Gate control.
b) Breakdown due to anode voltage momentarily too high.
c) Breakdown due to anode voltage rising too rapidly.
257 16.2 Sketches a typical main supply waveform showing possible transients with an indication of magnitudes.
257 16.3 Infers from 16.2 the points of possible breakdown.
257 16.4 Infers that a protection circuit which *a*) reduces the size of transients, *b*) slows down their rate of rise, is needed to be connected in a thyristor circuit.
257 16.5 Sketches circuit diagrams of the common protection arrangements with typical component values for a given load and supply.
16.6 Distinguishes between full and partial protection circuits.

Protection arrangements

258 (17) *Understands the basic requirements of fuses in thyristor and triac circuits and selects appropriate fuses for common designs.*

259 17.1 Lists typical surge current ratings against maximum normal current for a number of thyristor types.

258 17.2 Deduces that as a rule a thyristor surge current rating is of the order 10 × normal current rating.

258 17.3 States that maximum rated surge currents can only be conducted safely by a thyristor for 10 ms (half supply period).

258 17.4 Infers that under fault conditions the protective fuse must rupture during the first half cycle of fault current.

258 17.5 Defines the I^2t value for a fuse.

258 17.6 Calculates, for a given surge current rating for a thyristor, an appropriate I^2t value for a fuse.

258 17.7 Recognizes that the working temperature of the thyristor modifies the surge rating and hence the choice of fuse.

258 17.8 Prepares a table of typical I^2t ratings listed against normal thyristor current rating:

Normal thyristor current rating	I^2t

258 17.9 States that for a fuse to clear within the stated time under fault conditions its peak voltage rating must not be exceeded (i.e. danger of prolonging the arc).

258 17.10 States that a satisfactory fuse will have the I^2t rating appropriate to the thyristor and will have a peak voltage rating which lies between the peak supply voltage and the peak transient anode voltage rating for the thyristor.

17.11 Chooses appropriate fuses from a catalogue for some typical thyristor circuits and supplied.

(I) Sinusoidal Oscillators

151 (19) *Understands the maintenance requirements of sinusoidal oscillators.*

151 19.1 Derives the maintenance conditions required for steady state sinusoidal oscillation.

157,159 19.2 Draws the circuit diagram of a tuned collector and a tuned drain oscillator.

157, 159 19.3 Explains the operation of the circuits in 19.2.

157, 159 19.4 Derives the maintenance conditions for the circuits of 19.2.

155 19.5 Draws a circuit diagram of a Wien bridge oscillator.

155 19.6 Derives maintenance conditions for the Wien bridge oscillator.

161, 162 19.7 Draws circuit diagrams of a Hartley and a Colpitt's transistor oscillator.
161, 162 19.8 Explains the operation of the circuits of 19.7.

(20) *Evaluates the maintenance conditions in sinusoidal oscillators.*
151 20.1 Identifies the conditions to be satisfied for satisfactory oscillation.
157 20.2 Outlines the operating conditions of an $L-C$ oscillation.
151 20.3 Outlines the operating conditions of an $R-C$ oscillation.
164 20.4 Outlines the operating conditions of a crystal controlled oscillator.

(J) Non-sinusoidal Waveform Generation

170 (21) *Analyses the discrete component bistable multivibrator.*
170 21.1 Calculates the stable-state bias conditions in a cross-coupled transistor bistable multivibrator.
173 21.2 Calculates the loop-gain during switching action.
174 21.3 Determines trigger pulse amplitude.
174 21.4 States the effects of circuit capacitance on the trigger pulse-width.
174 21.5 Describes the factors which determine switching speeds.
175 21.6 Describes the effect of speed-up capacitance.
174 21.7 Discusses the need for trigger pulse commutation to obtain toggle action.
178 21.8 Draws the circuit and describes the operation of the Schmitt bistable multivibrator.
178 21.9 States the need for the loop-gain to be greater than unity.
179 21.10 Calculates the trigger levels and determines the hysteresis.
181 21.11 Deduces the transfer characteristic and appreciates the noise protection provided by hysteresis.
179, 184 21.12 Applies the Schmitt circuit to waveform squaring and level detection.
21.13 Designs and tests a discrete component bistable multivibrator.

184 (22) *Evaluates the discrete component monostable multivibrator.*
170, 174 22.1 Determines the stable state and triggering requirements of a cross-coupled transistor monostable circuit.
185 22.2 Calculates the period of relaxation.
186 22.3 Derives detailed waveforms for the circuit in 22.1.
175 22.4 Explains circuit modifications to provide improved waveshape of output.
219 22.5 Describes some typical applications.
22.6 Designs and tests a discrete component monostable multivibrator.

192 (23) *Evaluates the discrete component astable multivibrator.*
192 23.1 Calculates the two relaxation periods of a cross-coupled transistor circuit, and hence the mark-space ratio and pulse repetition rate.
193 23.2 Derives detailed waveforms for the circuit in 23.1.
175 23.3 Explains circuit modifications to provide improved waveshape of output.
193 23.4 Explains synchronization and frequency division using astable circuits.

	(24)	*Evaluates I.C. and hybrid circuit configuration multivibrator.*
182, 183, 188, 194, 195, 197	24.1	Uses operational amplifiers and integrated circuit NAND/NOR gates to provide bistable, monostable and astable circuits.
182, 188, 196	24.2	Calculates stable state levels and relaxation times, and derives waveforms.
	24.3	Designs and tests i) multivibrators based on the criteria in 24.1, and ii) multivibrators based on integrated circuit devices.
199	(25)	*Evaluates sawtooth and triangular waveform generators.*
199	25.1	Describe a sawtooth voltage generator using an e.m.f., a $C-R$ circuit and a switch.
200	25.2	Calculates the percentage deviation from linearity in terms of t/CR.
200	25.3	Calculates the sawtooth amplitude for a given non-linearity—hence understands the need for amplification.
200	25.4	Describes and analyses the Miller integrator and relates this to the requirements of section 25.1.
202	25.5	Appreciates the need for constant current charging of a capacitor to obtain a linear ramp and calculates voltage as It/C.
202	25.6	Analyses the Miller integrator as an approximate constant current charging method.
201, 202	25.7	Appreciates the need for high gain, but proves that the circuit action is independent of gain variation.
43	25.8	Predicts the use of a transistor as a constant current device.
204	25.9	States that the bootstrap principle is to provide a constant current source.
204	25.10	Analyses the bootstrap sawtooth generator including ramp rates, ramp amplitude and factors affecting linearity.
205	25.11	Points out the need for recovery to bring the Miller and bootstrap circuits back to the starting state of the ramp.
205	25.12	Outlines typical methods used in the two circuits and calculates recovery times.
206	25.13	Predicts the production of a triangular wave from a rectangular input.
207	25.14	Determines the voltage required to drive a linear ramp of current through *a*) a pure inductance, and *b*) an inductance with series resistance, such as a c.r.t. deflector coil.
208	25.15	Obtains the required ratio of step-slope.
209	25.16	Analyses the current waveform produced by an incorrect ratio of step/slope in the trapezoidal waveform.
214	(26)	*Understands the principles of clamping and clipping.*
214, 215	26.1	Explains why signals need to be *a*) clamped, *b*) clipped.
216	26.2	Draws diagrams and explains the action of i) a diode clamp and ii) a diode d.c. restorer.
215	26.3	Draws diagrams and explains the action of i) a diode clipper and ii) a zener diode clipper.

	(R)	**Linear Integrated Circuits**
48, 57, 79, 143, 209	(35)	*Understands the basic principles of linear I.C.s.*
119	35.1	Sketches the standard symbol for an operational amplifier.
119, 123	35.2	Explains the terms inverting and non-inverting inputs, offset voltage, open loop gain.
119	35.3	Recognises the amplifier responds to the difference in potential between the two inputs.
124, 132	35.4	Calculates for maximum output the magnitude of the p.d. between the two inputs.
124	35.5	Establishes the "virtual earth" concept.

Index